**Delmar Publishers Inc.
is proud to
support FFA
activities**

Environmental Science

for Agriculture and the Life Sciences

by

William G. Camp, Ph.D.

Professor of Agricultural Education

Virginia Polytechnic Institute and State University

and

Roy L. Donahue, Ph.D.

Professor Emeritus of Soil Science

Michigan State University

**Delmar
Publishers Inc.™**

I(T)P™

Freelance Project Editor: C. Denise Littleton
Cover Design: Brian Yacur
Cover photo credits: British Petroleum, NASA, and Brian Yacur

Delmar Staff

Publisher: Tim O'Leary
Senior Editor: Mark W. Huth
Associate Editor: Cathy L. Carter
Acting Developmental Editor: Gwen Ceruti-Vincent
Senior Project Editor: Laura V. Miller
Production Editor: Wendy A. Troeger
Production Coordinator: James Zayicek

For information, address Delmar Publishers Inc.
3 Columbia Circle, Box 15-015,
Albany, NY 12212-5015

Printed in the United States of America
Published simultaneously in Canada
by Nelson Canada,
a division of The Thomson Corporation

1 2 3 4 5 6 7 8 9 10 XXX 00 99 98 97 96 95 94

Library of Congress Cataloging-in-Publication Data
Camp, William G.

Environmental Science/by William G. Camp and Roy L. Donahue.
p.   cm.
Includes bibliographical references (p. 385) and index.
ISBN 0-8273-5025-2 (textbook)
1. Environmental sciences. 2. Human ecology. I. Donahue, Roy
Luther, 1908- . II. Title.
GE105.C36 1993
363.7—dc20   92-43120
                                    CIP

Contents

Dedication

This book is affectionately dedicated to our wives Betty and Lola for wonderfully Nonbiodegradable marriages, by **William G. Camp** and **Roy L. Donahue**

Preface

"Save the Earth"

How often have you heard that slogan? Well, it is a noble thought, but it is misdirected! The Earth is not in trouble and it does not need saving. Even if it did, there is absolutely nothing that people can do today to either save or destroy the Earth.

The Earth is a planet of the sun. It consists of a mass of solids, liquids, and gasses that occupy a very small corner of the universe. It provides a home for many billions of living organisms. The Earth was fine long before people came; it will be fine long after we are gone.

What we had better start being concerned about is how to:

"Save the Humans"

That is what this book is all about:

- How the environment works.

- How we humans relate to it.

- What we need to do today and in the future to have a sustainable and liveable world order.

To aid the student in identifying and understanding some important terms related to environmental science, at the begining of each chapter there appear "Terms to Look for and Learn," which relate to each chapter's subject matter. These terms appear boldfaced where their most complete definition is given, and not necessarily upon first usage in the chapter. A complete and alphabetized listing of these terms and their definitions can be found in Appendix C, Glossary of Environmetal Terms.

Please note that throughout this text are short references for sources that are provided in full in Appendix A, References and Bibliography.

CHAPTER 1
Ecosystem Earth

FAO photo courtesy Peyton Johnson.

Terms to Look for and Learn

Abiotic	Decomposer	Food Web
Albedo	Ecology	Homeothermic
Biome	Ecosystem	Muted Competition
Biosphere	Energy Transfer	Niche
Biotic	Environmental Science	Poikilothermic
Competition	Environmentalism	Producers
Competitive Exclusion	Full Competition	Solar Constant
Consumers	Food Chain	

Learning Objectives

After reading this chapter and participating in the activities, you should be able to:

- Explain the difference between environmentalism and environmental science.

- List and define some of the important environmental sciences.

- Discuss some of the major principles of ecology as an environmental science.

- Define and discuss the terms listed above.

Overview

The title of this book, *Environmental Science*, was chosen very carefully. You will notice that it is singular (science) rather than plural (sciences). It is important to understand that there is no such thing as a single, all-inclusive environmental science. Rather, there are many sciences that deal with various aspects of the environment. All of them have different perspectives and different specific interests. All of them reach different kinds of conclusions.

In the broadest sense, any science that deals with the environment could be considered an environmental science. Yet, in another sense, each of those individual branches of science deals with the same fundamental aspect of human survival: the environment. It is the purpose of this book to help the student explore the various sciences that deal with the environment and to make sense of them as they interact together—as environmental science.

A great many sciences have aspects that deal with the environment, and to that extent they could all be called environmental sciences. Some of the more important environmental sciences include (but certainly are not limited to) those listed in Figure 1-1.

Figure 1-1. Some Major Branches of Science that are Directly Related to the Environment.

Agronomy	Genetics
Botany	Marine Biology
Climatology	Meteorology
Ecology	Microbiology
Economics	Soil Science
Forestry	Wildlife and Fisheries
Geology	Zoology

This book will provide a glimpse of many of the sciences dealing with the environment. This book is for the student who is interested in an introduction to environmental science. It is by no means an attempt to present the detailed scientific content of so many varied fields of study.

Perhaps the one field of study that comes closest to being a unifying environmental science is ecology. For that reason, **ecology** will be the first environmental science considered. The remainder of this chapter and much of Chapter 2 will be devoted to an overview of the principles of the science of ecology.

Environmentalism and Science

Before environmental science is discussed further, there is a critical distinction to be made—that is the difference between environmentalism and environmental science. One need only read the newspaper or watch television to hear the terms "environmentalism" or "environmentalists."

Environmental Science

Environmental science refers to all of those branches of science that study the theory, mechanics, and interrelations of any aspects of the environment—the factors that surround and influence all organisms. This includes sciences dealing with the atmosphere, the Earth, humans, plants, and animals that inhabit the planet. It also includes natural resources management, air, soil, geologic formations, the management of soil and water to produce food and fiber, water, and many others.

Some scientists work to learn new things—to develop new knowledge merely for the sake of having that new knowledge. This is called basic research. Examples of basic research include studies to learn how plants use nutrients from the soil to function and grow, how animal genes work, and how acid rain is formed.

Other scientists seek solutions to real, specific, current problems. This is called applied research. Examples of applied research include studies to learn how to modify the soil to produce better crops, how to modify specific animal genes to produce longer wool fibers, and how to control acid-producing emissions. Whenever that research involves some aspect of the environment, whether it is basic or applied, then the research could be considered environmental science.

Environmentalism

Environmentalism is a political movement and not a science. Environmental activists are involved in political activities and not in science. It is true that many scientists who are involved in the fields of study listed earlier also consider themselves to be environmentalists. But being an ecologist or a forester or a soil scientist does not make one an environmentalist. By the same token, calling oneself an environmentalist does not mean that one is particularly knowledgeable about the environment. It merely means that one has a political agenda that deals with some environmental issue.

There are many special-interest groups that label themselves as "environmentalist." Some of the better known in this country are the Sierra Club, the Audubon Society, Greenpeace, People for the Ethical Treatment of Animals, the Environmental Defense Fund, and the National Wildlife Federation to name only a few.

Many other groups, such as Ducks Unlimited, American Forestry Association, and American Farm Bureau, are vitally interested in maintaining the quality of the environment, but do not refer to themselves as "environmentalists" because the term has come to be associated (probably unjustly) with groups considered extremist.

The Difference

In short, environmentalism is based on philosophy, emotion, and social considerations. It is driven by human values and value judgments. Environmental science is based on the realities of nature. It is driven by observations, careful calculations, and unemotional judgements. This book is about environmental science—not environmentalism. It is an attempt to separate reality from emotion regarding the environment in which we all must live.

Some Basics of Ecology

Ecology as a science is still fairly young, but the perspectives of the environment it offers us are very useful in understanding the environment. It is also quite useful in putting together a realistic philosophy regarding the environment.

Of equal importance is understanding how to apply ecology in assuring the livability of the planet into the future. In this chapter, Earth's environment will be explained from the perspective of the ecologist. Ecologists view the planet as a closed ecosystem functioning on the basis of energy transfer. Concepts to be built in this chapter include energy transfer, food webs, niches, competition, biosphere, and biome.

History of Ecology

The history of ecology (Greek *oikos* [house] and *logy* [science]) can only be traced back about a century. In fact, the first textbook on ecology was written in 1895 by Warming (1841–1924), a Danish botanist. It was first translated into English in 1909. The English title of that first "ecology" textbook was *Oecology of plants: An introduction to the study of plant communities*.

According to Botkin, ecology is the study of the relationships between living things and their environment. Colinvaux defined ecology as the study of the development, abundance, and distribution of species. From Botkin's perspective, the environment and organisms are considered fairly equally in terms of how they interact. From Colinvaux's perspective, organisms are the central focus. The environment is considered only as it relates to the development and survival of varying life-forms.

In reality, either definition is adequate. Using either definition, ecology can be used to provide the unifying framework for the environmental sciences. As an example of how ecology can serve as a unifying mechanism for the sciences dealing with the environment, we could consider the science of meteorology.

Meteorology looks at weather and climate. It looks at the formation and movement of atmospheric structures. One important outcome of meteorology is the ability to explain and predict weather. The only reason for studying meteorology is because of the relationships among atmospheric conditions and living organisms—a basic concern of ecology. By the same reasoning, **crop** and **soil sciences** (often called agronomy) seek to understand the processes and mechanics of plant production. The reason agronomists study plants is to make possible a reliable food and fiber supply for humanity—also a basic concern of ecology as you will soon see.

This does not mean that ecology is more important than the other sciences, or more scientific or more basic. It only means that ecology provides a broad framework with which to consider the other environmental sciences. The remainder of this chapter will discuss some of the major principles of ecology in order to provide the framework for the following chapters.

Ecosystem

One central concept in ecology is that of the **ecosystem**. An ecosystem is made up of a set of organisms, organic residues, physical and chemical components, and conditions. When all of those parts interact with each other within a single, definable space which is in some meaningful way separated from other spaces, that is an ecosystem.

In its broadest sense, the entire planet Earth could be considered an ecosystem. We will refer to that as the planetary ecosystem. All of the organisms on Earth occupy a single, definable space—the planet. It can be fairly clearly differentiated from other spaces by the limits of the atmosphere. The organisms all interact with one another in some way, given a long enough time period, even those on different continents. Thus, in that sense, Earth and all of its living and nonliving components can be thought of as an ecosystem. That is where the name of this chapter comes from—Ecosystem Earth.

In a more realistic way, smaller sets of organisms occupy smaller spaces and represent smaller ecosystems. Every ecosystem consists of **biotic** (living) and **abiotic** (nonliving) subsystems. An example of a biotic subsystem is the set of interrelated plants and animals that form the food web (food webs will be discussed later) that supports the bald eagles in a forest in Alaska. Components of that subsystem would include microorganisms, insects, spiders, shrews, moles, mice, rabbits, grasses, shrubs, forbs, trees, and eagles, among others. Sometimes we refer to all of the living organisms on the planet as the biosphere.

An example of an abiotic subsystem is the set of chemical and physical components that produce the salinity in an inland lake, such as the Great Salt Lake in Utah. Such a subsystem would include soil, parent material, rocks, minerals, rain water, snow, subsurface water, surface water, and the surrounding air—again, to name only part of the subsystem.

Within any ecosystem there will always be many subsystems operating. Just as importantly, every organism and physical part of a subsystem can belong to many different subsystems. For instance, a fish might belong in a biotic subsystem that includes sharks as the highest predator. At the same time, the fish could belong to another biotic subsystem that has no relationship to sharks.

Energy Transfer

There is one fundamental characteristic of any ecosystem, however. Energy transfer plays the central role in defining the bounds, character, and extent of the ecosystem. In fact, in its most basic sense, an ecosystem is an energy transfer system.

The **energy transfer** generally starts with the capture of heat and light energy from the sun by green plants. Solar energy is transformed into chemical energy in the form of food. As the food is used by plants and animals, that chemical energy is released in the form of biological activity or movement. For the energy transfer system to work, any complete ecosystem must have three basic kinds of living creatures present: producers, consumers, and decomposers.

Producers are organisms that convert free energy into stored energy by means of chemical processes. That statement can be interpreted to mean green plants that make sugar from CO_2 and H_2O do so by means of photosynthesis.

Consumers are organisms that use the stored chemical energy from the producers and transform it into some other form without producing any new stored food energy on their own. In other words, consumers feed on other organisms. Animals and people are consumers because they feed on plants and other animals.

Decomposers are organisms that take food energy from both of the other groups. The action of decomposers results in decay, rotting, and decomposition of the tissue they feed on. Typically, we think of fungi and bacteria as decomposers.

The three groups could be better understood by using an example you are probably familiar with. A potato plant is a green plant that **produces** a tuber, we call it a potato. The potato is food that you can **consume**. If you leave the potato too long in your kitchen, it will become mushy and start to smell bad. That happens when a fungus grows in the potato and causes it to **decompose**.

Potato Plant → 2 Potatoes (Producer)	Human Eats 1 Potato (Consumer)	Fungus Rots 1 Potato (Decomposer)

Energy Transformation

Energy exists in many forms, as we will see later. It can be transformed from one form to another. The First Law of Thermodynamics (also known as the Law of the Conservation of Energy) says that when energy is transformed from one form to another, no energy is lost.

That fundamental physical law involves many concepts well beyond the scope of this book, but, the essence of it is simple if described with an example. Suppose a certain amount of solar energy "falls" on a green plant. Part of the energy is stored in the conversion of liquid water in the leaves to water vapor in the air (transpiration). Part of it is radiated from the leaf in the form of infrared energy that tends to warm the surrounding air. Part of it is converted to food energy through photosynthesis. In the end, no energy is lost from the universe; it is simply converted into different forms as shown in Figure 1-2.

The Second Law of Thermodynamics says that every time a given amount of free energy, such as sunlight, undergoes a transformation, the amount of *free* energy is reduced. Part of the energy is lost to whatever system is doing the transforming. Part of it is lost as heat. In essence, that means that the usable energy within an ecosystem is reduced with every transformation.

Again, this complex concept can be explained with a simple example. The sun may shine for five months on an area one square foot in size to produce a plant that weighs one pound. It may take 100 one-pound plants to produce an ounce of seed. A mouse may have to eat ten ounces of seeds to produce an ounce of mouse flesh. A cat may have to eat twenty ounces of mice to produce an ounce of cat flesh. The area on which the sun must shine for 5 months to produce one ounce of cat flesh would be 20,000 square feet. Each time the energy was transformed, much of it was lost from the food web.

While this example may be a bit far-fetched (and the numbers are strictly fictitious), the point should be clear. Every time an energy transformation takes place, the energy available for biological production is reduced.

Figure 1-2. Energy flow begins with solar energy from the sun. As it reaches a producer (green plant), part of the solar energy is converted to chemical (food) energy, and part is radiated back into the atmosphere. In the end, the total amount tof energy remains the same.

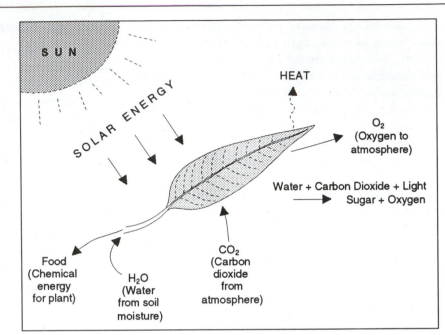

The unit of measure for solar energy is the Langley and the solar constant has been estimated as being equal to approximately 2 Langley units.

> *Note:* 1 Langley (ly) = 1 gram calorie/square centimeter

$$
\begin{array}{rl}
\times & 1 \text{ (ounce of cat flesh)} \\
\times & 20 \text{ (ounces of mouse flesh)} \\
\times & 10 \text{ (ounces of seeds)} \\
\times & 100 \text{ (one-pound plants)} \\
\times & 1 \text{ (square foot of space)} \\
\hline
= & 20,000
\end{array}
$$

The basic source of energy on Earth is the sun. There are other sources, such as geothermal, but the relative amount of usable energy from those sources is negligible. The amount of solar energy that reaches the upper layers of the atmosphere is very stable in terms of a human lifetime. Its intensity has been estimated for many years.

Physicists represent their estimate by use of the so-called **solar constant.** They define the solar constant as a measure of the solar energy that reaches the Earth per square centimeter (about 0.16 square inch) per minute.

When the vast area of the Earth's surface* is considered, and the fact that the flow of solar energy is continuous, the total energy reaching the Earth from the sun is staggering. Indeed, if all of the energy reached the surface of the planet, life as we know it would be quite impossible—it would be too hot. Only about two percent of that solar energy ever reaches the biotic subsystems of the ecosystem. The remain-

* The Earth's surface area of land plus water is approximately 196,949,971 square miles, of which 58,552,330 square miles (30 percent) is land.

der is reduced by reflection, scattering, and absorption (Colinvaux, 1986).

Reflection

Reflection refers to the effect of solar energy being returned to space by clouds, atmospheric dust, vegetation, and the surface of the Earth. Of those three reflectors, by far the most important is clouds. Clouds can reflect up to 70 percent of the light reaching them at a given time.

Ice cover can be found in the polar regions and snow-covered regions of the world. Ice cover is also very effective in reflecting solar energy. Freshly fallen snow reflects as much as 95 percent of the energy reaching it. That reflection is the reason for wearing sunglasses in snow fields when the sun is shining.

Scattering and Absorption

Scattering refers to the effect of light being refracted in the atmosphere. By scattering, the energy becomes unavailable to producers at the Earth's surface, yet is not reflected back into space. Absorption refers to the transformation of solar energy as heat within the atmosphere and on the Earth's surface. An example of absorption is a hot pavement on a sunny summer day. Some of the energy that is absorbed is later usable by organisms such as snakes to keep warm.

Only about two percent of the sun's energy reaches the biological surfaces of the Earth. Only part of that is converted into chemical (food) energy because of the inefficiencies described in the Second Law of Thermodynamics.

Albedo

Albedo refers to the ability of a planet or other celestial feature to reflect light. On Earth, the albedo of different surfaces varies greatly. Albedo is a function of the surface's color and texture, the angle of the surface to the sun, and the chemical makeup of the surface. Albedo is commonly expressed as a percentage—the percent of the sun's radiant energy reaching the sur-

face that is reflected. New-fallen snow has a very high albedo, largely because it is white. Tops of trees tend to have a very low albedo (see Table 1-1). The average albedo of clouds is about fifty-five percent. The overall albedo of the entire planet, given twentieth century conditions appears to be about 35%–37% (Lamb, 1972).

Food Chains/Food Webs

In recent years, marine biologists have discovered organisms in deep oceans that convert geothermal energy to chemical energy in the form of food. There are many isolated examples such as this, but in general, the only important producers in the biosphere are green plants using the process of photosynthesis. The characteristic green color comes from the presence of chlorophyll, which is the chemical that makes the transformations involved in photosynthesis possible.

Table 1-1. Albedos of Surfaces

Surface	% Reflection
New Snow	85
Old Snow	70
Salt deposits	50
White Chalk or Lime	45
Quartz Sand	34-35
Parched Grasslands	16-30
Green Grassland	8-27
Green Deciduous Forest	16-27
Grain Crops	10-25
Dry, Plowed Fields	12-20
Granite	12-18
Wet Fields, Not Plowed	5-14
High-level (Cirrus) Clouds	21
Cumulus Clouds	70
Water (Depending on Angle)	2-78

SOURCE: Lamb, H.H. (1972). *Climate: present, past and future,* Vol 1. New York: Barnes & Noble Books.

Figure 1-3. Photosynthesis is the single most critical chemical reaction in the process of life.

$$12\ H_2O + 6\ CO_2 + energy \xrightarrow{\text{chlorophyll}} C_6H_{12}O_6 + 6\ H_2O + 6\ O_2$$

In simplest terms, photosynthesis involves the combination of water and carbon dioxide into simple sugars (carbohydrates) and oxygen (gas). Literally translated, *photo* means "light" and *synthesis* means "put together." The significance of making sugars from CO_2 and H_2O is that the resulting carbohydrate compound is not as stable. It has a higher level of chemical energy than either carbon dioxide or water.

Chemical energy is transformed from solar energy in the process of photosynthesis. In general, practically all other biological processes in the biosphere are powered by that stored chemical (food) energy. The stored chemical energy is used by means of a series of chemical transformations beginning with the primary food (simple sugars). A simplified chemical equation for photosynthesis is shown in Figure 1-3.

Thus, the first basic energy transformation in most ecosystems is the production of hexose sugars $(C_6H_{12}O_6)$ during photosynthesis by green plants.

The Second Law of Thermodynamics implies that the amount of available chemical energy within the ecosystem is at its highest during photosynthesis. Each time there is another transformation from that point on, there is less food energy available. That is true because some of the energy is used to make the transformation possible and some is lost as heat. The use of food energy in biological processes results in the concepts of the food chain and the food web.

The producer (the green plant) produces $C_6H_{12}O_6$. The plant itself must convert that simple compound into other products in order to survive, grow, and reproduce. The chemicals in all of the plant cells are formed by chemical reactions, all powered by the release of chemical energy from those sugar molecules. One example of this process is the production of seeds during reproduction.

A **food chain** is a hierarchical order of organisms beginning with some producer, each of which feeds on the organism below it. A very simple example of a food chain might be visualized as shown in Figure 1-4.

In this oversimplified example, the oak tree leaves manufacture sugar during photosynthesis. The tree uses the energy from the sugars to produce the complex carbohydrates, fats, and proteins in its leaves and acorns. The acorns fall to the ground where they are eaten by the field mouse. The mouse releases part

Figure 1-4. A very simplified food chain is shown here as the oak tree converts the sun's energy to food energy and stores part of it in the acorn. The mouse consumes the acorn, then the hawk comsumes the mouse.

of the stored chemical energy in the acorn by means of the digestion process. It uses that energy to form the proteins and other compounds that make up so much of its own flesh. The mouse in turn is eaten by the hawk. The hawk then uses the stored chemical energy in the mouse's flesh to produce the work of its own metabolic processes. When the hawk dies its body chemicals are released for reuse in the ecosystem by the bacteria and fungi that grow in its tissues during decomposition.

The example just described was very much simplified. For instance, trees use the chemical energy to produce many more things than just acorns. Mice do not rely solely on acorns for food. Moreover, the food they eat is not all used to produce muscle tissue. Hawks certainly prey on more than just mice. And finally, when the hawk dies, fungi are not the only decomposers to attack its body. To recognize the actual complexity of such relationships, the concept of the **food web** was developed. A **food web** is a much broader concept than a food chain. The term food web refers to the complex series of overlapping food chains in an ecosystem. A typical food web might look something like the one depicted in Figure 1-5.

Competition

A critical part of the theory of population ecology is **competition**. In any natural setting there will be multiple species. When more than one species in an ecosystem uses a given resource, there is a potential for competition. Whenever that resource is in limited supply or one of the competitors harms the other or prevents it from obtaining the resource, competition occurs. Such competition can be either full or muted.

If more than one species must have a resource to survive, the competition will be *full.* What that means is that each species must compete without hesitation for the resource. If both species can use the resource, but can survive without it, can easily find an acceptable substitute, or the resource is not in limited supply, the competition may be *muted.*

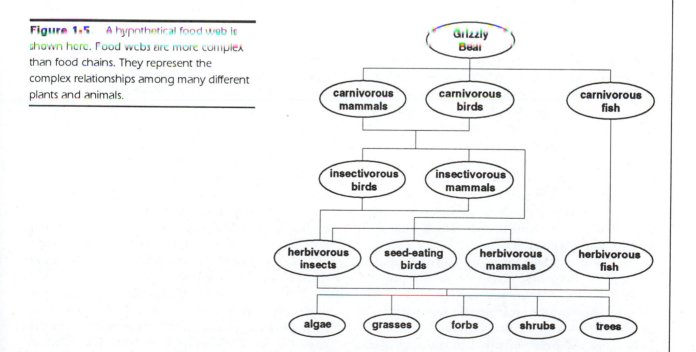

Figure 1-5.   A hypothetical food web is shown here. Food webs are more complex than food chains. They represent the complex relationships among many different plants and animals.

It is a general truism that no two things in nature are ever exactly alike. It follows that whenever there are two or more competing species for a scarce resource, one species will be more proficient (fit) at obtaining that resource or at preventing the other from obtaining it. Thus, full competition means that one species will eventually take all of the limited resource and muted competition means that multiple species can share a resource. In this sense, the old Darwinian concept of the "Survival of the Fittest" could be viewed in terms of the ability of one species to take a scarce food energy source to the exclusion of other species.

Full competition produces a situation in which one species can be expected to expand its use of the limited resource until it excludes the others from obtaining the resource. The result is the ecological principle of **competitive exclusion**. Competitive exclusion means that when full competition occurs between two species in an ecosystem, only one of the species will survive in that ecosystem. The other species will move, adapt, or become extinct. It will be excluded from the ecosystem by its competitor.

Niches

For any species to survive in nature, there must be a source of food energy for which that species has no full competition. Every species, therefore, must exist within a part of the ecosystem that is unique in one or more food-energy source characteristics. A unique subecosystem is called a **niche**.

For every niche there will always be a single species that is best suited to win in any full competition. More importantly, for every species there must exist a niche in which that species can dominate in any full competition. If that were not the case, then the species would be forced into extinction.

When we combine the concept of niche with the concept of competitive exclusion, a new principle emerges—that of "One Species, One Niche." This theory holds that for every species, there is a single niche within which it can survive and prosper. So long as that niche remains intact and no other species

emerges to engage in full competition for the limited food energy resources within that niche, the species can continue to survive. When that niche changes, or disappears, or another competitor emerges, the species must either adapt to a new niche or become extinct.

Biosphere

Within the Earth, there are living organisms of many kinds—some of which would be difficult to recognize as living. Indeed, one of the most fundamental problems in science is to define exactly what it means to be alive. At the very least, a living organism is capable of reproducing itself and expends energy in the process. Living organisms exist from far into the atmosphere to the depths of the oceans. In the most general sense, all living things on Earth can be thought of as a single biotic subsystem within the overall ecosystem.

Biosphere can be defined in two different ways. From the perspective of the physical sciences, the biosphere is that part of the Earth that contains living organisms, primarily the atmosphere, the surface crust (which includes soil), and water.

From the perspective of the ecologist, the biosphere is that part of the planetary ecosystem that is composed of living organisms. If we define all organisms as belonging to a single subsystem, that subsystem would be called the **biosphere**.

The biosphere can be considered to consist of any number of biotic subsystems, as was discussed earlier. The predominant organisms in any subsystem will always be plants. This is true for two reasons. First, producers (plants) must precede consumers in the flow of food energy because producers convert solar energy to food energy. Second, each transformation results in a decrease of available food energy. Green plants are the producers and provide the basis for almost all life. Plants dominate the ecological landscape.

Thus, major biotic subsystems have come to be described in terms of the climax plant life-form. Ecological succession and climax species will be discussed in detail in Chapter 2. For now, it is sufficient to define climax species as those plant species that would eventually tend to dominate a biotic subsystem if no natural or human intervention occurs. In other words, if equilibrium were reached within the subsystem, the dominant plants would be referred to as climax vegetation. As we will see in Chapter 2, this is a very arbitrary definition, but for now it will have to do.

Biome

It should be clear by now that a biotic subsystem can be as small as a drop of water or as large as the entire biosphere. For botanists and ecologists, a more encompassing classification of biotic subsystems was proposed in 1939 by Clements and Shelford. They introduced the term **biome**. Biome has since come to be understood as a biotic subsystem with a more or less uniform climax vegetation pattern, on a grand scale. Thus, a biome is a major biotic subsystem with its dominant vegetation being of limited species with similar environmental requirements.

With that definition of biome, it is possible to group many basically similar regions into large classifications. Ecologists have described eight great biomes: Tundra, Coniferous Forest, Temperate Forest, Tropical Rain Forest, Tropical Savanna, Temperate Grassland, Chaparral, and Desert.

As you can see, none of the great biomes exist in just one place. There are examples of tundra, for example in the northern regions of Asia, Europe, North America, Greenland, and so forth. There are examples of deserts, temperate forests, and temperate grasslands on most of the continents. The following discussions regarding the biomes are based on Colinvaux's *Ecology*.

Tundra

On large portions of the Earth's surface, the growing season is so short or the temperatures are so cold that large vegetation cannot survive. Yet, where such areas are not covered constantly by ice, there is a lush growth of small plants and many annuals produced from seeds each growing season. Interestingly, the real limiting factor on the survival of trees in this region is not the cold, it is the lack of available water. Trees can survive cold weather, but they require a supply of liquid soil water that is available year-round. When the ground is frozen most of the time, that water is simply not available to support trees.

Tundras are often rich in animal life, but those animals are adapted to the cold in some unusual ways. Many have heavy coats of fur or feathers. Many hibernate during the coldest seasons. Most of the tundras are located in far northern latitudes, mainly north of the Arctic Circle, but some examples of tundra exist in the far southern latitudes also. In addition, on high mountains, there are often so-called "alpine-tundra" areas above the tree line and below the permanent ice cover.

Coniferous Forest

Because coniferous trees have narrow leaves, they transpire less water than do broad-leafed trees. As a result of that, coniferous trees can survive in areas with less available water than broad-leafed trees. In very cold climates, much of the soil water is not available to plants because it is frozen for so much of the year. As a result, a broad band of coniferous forests exists along the more temperate (warmer) sides of the tundra regions. Some deciduous broad-leafed trees also live in this region, such as the maples, but the predominant species are coniferous. As with tundras, most coniferous forests are in the Northern Hemisphere. The Douglas fir of North America and the aspen forests of the northern continental United States, Canada, Alaska, and Europe are notable examples of this.

Temperate Forest

Further toward the equator, where the growing seasons are longer and the temperatures are warmer, broad-leafed deciduous trees make up the climax vegetation. Because of the mild summer and the normally adequate rainfall of these regions, deciduous trees such as oaks, hickory, and sugar maple grow rapidly. The winters are milder than in the coniferous biome. The lack of available soil water during winter is not a limiting factor on their survival. Temperate forests cover only a small portion of the Earth's surface compared to coniferous forests.

Tropical Rain Forest

Closer still to the equator the summers are longer and hotter and the winters milder and shorter. Within that band of latitudes, areas that receive higher-than-average rainfall on a year-round basis produce tropical rain forests. The trees are usually broad-leafed and remain green year-round. In contrast to the other forest biomes, tropical rain forests rely more heavily on pollination by animals rather than wind, so flowering plants are more common. Because the soil water is available year-round, growth is very lush. There is a great diversity of species, both of plants and animals.

Tropical Savanna

In the latitudes close to the equator winters are warm. In those regions where rainfall is extremely limited during one part of the year, trees cannot survive as readily. Trees that do survive are scattered and stunted in size. An extended annual dry season with frequent droughts is characteristic of this region. The dominant vegetation is grasses.

Perennial grasses die to the ground during the droughts, yet they can retain enough moisture in the root system to survive the dry season. Annual grasses die completely with the dry season but leave dormant seeds on the soil surface. As soon as the rains come again, the perennial grasses sprout and the seeds germinate. This is followed by a rapid period of growth and new seed formation in preparation for the next dry season. This is particularly interesting because the cycle is basically the same as that of the tundra, and for the same reason—lack of available soil water for a part of the year.

Temperate Grassland

In temperate regions with low annual rainfall amounts, the limiting factor on tree growth is again available soil water. In those regions, as in the savanna, the climax vegetation is grasses. Historically, these areas, also called ranges, prairies, and steppes, have been among the most important for human development.

Forests are particularly productive of small herbivorous (plant-eating) and carnivorous (meat-eating) animals. But the grasslands are very productive of large herbivorous animals. The huge herds of bison on the American plains were one example. This biome is dominated by grasses of heights ranging from a few inches in dryer regions to six to eight feet in areas with plentiful rainfall.

Often our stereotype of early humans is that of cave dwellers in forest regions. However, the reality is far different. Taken together, the savannas and the temperate grasslands have probably contributed more than all of the other biomes combined to the survival and development of humans as a species.

Chaparral

Chaparrals are characterized by hot, dry summers with cool, moist winters. They typically lie near oceans, which influence the level of humidity. There is enough rainfall to support a taller shrub and brush growth that overtops and shades out grasses. But there is not a reliable enough soil water supply to support full-sized trees. Chaparrals experience very dry periods with high incidence of wildfires. The hills in Southern California and inland along the Mediterranean are classic examples of chaparral.

Deserts

In its most extreme form, the desert is an area with essentially no available soil water. Such a place would exist in a totally barren state with only sand and rocks. Such a place is almost totally devoid of life and fortunately quite rare. The desert biome as described by ecologists is any region lacking adequate rainfall to support savanna or temperate grassland yet having enough soil moisture to support specially adapted vegetation and animals to survive.

Typical desert climax vegetation includes plants adapted to low levels of transpiration. Examples are succulents with defenses against herbivores like spines and thorns, such as cactus plants. Animals in deserts are generally adapted to low energy expenditures so that heavy consumption of food is unnecessary. That means that **homeothermic** (warm-blooded) animals are often displaced by **poikilothermic** (cold-blooded) animals as the predominant species.

Summary

There are many sciences that deal with the environment. They include **meteorology**, **geology**, **agronomy**, **botany**, **marine science**, **zoology**, and **forestry** to name only a few. All of these can be accurately referred to as environmental sciences. It is important to differentiate **Environmentalism** and **Environmentalists** from **Environmental Science** and **Environmental Scientists**.

■ Environmentalism is a movement based on emotion and politics. Environmental sciences are those sciences that study various aspects of the environment.

■ Environmentalists operate from a social agenda and environmental scientists operate from a scientific agenda.

■ Environmental scientists may, or may not be active in so-called environmental causes. As we will see in a later chapter, many environmental causes are not based on scientific realities.

One of the basic environmental sciences is that of ecology. Ecology is a fairly young science. It deals with the study of the relationships among living things and their environment. A basic assumption of ecology is that all organisms exist within a planetary ecosystem. The planetary ecosystem consists of biotic and abiotic components which interact in predictable ways. The planetary ecosystem consists of many smaller ecosystems, ranging in size from nearly continental to smaller than a drop of water. Ecosystems can be defined as almost any size and in terms of almost any organization of biotic and abiotic components.

A fundamental characteristic that can be used to define a particular ecosystem is the energy transfer among life-forms. Energy for organisms originates primarily as sunlight and is converted to food energy by the process of photosynthesis in green plants. Plants that convert external energy into food energy stored within chemical compounds are called producers. Organisms that do not convert external energy into stored food energy must rely on producers for their food energy. Such plants and animals are called consumers. All organisms are members of at least one food chain/food web.

A principle of ecology that explains the diversity of life on Earth is competition. Competition occurs whenever (a) individual organisms or species must rely on a scarce source of food energy or (b) individuals or species obtain food energy at the harmful expense of others. Within any ecosystem, each species must have its own niche in order to survive in the long run. That is based on the assumption that in any situation there must always be one species that reproduces faster or is more efficient at obtaining food energy from a given source than its competitors. In this sense, niche is defined as a unique combination of food energy sources.

The biosphere consists of all of the living organisms within the planetary ecosystem. Biomes are major biotic systems within the biosphere. Similar biotic subsystems can be grouped into the great biomes or biome types as follows: Tundra, Coniferous Forest, Temperate Forest, Tropical Rain Forest, Tropical Savanna, Temperate Grassland, Chaparral, and Desert.

DISCUSSION QUESTIONS

1. Why is there no single environmental science? What are environmental sciences?

2. What is an ecosystem? How can a single organism be in many ecosystems?

3. What is a food chain? A food web? Why is every organism part of a food web?

4. What does it mean to say that all of life is based on energy transfer?

5. What is a biome? How are the concepts of biome and energy transfer so closely related?

6. What are the great biomes? What are the major characteristics of each?

ADDITIONAL ACTIVITIES

1. Much of this chapter was based on the book *Ecology*, copyright © 1986, by Dr. Paul Colinvaux of The Ohio State University. For those students with a desire to learn more about the intriguing science of ecology, reading all, or even portions of that book would be an excellent experience.

2. Determine within which of the great biomes you live. Find some examples of vegetation within or around your community that ARE NOT characteristic of that biome. Find some examples that ARE characteristic of that biome.

3. Collect newspaper and magazine articles dealing with environmental issues for class discussion. Try to determine which of the issues are based on scientific fact and which are based on other things such as emotion or political agendas.

CHAPTER 2

The "Balance of Nature"—A Great Myth

Photo courtesy L.G. Kesterloo, Virginia Commission on Game and Inland Fisheries.

Terms to Look for and Learn

Carrying Capacity
Climax Vegetation
Divine Order
Equilibrium
Exponential Growth
Gause Curves
Logistic Curve

Lotka-Volterra Curves
Maximum Growth Rate
Maximum Sustainable Harvest
Natural Succession
Primary Succession
Secondary Succession
Stability

Learning Objectives

After reading this chapter and participating in the activities, you should be able to:

■ Explain the concept of nature as "divinely ordered."

■ Discuss the dynamic processes involved in ecological succession.

■ Define and give examples of climax vegetation.

■ Explain the classical theories of ecologists that have been used to describe the process of population equilibrium.

■ Discuss why you believe that the "Balance of Nature" is a myth or that it is a reality.

Overview

The concept of a balance in nature has been long accepted by most people, as it was by early scientists. But it is discounted as a harmful myth by most scientists today. One cannot seriously study environmental science without first understanding that change rather than balance is normal in the environment.

Both living and nonliving subsystems are dynamic rather than stable. The concept of a balanced nature is simply false. In this chapter we will first consider the philosophical and theoretical bases on which the concept of **equilibrium** (i.e., balance) in nature was built. We will then examine the dynamics of change in the ecosystem. Finally, we will learn why there can never be a balance in nature so long as there are living things.

Historical Perspectives: Why Mankind Needed to Believe in a Balance of Nature

From the very beginnings of civilization, people have sought to understand nature and humanity's place in it. The earliest people developed superstitions and mythologies to explain nature. Early humans held a mystical view of nature. We still refer to "Mother Nature." Early humans sought guidance in their decisions from the spirits of animals and the various elements of the environment.

To early humans, security was the most important and immediate primary need. Security was even more important than food and water. Given the choice between finding food and running from a hunting lion, a Neanderthal man would always run. He could search for food and water later, assuming he survived the lion.

The driving needs for safety, shelter, food, **sex**, and water came to be translated as a need for **stability** and predictability. In order to feel safe in a world that was not always hospitable, early people needed to have a stable and predictable world—an ordered system.

Divine Order

With no scientific knowledge, early thinkers gave credit for life and the order of nature to the gods, called **divine order**. For instance, they believed that if there was a sun in the daytime sky, then there must be some supernatural explanation. There must be a being who lights a magnificent torch and carries it across the sky—after all, torches produce light.

The ancient philosophers sought truth in their gods and in the order of nature. They believed that without humans, the world would have been a perfect place of tranquility and order. This view of nature saw stability and order as natural and disorder as human-caused. The purpose of organized philosophy, which preceded true science by several thousand years in the evolution of human thought, was to discover truth and good. The early truth was that nature and the environment were inherently stable and good. They believed that stability was based on divine decisions and supernatural powers.

By that reasoning there must be a perfect order that is divinely determined. Thus, any attempt to explain nature or to examine the order of nature outside of accepted religious teachings was heresy. To suggest that Earth was not the center of the universe, that the sun did not indeed move around the Earth, was an offense against the gods that might even result in the heretic's death. By the same reasoning, the concept of evolution was blasphemous and an attack on religious teachings.

Yet, they saw clearly all about them pain and disorder. They reasoned that if divine order were missing from the world, it must be missing for some reason. There have been two different explanations for pain and disorder:

a. actions taken by mankind, or

b. the failure of mankind to act.

Actions Taken by Mankind

As early as Pliny the Elder (AD 23 or 24–79), philosophers had accused mankind of damaging the natural order. In the last decade of the twentieth century, every day someone else charges people with interfering with the natural world. The Christian *Bible* holds that humanity was cast out of the perfectly ordered Garden of Eden because of sin. By this reasoning, we humans were at fault for the disorder in the world because we are "spoilers" (Botkin, 1990).

Failure of Mankind to Act

Cicero (106–43 BC), a Roman statesman, concluded that nature was so wonderful that it must have been created for the benefit of mankind. Trees cannot enjoy a lovely garden, they have no conscious understanding. Animals cannot contemplate and understand nature. Why else would nature have been created if not for the benefit of mankind? Philosophers for thousands of years have proposed that mankind was godlike or created in the image of God.

Or conversely, they created their gods with human characteristics and thought processes. They reasoned that it was mankind's role to dominate, shape, and use nature. They viewed nature as untamed, hostile, and forbidding. It was mankind's role to create the divine order from chaos. It was mankind who cut back the thorns and wilderness and replaced them with flowers and gardens. It was mankind who removed the wolf and replaced it with sheep. It was mankind who drained the putrid swamp and replaced it with rich and productive farm land. Thus, if there were environmental problems, it must be as a result of mankind's failure to properly husband nature.

Descriptions of "Mother Nature" and the "Living Earth" have been around since ancient times. The early Egyptians saw the Nile as the great mother. Prehistoric peoples of Eastern Asia and Europe thought of nature as the "Great Mother" and felt a spiritual connection with nature and Earth.

Later Views of Nature

We have seen that the earliest philosophers viewed Earth as "divinely ordered." No question of nature as being dynamic rather than divinely ordered was religiously acceptable. It was not until the 1500s that scientists and philosophers could begin to question the description of a balanced, ordered nature.

Sir Isaac Newton's (1642–1727) ideas about the physical sciences, along with work by other early scientists, led to a mechanical view of nature. Nature could be described in a mechanical flowing sense. The view of Earth as a mechanical system (as described by many ecologists today) developed in the seventeenth century, expanded in the eighteenth century, "blossomed" in the nineteenth century, and still persists today (Botkin, 1990).

It was during this time that scientists began to seek explanations for nature—explanations that were observable and verifiable. If science could unravel the cause-effect relationships in nature, we would be able to reshape the world in ways to benefit mankind. The logic was that if nature's basic relationships (the natural order) could be discovered by science, then mankind could intervene to modify the natural order of things to our advantage.

Even today, people speak of the "delicate balance of nature" and of the "virgin forests" with reverence. This perspective assumes that without mankind's interference and destructiveness, nature would be serene, stable, and presumably pleasant. There would be an ecological community with plants and animals living in harmony and equilibrium. But, as we will see, nothing could be further from reality.

Ecological Succession

As we learned in Chapter 1, the dominant life-form in any ecosystem must be plants. That is true because it is plants that transform solar energy into food (chemical) energy. Under certain circumstances animals may temporarily dominate a given landscape. An example is when a swarm of locusts (grasshoppers) strip all the vegetation from an area. But, such a situation cannot hold for long. Eventually the swarm runs out of food and must move on or die.

As we also learned, in any given setting where there is biological competition, there will always be a dominant species. Thus, in a given location and under a given set of environmental conditions, the most important forms of plant life tend to set the biological character of the ecosystem. It is for that reason that the great biomes exist, with each being dominated by a set of plants and animals with similar biological requirements.

Botanists speak of "**natural** or **ecological succession.**" This is the tendency of an area to go through a predictable sequence of dominant plants. A cleared area will first be dominated by quick-growing vegetation. Later will come stronger and more slowly-growing vegetation. Finally, still other forms of vegetation will dominate. In theory this would eventually produce a stable environment in any locality in which there is no external interference—as by mankind. Thus, ecological succession involves the sequential dominance of a given ecosystem by a fairly predictable series of plant species.

But, the process of succession often changes the very make-up of the ecosystem involved. The result is that the plant life the ecosystem is capable of supporting varies greatly across time. Plants and animals living in and on a soil change its character. The characteristics of the ecosystem are shaped by the plant species as much as by its actual physical characteristics at a given point in time. In this sense, ecological succession can be thought of as ecosystem development. Colinvaux defined ecological succession as ". . . the gradual change that occurs in an ecosystem of a given area of the Earth's surface on which populations succeed each other."

Successive Species

As we have seen, the dominant life-form in any ecosystem over an extended period of time must be plants. Thus, ecological succession is generally described in terms of plant species. In any given ecosystem, at any point in time, there is probably a wide variety of plant species present. Also, as was explained earlier in this chapter, no ecosystem is stable over extended time periods. Thus, one species of plant after another is replaced in the ecosystem as conditions change and as the ecosystem matures.

A hypothetical example will help to make this process clearer. Assume that there is a small natural pond near where you live now. In that pond there may be many types of plant life: algae, grasses, water lilies, and many others. Around the pond there are probably shrubs and trees. The pond has fish, and perhaps frogs, insects, and many other animals. Left alone, the pond will someday become filled with sediment from water that drains into it from the surrounding countryside. As that happens, the water will become shallower and the types of vegetation will gradually change to plants that grow in bogs.

Eventually the pond will become completely dry and the plant life will change to grasses and forbs. These will come first because they grow quickly and their seeds are spread in a number of ways. In a short time, the grasses will be overshadowed by short, rapidly growing shrubs and bushes. Eventually, the low growing plants will be overshadowed by trees.

If you live in a temperate climate, the first trees to dominate the area will probably be tall coniferous trees that grow quickly. Deciduous trees frequently grow in height more slowly, and so the coniferous trees will probably dominate the new forest for a long time.

A common characteristic of deciduous trees is that many are shade-tolerant whereas many coniferous trees are not. That means that in the coniferous forest there will be many smaller broad-leafed trees growing in the partial sunlight reaching the forest floor, but there will be few coniferous saplings. As the mature coniferous trees grow old and die, the next generation of dominant trees will be the deciduous saplings that survived in the shade of the faster growing coniferous trees.

This particular example would be referred to as a **primary succession**. It involves the ecological development of an area that has been bare for some time or that for some reason has become newly available for vegetation. Primary succession occurs on such land as filled-in lakes, sand dunes, volcanic ash, mine spoils, and glacial deposits. If the area undergoing ecological succession had been simply disturbed by plowing, fire, flood, or other catastrophe, the process would be referred to as a **secondary succession.**

In an area where the vegetation is killed by fire, a typical succession might be quicker. It might involve grasses immediately, shrubs soon, pines or cedars later, maples later still, and oaks finally overshading all and preventing understory growth. To a degree, this succession can be explained by the nature of the seed dispersal mechanism, germination rates, and shade tolerance.

Climax Vegetation

The example of the pond is merely one hypothetical case of ecological succession. In that particular instance, the process could take many thousands of years or it could be completed in a few hundred years. How long it would take depends partly on how soon the pond is filled by sedimentation.

Regardless, the sequence of plant species succession, while not precise, is fairly predictable. The range of possibilities is determined by the climate, but also modified by seed sources and soil conditions. Within that range, the precise species of plants that will dominate at a given point in the succession is not very predictable, and chance plays a major part.

If nothing catastrophic were to happen during the process of succession, (an ice age or a lethal disease for example) eventually one species or a group of similar species would dominate the ecosystem and not be replaced by other species. When the succession has reached equilibrium, the dominant plants would be said to be the climax species or **climax vegetation.**

Thus, a realistic definition of a climax community is "a community similar to contemporary communities in comparable environments that are free from physical disturbance and in which important species can persist for many generations" (Colinvaux, 1986, 586). This gives rise to the concept of the biome, which was discussed in Chapter 1. As you will

recall, a biome is defined by the character of its climax vegetation.

If an ecosystem were to go through an undisturbed ecological succession, it would someday reach equilibrium. At that point nature would be "in balance." During the process, it would have been proceeding in an orderly way toward balance.

Unfortunately, long-term equilibrium never occurs in nature because ecosystems are never undisturbed for long. There is no stable environment. Fire is natural, old trees die, insects and diseases produce change, new diseases result from genetic mutations of old diseases, to name only a few things that act to change an ecosystem. Thus, even so-called climax forests are not uniform and are not stable. They consist of smaller patches of vegetation in varying stages of succession from new growth to mature, dominating trees.

The last great Ice Age produced an ice sheet as deep as four thousand feet. It covered as much as 15 million square miles of what are now North America, Europe, and Asia. In this country, moraines indicate that the ice moved as far south as central Ohio, Indiana, and Illinois. They moved westward to the Rocky Mountains and eastward into the edge of the ocean. Areas that are now spruce climax forests were then tundra or underwater or even covered by ice. Many millions of years ago, between the ice ages, the central part of the North American continent was under a shallow sea. The ecosystems have always been changing and will continue to change as long as life continues on the planet.

Constancy of Change

In the past few thousand years, an additional factor has begun to affect the stability of the environment— in an ever-increasing way. Human activity has come to be important in this regard. We clear land for farming, highways, and cities. We plant trees that would not grow naturally in an area. We divert rivers and drain wetlands. As human technology has advanced, our impact on ecological succession in the places we live has increased dramatically. But the point is that with or without human activity, the environment is constantly changing—it is never stable.

Nothing on Earth is more certain than change. The range of possible plants and animals in any ecosystem is determined almost totally by environmental conditions. The most basic condition is climate, primarily temperature and moisture. Within that environmental framework, the exact mix of living organisms at any given time is largely determined by chance.

Climatic Change

The amount of solar energy reaching the Earth's atmosphere has been fairly constant over the past several million years. But no one knows how much, or even if it has varied over geological time beyond 100 million years. It is certain, however, that the Earth's atmospheric and surface temperatures have varied greatly over time.

The Earth's orbit around the sun is not perfectly stable. Even if it were, the alignment of the planet's axis wobbles slightly. Thus, the climate within a given area may vary enormously across the centuries. As discussed later in Chapter 20, global temperature change has been observed throughout Earth's history, and is not uncommon.

Changing climates drastically affect the biotic components of ecosystems. A shift of only a few degrees can completely change the dominant forms of plant life in an area. During the great ice ages for instance, as the temperatures of the planet's surface cooled, the size and thickness of the ice packs increased. As a result of the transformation of surface water to ice, levels of the Earth's oceans fell by as much as 110 yards below current levels. The ice cover moved across much of Europe, Asia, and North America.

The result was a complete change in the locations of the great biomes in regions that are now North America, Europe, and Asia. What had been tundra was covered by glacial ice. What had been coniferous forest became tundra. What had been temperate regions covered by deciduous forest became coniferous forests, and so on. Recent discoveries indicate that there may have been many ice ages in the Earth's history. Each successive ice age meant massive changes in the biosphere.

But, the ice ages are only one example of this constancy of change. The eruption of Mount Pinatubo in the Philippines during 1991 forced so much volcanic dust into the atmosphere that the albedo of the entire Earth's atmosphere was changed and predicted to remain altered for several years. You may recall from Chapter 1 that the amount of solar energy available for plants to convert to food energy by photosynthesis is controlled in large measure by the reflectivity of the atmosphere. Dust in the atmosphere, like cloud cover, reduces that level of available energy significantly.

Every such environmental change means countless and unforeseeable shifts in the pattern of plants and animals that will dominate and survive within a given ecosystem. Please recall the concepts of competition and niche from Chapter 1 to understand why this is true.

The Butterfly Paradox

Even the climax vegetation in an ecological community is not truly constant. Though it may remain stable for generations, it will eventually change. There are many reasons for this, but pure chance is an important one. Take for instance the so-called "butterfly paradox." A physicist once proposed a set of circumstances under which even as minor an event as the flapping of a butterfly's wings in South America could lead to a tornado in Kansas. The concept is simple but the physics are difficult, so let's look at this concept from the standpoint of a biological example.

Let's assume that a honeybee in the desert of Australia once had a choice of landing on two different plants, one with a white flower and one with a yellow flower. By chance, it selected the yellow flower. That plant became pollinated because the honeybee had just left a male plant of the same or a closely related species. The result was a new hybrid that had new properties.

Perhaps one of the properties would be the ability to survive in a drier climate than either of its parents.

The resulting offspring could eventually populate an area that would otherwise have not had any substantial plant cover. Suppose that the new plant hybrid provided an ideal food for a particular insect, which in turn provided a major food source for a insectivorous bird, which in turn became prey for a carnivorous mammal. Suppose that in addition, the plant's root system were very effective in holding soil particles together and in retaining soil moisture. Eventually, the entire ecosystem in the region would be affected. It could be that someday the entire biosphere would be affected in some way by the 50:50 chance that the honeybee landed on and pollinated the yellow flower rather than the white flower.

This example may seem trivial, but consider that the entire theory of evolution is based on just such random events and the cumulative effects of their outcomes. A genetic mutation is no more trivial and no less the result of chance than the random cross-fertilization of different plant species or the chance cross-breeding of different animal species. Thus, change is a constant in biotic systems as well as abiotic systems.

Population Limits

For any organism in any given setting, there will always be a limiting factor on its population. The limiting factor(s) for a species may be available food energy, water, shelter, breeding range, territory, predators, diseases, or some other factor. Regardless, the limiting factor determines the upper limit of the population for that specific species in that specific ecosystem. If that limiting factor were removed as a limitation, the second most limiting factor will replace it.

For instance, the availability of prey for food energy may be the limiting factor on numbers of red winged hawks in a given community. Then, if a new rapidly reproducing rodent were introduced into the ecosystem, food energy might become more plentiful. Food might no longer be the limiting factor on the red winged hawk population. Food might then be replaced by the availability of nesting sites as the limiting factor.

One biologist, Daniel Botkin, who studied moose populations on an island concluded that sodium in the diet was the limiting factor on the population of that large mammal in the relatively closed and protected ecosystem. There will always be some limiting factor on every species in every ecosystem.

Biologists have long studied population limits. Their studies contributed much to the formation of the science of ecology in the last century. The upper limit of the population of a given species in a given ecosystem is referred to as the **carrying capacity** of the ecosystem for that species. As long as a given species does not exceed its carrying capacity in the ecosystem and there is a stable environment, the population should increase until some limiting factor prevents further growth. According to this theory, the population of any species within a given ecosystem eventually should reach equilibrium at the level of its limit.

Logistic Curve

The **logistic curve** is generally known in biology as the population curve. The concept developed as a result of experiments by a number of biologists and ecologists. Various organisms were placed in closed containers and allowed to reproduce freely. The experiments were typically performed in sterile, nutrient-rich, single-organism communities. In such an environment, the total population of the organism over time would increase along an **S**-shaped curve until the population limit was reached.

An analysis of this curve shows that the population of an organism (for instance an amoeba) in a closed system with no limiting factor except space, increases slowly at first. Then it begins to increase more rapidly until eventually the space begins to fill. When that happens, the population growth begins to slow, and eventually level off at the limit of the system. In geometric terms, the population increases first at an increasing rate and then at a decreasing rate until the limit is reached.

Figure 2-1. The logistic curve has been used to explain growth patterns in populations of all kinds of plants and animals.

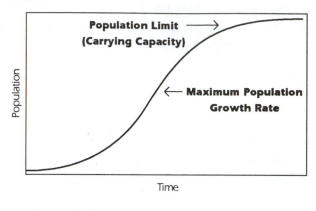

From this, it is clear that the population of a single organism should have a rate of growth that reaches its maximum at some point. That point is somewhere about halfway between the minimum number necessary for reproduction to occur and the maximum population limit for the ecosystem. According to the logistic curve, that point of maximum growth rate occurs at exactly half of the population limit. It is there that the curve's rate of increase changes from increasing to decreasing. It is there that the rate of reproduction growth is at its highest.

Consider how this theory might work in a real-world example. Imagine an island in the middle of a large river with no grazing animals and no carnivo-rous mammals. Then imagine a very cold winter in the year 1900 that froze the water's surface. Suppose a few moose crossed the ice and became stranded on the island. We will call our island "Isle Royale." Let's assume it has an area of 842 square miles and is located in Lake Superior. In the absence of a preda-tor, the moose herd would multiply by the logistic until its population limit is reached. At first the food and shelter would be plentiful and the Isle Royale herd would be healthy. The island would appear to be a paradise for the moose.

At first the only limiting factor on the moose population would be its reproductive rate. At some point the population growth would be limited by some other factor. In this case the limiting factor would probably be food resources or perhaps some specific nutrient such as sodium. At that point the moose would begin to overgraze the plant cover but there would still be many young moose and pregnant females.

The first winters after the overgrazing began would be rough ones. Many young, old, and sick moose would starve and others would be undernour-ished. Eventually, the herd numbers would tend to level off at the population limit and the result would be a moose herd characterized by an ongoing state of near famine. During years with good weather the herd would grow, only to be trimmed by disease, starvation, and harsh winters. In general, the vegeta-tion would be so overgrazed that the original popula-tion limit would be reduced for the malnourished herd. Isle Royale would become more a living hell than a paradise for the moose.

Isle Royale is not a hypothetical example. It really exists in Lake Superior, just off the state of Michigan shore, where it has been designated Isle Royale Na-tional Park. The description of the development of the moose herd after its introduction in 1900 was a true story. By 1930, the impact of the moose on the island's vegetation was so great that the entire herd was threatened with starvation. Wolves were intro-duced in 1946 as predators to help keep the moose population in check, and the populations of both animals vary widely to this day.

It is important to recognize that the theory of the logistic curve fits the human need for a stable nature. Using this theory, there is indeed a point where the population of a single species would be at a balance.

But there are no closed ecosystems on Earth. There are no sterile environments. There are no competition-free ecological communities. Everywhere we look in nature there is competition, predation, and disease. This reality meant that early biologists had to look beyond the oversimplified explanations of a simple logistic curve. That gave rise to another theory to account for the effects of a pair of organisms in an otherwise closed system.

Lotka-Volterra Curves

Carrying capacity ignores predation, so a second theory was needed to account for that. An American scientist, Alfred Lotka, and an Italian scientist, Vito Volterra attempted to explain population limits in an environment where predation was allowed.

The theory, using the **Lotka-Volterra Curves**, describes a predator and a prey with population levels that are symmetric, out-of-phase cycles like the trigonometry concepts of sine and cosine. As preda-

tor numbers rise because of plentiful prey, prey population decreases to the point that predator numbers exceed carrying capacity, causing decline in predators, which results in lower pressure on prey, and so on.

In the absence of substantial predator numbers, prey population would increase exponentially (by the logistic curve). For students who would like a more thorough discussion of **exponential growth**, see Chapter 11. In the presence of abundant prey, predators increase by the logistic curve. Two kinds of balance would be possible using the Lotka-Volterra theory: (1) continuing opposing oscillations, or (2) dampened oscillations leading to decreasing swings and eventually to a static population for both predator and prey "balancing" each other.

Figure 2-2A. As the population of a species nears it carrying capacity, it becomes unhealthy. A healthy population exists only below that level. The healthy deer shown in Figure 2-2A have plenty of food. (Photo courtesy L.G. Kesterloo, Virginia Commission of Game and Inland Fisheries.)

Figure 2-2B. This photo shows the result of overpopulation. This deer feeding area has been overgrazed and the herd has been forced to find food elsewhere. (Photo courtesy L.G. Kesterloo, Virginia Commission of Game and Inland Fisheries.)

Again, this theory has impacted heavily on the fields of ecology and natural resources management over the past decades. An example of the use of this theory should make it clear. From the example of Isle Royale it should be evident that in order to have a healthy population of moose in a forest, it is necessary that some predator be present. So, a benevolent ecologist would put a pair of wolves onto the island to provide predation.

Some may think predation is cruel. Not really—in nature, there is no cruelty—there is only reality. The goal of the ecologist would be a population of healthy moose. In order to achieve that goal, some of the younger and weaker animals must be killed so the others can have a healthy life. In essence, the wolves would be introduced to "harvest" the excess moose for the good of the overall herd. It is important to understand that the life of a single plant or animal is not the question. The question is will the species continue to survive and flourish? True environmentalists are concerned with populations rather than individuals.

According to the theory based on the Lotka-Volterra Curves, the wolf population would grow because of the plentiful food supply (moose) until eventually the two populations were in "balance." This fits the ancient belief that nature should be stable—that there should be a "Balance of Nature."

Gause Curves

Over the past half-century, repeated scientific experiments have failed to confirm the Lotka-Volterra theory. The most famous of the scientists to test the theory was G.F. Gause who experimented with microbes in cultures. The **Gause Curves** showed that in the absence of predators and given no population-limiting factors, prey species multiplied logistically. As soon as predators were introduced, prey populations began to decline immediately. But, without some other limiting factor (interference) the prey soon was totally extinct and the predator became extinct shortly thereafter.

Figure 2-3. The Lotka-Volterra Curves with constant cycles are shown here. The theory represented by the Lotka-Volterra Curves says that as predator numbers go up, prey numbers will go down. Then, when prey numbers get low, the predators will begin to die. That in turn will allow the prey numbers to rise again.

Figure 2-4. The Lotka-Volterra Curves reaching a stable equilibrium are shown here. A variation of the Lotka-Volterra theory says that the predator and prey populations could eventually stabilize at an equilibrium point.

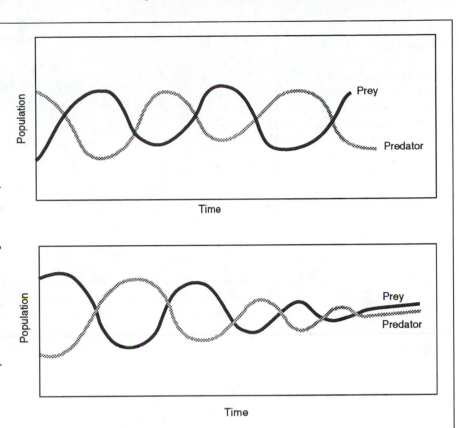

The implications of the Gause experiments are clear. A balance of predator and prey populations as envisioned by Lotka and Volterra cannot be expected in nature. Indeed, in a truly self-contained ecosystem with only two species of organisms, one prey and the other predator, the survival of either would be impossible.

The Contradictions

Clearly, the logistic curve is too much affected by context to prove useful in the real world. No plant or animal species of importance is ever in a nutrient-rich, competition-free, sterile environment. In reality, once a population approaches its limit in an actual ecosystem, its feeding damages its food source and then numbers must fall, often dramatically. Gause demonstrated the hopelessness of trying to find a stable population balance between two species associated by predation.

Botkin argued that the limitation of the findings of Gause are based on the extreme controls on the experiments. The cultures were nutrient-rich and in ideal growing conditions with no changes except for introduction of the prey/predator. The experiments allowed only two species in fixed and enclosed containers. Multiple species experiments have never duplicated these results.

Carrying Capacity

The ecological theories arising from the logistic curve, the Lotka-Volterra, and the Gause models have serious limitations. Yet, it is still possible to estimate the numbers of a given plant or animal species that should be able to survive in a given ecosystem over an extended period of time.

As was discussed earlier, carrying capacity refers to the number of plants, or more often of animals, of a

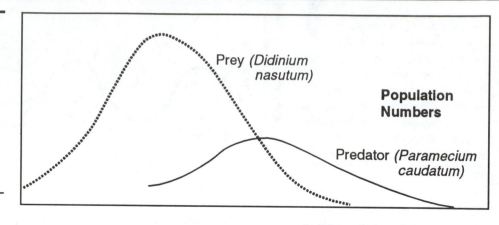

Figure 2-5. The Gause Curves are shown here. According to the experiments of Professor Gause, when a predator species and a prey species are put together in a closed environment, the prey will all be eaten and the predators will all starve—disaster for both.

particular species that can live in a given ecosystem on an ongoing basis. According to the logistic curve, carrying capacity is defined as the point on the curve when the total number of the given species is at its maximum.

Maximum Sustainable Harvest

According to the logistic curve formula, the maximum rate of population increase, or **maximum growth rate**, occurs at the point where the rate of increase changes from increasing to decreasing, as shown in Figure 2-1. Because of the nature of the formula, that point is exactly halfway between the carrying capacity and zero.

Because the population of the species is growing fastest at this point, this is also the point of **maximum sustainable harvest** for game and food species. The maximum sustainable harvest is the highest rate at which individuals of a species can be taken for human use without causing a decrease in the population of the species.

Assumptions and Contradictions

The use of these technologies in environmental management assumes several things. A basic assumption is that nature tends toward equilibrium within a given ecosystem. A second assumption is that stability is automatically good and that change is automatically bad. A third assumption is that it is possible to determine the carrying capacity for a given plant or animal within a given ecosystem. A fourth assumption is that we can determine where along the logistic curve the plant or animal population is at any given point in time.

Sadly, all of those assumptions are wrong. Yet essentially they have shaped most of the decisions being made today by conservationists and environmentalists alike.

Reshaping the Way We Look at the World

Nature before human intervention was neither good nor bad. More importantly, it has never been stable or ordered. For all of planetary history, the norm on Earth has been change, rather than stability; transition, rather than order.

If one would question this assertion, then one must point to a time and a condition that was ordered, stable, and natural. One must find a time and place of equilibrium in either human or planetary history. But no such condition has ever existed on Earth. Without human intervention, the ice ages have come and gone, the continents have formed and drifted apart, life has begun, and millions of new species have come and gone from the environment. And, more importantly, with or without human intervention, all of those great forces of nature likely will continue to mold, change, and remold the planet into the foreseeable future.

By the same token, there is no simple mechanical explanation for environmental relationships that will allow for a stable modeling of natural processes. Environmental truths change over time, so that even if we do discover cause-effect relationships, they are true only within limited ranges of conditions in an environment where those conditions are constantly changing. This concept has given rise in recent years to the emerging philosophy (perhaps science) of chaos. In this way of thinking, transition and instability are the only fundamental truths in nature.

Thus, if humanity would seek to understand nature, we must first accept one basic truth. *The truth is that there is not now, and there has never been, a stable order to nature.* The environment is in a state of flux that has existed since before the planet was formed and which is predicted to continue as long as the Earth exists.

The concept of a "balance in nature" is a myth that has proven to be counterproductive of human efforts to understand nature. A "return to nature" in the sense that there is some naturally correct way of doing things is an illusion—there simply has never been a naturally correct and stable order. The authors predict that there never will be.

This does not mean that mankind is free to do anything to the environment in order to promote economic or social growth—*far from it.* It only means that attempts by humans to reestablish the natural order of things are doomed to failure. In fact, they are based on a false premise. Even if all of the complex and changing relationships in the environment were fully understood, there is no such thing as equilibrium— as a "balanced nature."

There is no stability; there is only change. There is no "natural order"; there is only transition. We cannot "return to nature"; however, we can coexist with natural processes in a way to minimize the impact of our activities on the ecosystem. And we can learn to accept the fact that humanity has never been, and will never be, able to exist on the Earth without altering the biosphere in very fundamental ways.

Summary

The assumption that nature, without human interference, is orderly and benign is as old as human philosophy. A belief that the environment was orderly and stable was necessary for people to feel secure. They could not accept the nature of dynamic change within the environment. Change meant a lack of stability and therefore a lack of safety. Early peoples developed mythologies to explain nature based on divine decisions—to describe nature as divinely ordered. In various ways, mankind was held responsible for all evil (defined as disorder) because they either had failed to do what the gods had intended or because they had taken the perfectly ordered nature and damaged it somehow.

Even though there is not a perfectly ordered environment, many patterns of ecological development can be predicted with some accuracy. For a given ecosystem, the range of biological possibilities is set largely by temperature, available moisture, and other environmental factors. Within that framework there is a broad range of possible organisms that could survive and prosper. In general, the organisms in the ecosystem can be expected to progress through a fairly predictable sequence that is known as ecological succession. The exact mix of organisms at any point in time is very much based on chance, but the general sequence is reasonably predictable.

Eventually, an ecological community can be expected to reach a point at which the rate of change in the ecological succession slows. The dominant vegetation in the ecosystem at that point is referred to as climax vegetation.

In earlier years, climax vegetation was thought to be a permanent state of affairs that could be predicted exactly for a given ecosystem. Indeed, that predictable climax vegetation was thought to be the norm toward which all ecological succession was directed. That reinforced the concept of nature in balance. It is now known that there can never be a true climax vegetation with permanent stability. It is also known that dynamic development rather than stability is the norm in nature. Change is the only true constant in the environment.

For any given species of plant or animal, there is a maximum population that can live within a given ecosystem. Beyond that number, some limiting factor will eventually prevent further population growth. The limiting factor may be appropriate nesting space, water, food, shelter, compatible breeding, or simply room to move about. That maximum population is referred to as the carrying capacity for that species in that ecosystem at that particular time.

Without interference or predation, plants and animals tend to reproduce exponentially until they begin to tax the carrying capacity of the system. The theory of the logistic curve was developed to describe the rate of population growth of a single species in such an environment. Because the logistic model assumes no interference, the Lotka-Volterra model was developed to explain the effect of predation on population growth. It was a good theory but experiments by Gause showed that its assumptions were incorrect. Sadly, the Gause experiments too were limited by the rigorous controls and artificial conditions. Regardless of their flaws in the real world, the three theories have been, and remain, very influential in the decision- and policy-making processes among conservationists and environmentalists today.

As a major player in the ecosystem, humankind affects ecological succession of all of the biosphere. That in itself is neither good nor bad. The problem develops when the effect is to damage the long-term health of the ecosystem. From a human perspective, that means the long-term carrying capacity of Earth for humans. Humanity is subject to the same ecological processes as all of the other biotic players.

DISCUSSION QUESTIONS

1. If there is actually no such thing as permanence and stability in nature, why did intelligent, thoughtful people come to believe in a stable nature in equilibrium?

2. What would a "Balance of Nature" really mean?

3. What environmental factors prevent the establishment of true stability in the natural world?

4. Is the impact of mankind on the environment inherently good, bad, or neutral? Why?

5. What is the theory of the logistic curve? Why has it had such a major impact on policymaking in terms of environmental management?

6. What are the inherent flaws in the three major theories that have shaped environmental decisions over the past century?

ADDITIONAL ACTIVITIES

1. For those with an interest in further exploration of the emerging concepts of dynamic and constant change as the basis for the ecology of the future, read Daniel Botkin's fascinating book, *Discordant Harmonies*, copyright ©1990, published by Oxford University Press.

2. Organize a class debate on the topic: "Resolved: That the population of every species of plant and animal should be left alone by mankind, and allowed to seek its natural population limit."

3. Invite a game biologist to speak to your class on the topic of managing game or fish population levels.

4. Read the article on elephants in the July, 1991 issue of *National Geographic* magazine. Discuss in class how the concepts of population growth and limits apply in such a case.

CHAPTER 3
Concepts in Environmental Management

Photo courtesy Bill Camp.

Terms to Look for and Learn

Amorality	Multiple Use
Anthropocentricity	Natural Resource
Common Properties	Nonrenewable Resource
Conservation	Opportunity Cost
Development	Preservation
Exhaustible	Renewable Resource
Exploitation	Sustainable Harvest
Inexhaustible	Utility
Land Use Planning	

Learning Objectives

After reading this chapter and participating in the activities, you should be able to:

■ Explain what makes something a *natural* resource.

■ Discuss the similarities and differences among the philosophies of exploitation, conservation, and preservation.

■ Differentiate between renewable and nonrenewable natural resources.

■ Explain the concept of common properties and its implications.

■ Discuss economic concepts that apply to environmental decision-making.

Overview

Managing our natural resources and maintaining a livable environment are not simple tasks. In even the smallest and most simple ecosystem, there are many complex relationships. The food webs in a small pond are numerous and complicated. They include bacteria, algae, other small plants, insect larvae and adults, perhaps tadpoles, and even fish. There are birds that drink from the pond and feed on the plants and animals of the pond. In a small stream it becomes far more complex because many more plant and animal species live there. Also, the water moves along, making the relationships even more complicated. A large, stream-fed lake becomes too complex an ecosystem to really describe. Its interrelationships are too complex even to catalog, yet that is unbelievably simple compared to the global ecosystem.

Defining Natural Resource

Natural resources have been defined in many different ways. From an ecological standpoint, one definition of a natural resource could be anything that occurs in nature that humans use to provide any form of energy. From an economic standpoint, a natural resource is anything that occurs in nature from which humans gain any form of utility. From a political standpoint, a natural resource is anything that occurs in nature that one human community uses to gain advantage, either relative to the previous situation or relative to some other human community.

Even though the previous two chapters are heavily concerned with ecology and the principles of that science, it seems that the economic definition is more all-encompassing in this case. From the perspective of this book, natural resource can be defined as anything occurring in nature that can provide utility or that can be transformed into something that provides utility to people.

Definitions Change

Although they probably did not think of it in that way, early humans had a very practical definition of natural resources. They looked upon the animals whose flesh they used for food, bones they used for tools, and hides they used for clothing and shelter as a natural resource. They looked upon the land and the waters as natural resources. They learned to use fire and looked upon that as a natural resource.

What was an important natural resource in prehistoric times would hardly be valued today except to historians. As an example, early humans searched for supernatural explanations and artifacts to assist them in their most difficult task—survival. A small piece of rock that happened to be in a shape resembling an animal could be taken as a totem (symbol). We would find it hard to value today, but such a totem was of great importance to our ancestors. They might shape their very lives around their belief in the power of a small piece of rock.

Today we use silicon to manufacture computer chips that did not exist a half-century before. A silicon chip would not have been considered an important resource in 1940. Today, computers rely totally on silicon chips to work. As technology evolves and value systems change, the things that are useful to people change also.

Utility

Utility is an important concept from economics. It involves the notion of usefulness and is based on what people value. If something is useful to people, they will value it. They will seek to secure it and to use it. That willingness of people to seek an item is what gives it its value. The usefulness of the item is known as its **utility**.

Natural resources are of no utility, in themselves. Gold has no real, inherent value. But gold can be used to make something of beauty—something that we value. Thus, gold has utility. We do not value oil, just as crude oil. After all crude oil as it comes from the ground is a pretty messy material. It is smelly and slippery. But, in the mid-1800s, oil became valuable because it could be burned to provide light and heat. Later, people learned that crude oil can be distilled into several products that possess utility. Gasoline can be burned to power an internal combustion engine. Grease can be used to lubricate moving parts. Other useful products such as plastics and synthetic fabrics can be manufactured from oil. Crude oil was once just something that was messy, polluted soil, and killed plants if it surfaced. At that time it had no utility. With changes in technology, it became valuable because it gained utility.

Air is not of value to us, until we need to take a breath. Then its utility becomes very high. Clean water only has utility when we need to use it. A tree becomes a natural resource when we value what we can get from it. Using this definition, a natural resource is important for its utility, and utility has many different forms.

For example a tree may have many different forms of utility. If it is a sugar maple tree in Vermont, it may have utility as a provider of sweet sap which can be converted to maple sugar or maple syrup. If it is a pine tree on a tree farm in Georgia, it may have utility as a source of pulp from which paper is manufactured for books on environmental science. If it is a Douglas fir tree in Oregon, it may have utility as a source of lumber. If it is a giant redwood in California, it may have utility in its beauty. If it is a bristlecone pine on the Pacific coast of the United States, it may have utility simply because it is over 4,700 years old—the oldest living thing known.

As we saw earlier, utility is the property of being valued by people. And people value things for many different reasons. A piece of hard, rocklike material may be simply a rock to be moved out of a field; or it may be a piece of the Berlin Wall; or it may be a "pet rock"; or it may contain a diamond. In any of those cases, it would have utility.

Utility gained from the direct or indirect use of a resource, as in manufacturing a product, can be referred to as extrinsic utility. Utility gained from an

Figure 3-1. The utility of this tree may be in its wood, in its sap, in its ability to hold soil particles or moisture, in its ability to manufacture oxygen for us to breathe, in its cooling effect on the surrounding air, or simply in its beauty. It is of no value to people unless they can utilize it. (Photo courtesy Bill Camp.)

intangible use of a product is intrinsic utility. The maple syrup, paper, wood, and even the oxygen manufactured by the tree discussed before provide extrinsic utility. People value things with extrinsic utility because they use the resource.

The cooling effect and beauty and historical significance provide intrinsic utility. People value things with intrinsic utility because they like them. Both forms of utility are valid because they are both valued by people. The only property that is essential for utility to exist is the willingness of people to seek it and to give up other things in exchange for it.

It is this changing and changeable nature of utility that leads to all of the disagreements over environmental issues. A stockholder in a lumber company may value the trees in a forest for a completely different reason than a naturalist, or a soil conservationist, or a poet does. It is important to remember this concept as you make value judgements. Things that you value may not be important to others and things that other people value may not be important to you. That does not mean that your values are necessarily right and that theirs are wrong. It only means that they are different. Utility is a very personal thing and its interpretation is never the same among two or more people.

Anthropocentricity

Anthropocentricity can be defined as humans' approach to nature and natural resources from a "human-centered" approach. This anthropocentric perspective traces back to our definition of natural resource as anything occurring in nature that has utility for humans. As discussed earlier, utility means different things to different people.

Natural resources are classified according to how they relate to human use. The following sections provide some important anthropocentric definitions and discussions of our world's natural resources.

Inexhaustible Resources

Sincere and well-meaning people can, and often do, disagree over what is important, even when they all have access to the same facts. Unfortunately, most environmental issues are based on questions that are beyond the simple interpretation of factual data.

Natural resources that are essentially without practical limit are said to be **inexhaustible**. That does not mean that there is an infinite amount of the resource, it only means that there is no practical limit. An example of an inexhaustible resource is sunlight. Surely, it would be possible to determine how much total light can be generated by the sun. Using the concept of infinity in a mathematical sense, the amount of sunlight would surely be finite; but it just keeps happening. If the sun were to stop shining, all life on Earth would cease very soon anyway. So for all practical purposes, there is no limit to sunlight. Solar energy is inexhaustible.

Rainfall is an inexhaustible resource. If it were possible to catch all of the rainfall that reaches the Earth on a rainy day and to bottle it, that would not be the end of rain. The next day, there would simply be more rain. This is not to imply that there will not be a shortage of available water for the city of Los Angeles in the near future. It simply means that however much rain falls today on Earth, there will be more rain tomorrow, and it will continue without exhaustion. If the rain were to stop worldwide, human life as we know it would soon cease. So, for all practical purposes, water is an inexhaustible resource.

Exhaustible Resources

Natural resources that have a practical limit are said to be **exhaustible**. There is a practical limit to the amount of crude oil in ground deposits that can be economically recovered at present prices and using present technology. Nobody would argue against that statement. We do not know what that limit is because we have not found all of the oil fields in the world, but surely there must be a practical limit, given those conditions.

There is a limited number of elephants in the world. If their tusks are valued (extrinsic utility) more than their lives (intrinsic utility), we may soon exhaust the supply.

Although surface water, in the form of rain, is for all practical purposes inexhaustible, groundwater in defined aquifers is certainly exhaustible, as we will discover in several chapters later in this book. Moreover, if rainfall is redefined as drinkable surface runoff in a specific watershed during a specific period of time, there can easily be a practical limit. For instance, there is no practical limit to the rainfall that will eventually occur in the watershed supplying water to the city of Miami, Florida; but, there is a practical limit to the water available for that city in a given year. Thus, available water is seen as an exhaustible resource when taken in a specific geographic area and during a specific timeframe.

Renewable Resources

An exhaustible resource can be either renewable or nonrenewable. A **renewable resource** is one that replenishes itself. It does not necessarily require recycling for a perpetual supply to exist. As a rule, renewable resources are those that involve living organisms. Examples of renewable resources include wild animals, fish, trees, grasses, birds, and many others.

A resource that is renewable can be used at some rate without reducing the amount of the resource remaining for future use. Mushroom collectors in Indiana can go to the woods every fall and collect the edible fungi. There will be more mushrooms produced to replace those taken. In a sense, this category is somewhat like inexhaustible resources. The difference is that for renewable resources, if all of the supply is once used, the resource will disappear. In living populations, once an animal or plant is extinct, it is gone forever. Organisms like this are now protected by listing them as threatened or endangered.

Perhaps the most notable example of the loss of a renewable resource in recent history is the passenger pigeon. At one time passenger pigeons inhabited North America in such numbers that John Audubon once estimated a single flock at over one billion birds. The sixteen- to eighteen-inch-long bird travelled in such dense flocks that they would literally strip the vegetation from an area where they landed and leave the land bare. They were such a nuisance that Americans practically declared war on the birds. They killed the pigeons in massive numbers, sometimes for

food, other times just to be rid of them. Eventually, the large flocks disappeared. By the time naturalists realized it, the birds' numbers were below that required to maintain a breeding population. The last known passenger pigeon died in 1914 in a zoo in Cincinnati. Her body is preserved in the Smithsonian Institution Museum in Washington, D.C. shown in Figure 3-2.

Figure 3-2. The last known passenger pigeon was mounted for display at the Smithsonian Institution Museum in Washington, D.C. (Photo courtesy Ron Garrison, Zoological Society of San Diego.)

Another important example of a renewable resource is the forest. When the first Native Americans settled what is now the United States, the land was covered mostly with either forests or grasslands. Camp and Daugherty discussed the extent of the forests in this country prior to European settlement.

> In 1607, at least half of our land area was forested. This amounted to over a billion acres. . . . Since colonial times, United States forests have produced about 2,700 billion board feet of timber. Each decade we take more wood from our forests than the decade before, yet each year the trees in our forests grow more wood than we harvest. Our 760 million acres (of U.S. forest today) produce more wood each year than our over one billion acres did in 1607. (Camp & Daugherty 1988, 5)

Nonrenewable Resources

Nonrenewable resources are those exhaustible resources that do not regenerate themselves. This does not mean that the resources are in short supply. It simply means that there is a fixed supply with a practical limit. There is so much iron ore in known deposits that at least a 117-year supply is currently available. Moreover, every year it seems that more iron deposits are being discovered. Between 1950 and 1970 the annual rate at which iron was being used worldwide actually increased. During the same period, the supply of iron in known deposits increased by 1,221% (Simon 1981). Nevertheless, by definition iron is a nonrenewable resource.

Trade-Offs

It is literally impossible to satisfy all human needs and desires. Our appetite for utility is limitless but our resources are limited. Just as importantly, we cannot simply take everything we want from our environment without damaging it. Because we cannot have everything we could ever imagine and desire, people must make choices. The choices we make as individuals and as a society will be made on the basis of our wants and needs. But those choices must also take into account that the environment needs protection, maintenance, and often improvement. Finally, those choices must take into account that future generations of humans will also have needs and wants.

Opportunity Cost

For every decision made regarding the environment, something should be gained—some utility. But, for everything gained, something must also be given up. The utility given up to gain something else is called **opportunity cost**.

To drink a glass of orange juice, we must give up the money necessary to purchase it. To get that money, we, or our parents, or at least someone had to give up time and energy in the form of work.

From an environmental management perspective, opportunity cost is a critical concept. If Americans want to have houses built from wood, they must give up some trees. If we want to have beautiful mature forests, we must be willing to face higher lumber prices.

Agendas

An agenda is a list of things to do. A meeting should have an agenda to keep it focused and on track. A political party publishes its agenda and calls it a platform. A politician makes promises to the voters. Those promises are designed to please potential constituents because of the politician's perceptions of the voters' agendas.

There are basically two kinds of agendas—public agendas and hidden agendas. Public agendas are the openly-stated reasons for doing something. Hidden agendas are the underlying, normally unspoken reasons for doing it.

During the late 1980s some environmentalists protested against the harvesting of mature forests in

Oregon. The hidden agenda was to preserve the forests from clearcutting. But there is no law against clearcutting forests and so that argument would not likely be successful in the courts. There is a small bird, the Northern spotted owl, that is on the endangered species list and that lives in the mature forests of the Pacific Northwest. The environmentalists' argument for leaving the forests untouched was to preserve the nesting habitat for this endangered species. Loggers protested that their jobs are more important that the spotted owl.

As a result of this debate, one million acres of forest in the Pacific northwest were set aside from logging. The irony is that neither side was satisfied.

The logging interests argued that preserving the forests cost jobs and increased the cost of lumber in the United States. Consider that "logging interests" include more than just loggers. Logging interests also include the merchants and business people who provide goods and services to the loggers and their families. As well, logging interests include the large corporations that process the logs for consumption and the stockholders of those corporations as well. A person with a logging interest may live in New York City as well as in Oregon.

On the other side, environmentalists argued that a million acres was not enough. And remember that the environmentalists concerned with this issue were not just those living in the immediate area. The issue was an important precedent for environmental activists everywhere. If the federal government could be forced to set aside forest in the Pacific northwest for the Northern spotted owl, then similar actions elsewhere would be easier to promote. Beyond that, people whose homes and farms would be affected by the increased runoff and soil erosion resulting from clearcutting that million acres of forest were "environmentalists" on this issue.

The questions raised by the Northern spotted owl controversy in Oregon and Washington were truly complex. They involved many opposing groups of people. They involved politics and emotions more than scientific facts. That is not unusual in the arena of environmental concerns.

In order to keep owls alive in the woods of Oregon, then as a society we must be willing to give up the jobs of some of the many loggers whose work is in the forests of those states. That is the opportunity cost of maintaining the habitat for one small species of bird. On the other side, in order to maintain the economic well-being of the families whose primary source of income is from logging, we must be willing to give up the mature forests in the region. Society must be willing to accept the increased runoff and soil erosion that will occur in the streams until the forests return to protect the soil and to hold the soil in place. These are not simple decisions and they are not without cost. The things that are given up must always be taken into account as environmental decisions are made.

Environmental Management

Multiple Use

Multiple use is a concept that was originally developed by foresters. It is becoming more important in the entire environmental movement.

A forest can be used to grow trees for wood. At the same time, it can also be used for hiking and other recreation. It can be used for wildlife habitat. It can be used for hunting as well. All of these multiple uses can be done at the same time.

A lake can be used as a source of water for a city. If it is stocked with fish, it can be used for fishing. It can be used for water skiing and it can provide other recreational facilities as well. Multiple use means that the same resource is put to more that one application at the same time.

Multiple use is a very productive concept, because it means more utility from the same resource. Careful management is needed to prevent one use from damaging the resource so that it is no longer useful in other ways.

Sustainable Harvest

Sustainable harvest is defined as the number of a given species that can be taken from a given ecosystem on a continuing basis without reducing the total productivity and utility of that species. That may be in the form of catching fish by sports fishermen, harvesting fish in nets by commercial fishermen, trapping fur animals, or hunting, just to give some examples.

Estimates of sustainable harvest rates for fish and wildlife have historically been based on the logistic curve. As illustrated earlier, the logistic theory proposes that the population of an animal in a given system increases until it reaches its limit. Theoretically, once the animal reaches its population limit it can remain at that level indefinitely, so long as conditions remain unchanged. Population increase starts slowly. It then grows at an increasing rate. Finally, it increases at a decreasing rate until the population reaches its carrying capacity.

One reason for studying sustainable harvest is to determine how many fish could be taken from a given fishery or how many animals can be taken from a given area on a continuing basis without destroying the fish or animal population. Let us use the population of grouper (a large edible fish) in a fishery around the large reef just south of the Florida Keys to illustrate.

The grouper population growth rate affects how many of the fish can be removed from the ecosystem during a given time period. At the level of the carrying capacity, the fish would not be reproducing very rapidly. If the population of groupers were to remain near its upper limit, then the sustainable harvest rate would be quite low. But, if the grouper population is forced to remain lower than its limit, a larger sustainable harvest would be possible. Not only that, but the surviving grouper would be healthier and would grow faster.

The question is how much below the carrying capacity should the population be forced to remain. According to the logistic curve, the maximum sustainable yield for grouper in the fishery along the reef off the Florida Keys could be determined by estimating the carrying capacity and dividing by two. The allowable harvest rate for grouper could then be set to equal the reproduction rate at the point of maximum growth.

Thus, according to the theory of the logistic curve, the healthiest population of an organism should be at about half of its population limit for a given ecosystem. In the case of the moose herd on Isle Royale, it would have been at that point that reproduction was at its highest and just before the food resources began to be in limited supply. All one would need to do is to determine the population limit, divide that number by two, and the ideal population level for a given organism is the result.

That point is also the maximum growth rate, as well as the maximum sustainable harvest rate for an organism. Much of the decision making in the field of natural resources management has been based on this theory.

Population Controls

Hunting for deer or other animals has become the target of a number of animal rights groups. At the same time, wolves and other large predators have become quite rare in most of the United States. Without predators to reduce their numbers, the deer populations in many Eastern states has grown rapidly. As a result, accidents involving deer being hit by automobiles have become quite common on rural roads. Deer are beginning to present a more serious problem for farmers and other home owners. If allowed to continue to multiply, they will with certainty begin to damage their own food supply and the result will be massive die-offs of deer in the wild.

The question is what is more desirable. Introducing predators into the wild and allowing them to survive would present a "natural" population control on the deer. The trade-off is that we must be willing to have large carnivorous predators in the forests before this solution will work.

Allowing increased hunting for deer would also provide a degree of population control. To accomplish this, hunting must be encouraged rather than discouraged. The trade-off is that many people be-

Figure 3-3. Uncontrolled populations of deer and other animals mean problems for people as well as for the animals. This deer was killed when it ran in front of a pickup truck on the highway. The driver was lucky not to have been injured. (Photo courtesy L.G. Kesterloo, Virginia Commission of Game and Inland Fisheries.)

lieve that hunting is an inhumane activity that should not be encouraged at any cost.

The third solution is to allow the deer population to grow out of control. The trade-off will be the resulting damage to crops and gardens, increased accidents, and occasional massive deer die-offs from starvation and disease.

For every decision there are opportunity costs. There are always trade-offs in environmental issues.

Philosophies of Resource Use

All living things depend on their ecosystem for survival, and humans are no exception. For the cave dwellers of the Neanderthal age, their ecosystem was limited to the area immediately surrounding them. But for modern humans, the entire planet is the ecosystem. We derive our livelihood from the global ecosystem. Over the past 300 thousand years or so since *Homo sapiens* emerged, humans have thrived. There was a human population in excess of 5.3 billion by the end of 1990 and a projected population of 6.2 billion by the end of the twentieth century.

With that many humans, it is literally impossible to survive at any level without severely disrupting the ecosystem. On the other hand, the near-term well-being of individual humans or even social groups cannot be allowed to threaten our long-term survival. The abuse or misuse of our natural resources and of our ecosystem cannot be allowed.

Humans must eat. Feeding over five billion humans requires large-scale, scientific agriculture. At the present state of technology, it would literally be impossible to produce enough food without chemical fertilizers and pesticides. Yet, some chemicals are slowly damaging the soil and water supplies. Alternatives must be found that will be more environmentally friendly. Genetic engineering and other research being done now promise solutions, but in the meantime we must continue to degrade the carrying capacity of the planet. We have no choice. The long-term survival of humanity demands that agriculture become less damaging to the environment. But the near-term survival of people alive today forces us to continue to stress the environment.

Our homes and factories produce vast quantities of wastes: solid, liquid, and gas. In the past, that was simply discarded into the environment. Liquid wastes were put into lakes, streams, rivers, and the oceans. Gaseous wastes were put into the atmosphere. Solid wastes were dumped onto the ground or buried in it. That solution cannot be used forever. The damage to the ecosystem is becoming too great.

As we make the decisions that will allow for the survival of our species as a whole and of its individual members, there are at least four competing philosophies: exploitation, development, preservation, and conservation. We can think of these as being on a continuum with exploitation and preservation at the extremes and conservation between them.

Figure 3-4.   A continuum of environmental philosophies states that people's attitudes toward any given environmental question can be characterized as preservationist, conservationist, development-oriented, or exploitationist.

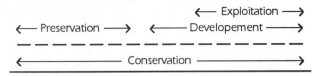

Exploitation

A philosophy of **exploitation** means that nature is regarded as something to be used. In its most general form, exploitation is simply the act of taking something directly from nature then using, reshaping, or moving it for profit.

Exploitation has a harsh sound, but without it, our society simply could not have moved to its current level—in fact it could not even exist. We exploit nature when we drill for oil and burn gasoline. We exploit our natural resources when we clear a forest and build a house for our family to live in. We exploit our water supply every time we take a drink of water or irrigate our crops.

Exploiters can be very sensitive to the environment or they can be totally indifferent. It is possible to exploit natural resources with very little change to the ecosystem. In its extreme form, exploitation may be done with no regard for the effect on the environment. Extreme exploiters would drain a swamp and harvest the trees in it for the exotic wood.

In the United States, such exploitation was once the norm. The passenger pigeon was exploited into extinction, and we almost did the same to the bison. Fortunately, this nation is becoming more aware of the environmental impact of the lifestyle created by exploitation.

Development

Development involves changing some aspect of nature to make it more useful to people. A developer would drain the swamp and build a highway across it so that people could drive more quickly between two cities.

Developers can be totally insensitive to the environment or they can be very sensitive. It is possible to minimize the impact of development on the environment. On the other hand, many developments can have extreme environmental impacts and they must be regulated.

Preservation

Preservation is the exact opposite of exploitation and development. A preservationist advocates keeping things as they are or returning them to a "natural state." As in exploitation, there are many degrees of preservationism. People may be preservationists with regard to selected things and not to others. An avid hunter can oppose poaching of elephants. A dedicated fisherman can be a preservationist with regard to whales. A tree grower can be an advocate of maintaining wilderness areas intact.

Extreme preservationists could halt the construction of a needed highway because it would require that a swamp be drained. They would then set aside the wetland swamp as a wildlife refuge and refuse to allow trails or pathways to be built. They could then discourage people from entering the area. Preservationists sometimes seem to want to make a museum of the whole world.

Conservation

Between the two extremes lies the philosophy of conservation. Conservation is defined as the wise use of resources. It involves planned exploitation along with planned preservation. Conservationists advocate the management of natural resources to allow for their use without inflicting unacceptable damage to the environment.

Conservationists would probably allow the road to be built in the example used before. But before building it, they would require that an environmental impact study be done. Then they would route it around parts of the swamp, leaving adequate habitat for the endangered species. They would provide trails into parts of the remaining swamp area and encourage school tours and limited tourism as a means of educating people about the environment and providing income to help maintain the habitat. Multiple use is a conservationist's concept of natural resources.

Reconciling the Differences

As with so many other aspects of environmental issues, there is no clear right and wrong in this regard. Clearly there are some things in nature that are worth keeping (preservation). On the other hand, people cannot survive without exploiting nature. Indeed, exploitation is probably the most natural act possible. We all **must exploit** (use) natural resources from the ecosystem in order to live.

Human society as we know it could not exist without some sort of development. People need shelter and there are only so many caves. Housing in some form must be developed. The middle ground is conservation. A philosophy of conservation recognizes the needs of humans as well as their wants. It advocates meeting as many of the needs and wants as possible, within the framework of environmental responsibility.

Again, the fundamental concept of opportunity cost becomes clear. Humans cannot survive in the near-term without altering the ecosystem, yet, cannot continue to alter the ecosystem recklessly without endangering long-term survival. Human population growth and advanced technology have so vastly increased our ability to modify the environment that we cannot continue our present direction indefinitely.

Land use planning for the future can help balance out some of these contradictions. Suggested alternatives are discussed in the following paragraphs.

Amorality of Science

There is a fundamental difference between such philosophical considerations as those just discussed and science. Science is in the business of examining and interpreting evidence. Scientists seek answers based on observation and evidence.

This means that science, including environmental science, is not concerned with what is *morally right* but rather with what is *technically accurate*. Because science is not concerned with moral judgements, it can be said that science is *amoral*.

The **amorality** of science means that questions of what is right and wrong, of what is good and bad cannot be answered by science. An environmental scientist can legitimately determine the impact of certain behaviors of people on the environment, and make recommendations based on those findings. But when a person makes a value judgement that those human behaviors are bad or good, that person has gone beyond science and entered philosophy. This is an important point to remember while reading the remainder of this book and reading further about the environment.

The Paradox of the Commons

As you walk or ride down the street, you will see many different kinds of property. Some belong to specific people. For instance, there will be houses and land that are the property of individuals. There will be streets and sidewalks that belong to the federal, state, county, city, town, or some other form of government. If you look up you will see clouds. There may be streams or lakes. The things that do not actually belong to some individual or business are so-called **common properties**.

Common properties may be claimed by a government, but in reality when everybody owns something, nobody really owns it. As you drive along an interstate highway, you may stop at a rest stop. Such rest stops belong to the government. But the government is the people. Therefore, as a citizen, you are part owner of those common properties. At the same time, you would be well advised not to try to take any of "your property" with you when you leave the rest stop. In a sense, everyone owns them, but no single person can make management decisions regarding common properties.

The paradox of the commons is that if everyone owns the common properties, then everyone should be responsible for their upkeep. Yet, it makes no economic sense for anyone to contribute to their upkeep. If everyone has equal access to common properties, then everyone should be responsible for their wise use and management. Yet, it makes economic sense for everyone to use as much of them as possible. In fact, it is completely logical, from an economic viewpoint, for any given individual to take from the commons as long as it is profitable to do so, even at the expense of damage to the environment.

For an example, suppose that there is a fisherman named Juan who lives in a fishing village in Portugal. He owns a small boat and earns his family's income from his catch. The more fish he catches the more income he makes and the more useful things he can buy for himself and his family.

Further, suppose there are several hundred fishermen in the same village. The fish do not belong to anybody in particular—they are a common property. In most years, the fish would be plentiful and would provide a reliable income. Then suppose that one year the fish are not so plentiful, that there is a storm or a disease that decimates the fish population. If everyone continues to fish at the same level, the fishery may suffer profound damage and it might take years to recover. Juan may know this, but there is little that he alone can do. If he stops fishing at half his normal catch, the next fisherman will simply catch more fish.

From Juan's perspective, the only logical decision is to continue to fish as long as it is profitable to do so because his conservation efforts will not affect the overall fishery population. The only viable solution for the fishermen is to organize a conservation effort that would involve all of them. But that would rely on voluntary compliance and that seldom works. Now, suppose that half of the fishermen comply with voluntary limits. The fish prices would be higher. Then the half who did not comply would profit from the increased prices of the fish that would result from the shortage. Juan would not be able to feed his family and his neighbors would grow richer.

Figure 3-5. Common properties belong to everyone, yet they do not really belong to anyone; that produces the paradox of the commons. (Photo courtesy Bill Camp.)

The same logic holds true in the case of almost all common properties. A classic example of abuse of the commons is the acidification of the air from industrial emissions in the United States. The emissions in one state drift over another state or over Canada. Most people agree that it is a bad thing. But without governmental regulations and enforcement, it makes no economic sense for any single industrialist to remove the acidifying emissions from his or her factory discharge.

The ownership of resources has major implications for environmental management decisions. Private domain resources are likely to be exploited to the maximum economic potential. To do anything else for the individual makes little economic sense. But, at least the individual manager or owner can foresee the long-term effects of his or her decisions. Because of that, it makes economic sense to manage private domain resources conservatively, with an eye to the future.

On the other hand, public domain resources are not subject to the same kinds of foresighted planning on the part of individuals. Management of public domain resources cannot be done in a reliable way without governmental regulation and enforcement.

Summary

A natural resource is anything that occurs in nature that provides utility or that can be transformed into something that provides utility to people. Utility is the property of usefulness. An item's utility can be determined if people will seek it out and will give up something else to obtain it. Natural resources can include anything that has value to us only in terms of its utility. A resource is of no inherent value until some use from it can be determined.

There are many forms of utility. Two forms that were discussed in this chapter are extrinsic and intrinsic. Extrinsic utility is value gained by the actual tangible use of a resource. Intrinsic utility is value based on enjoyment or other forms of nontangible human response.

Natural resources are classified and managed in an anthropocentric manner. This results from the unavoidable fact that human thinking and management is based on human values. From the perspective of society and humans, natural resources may be exhaustible or inexhaustible. Of the exhaustible resources, some are renewable and others are nonrenewable. How we deal with environmental resources is largely determined by whether they are inexhaustible, renewable, or nonrenewable.

At the same time, a resource can be inexhaustible and still be subject to damage. A resource can be theoretically exhaustible and still subject to shortages. On the other hand, a nonrenewable resource can be in such great supply that its conservation is not necessary.

In all decisions regarding environmental management, there are trade-offs. Faced with any decision, environmental policy-makers must weigh the pros and cons of each possible alternative. The opportunity costs of any decision are the other alternatives that must be foregone in favor of the alternative chosen. Another complication in environmental policy-making is the conflicting agendas of various groups of people. In fact, it is often a person's or group's hidden agenda that motivates behavior rather than its public agenda.

Multiple use is a particularly powerful tool in environmental management. It allows the same resource to be used in more than one way at the same time. This is a way of increasing the total utility of a resource and of serving the agendas of more than one group of people at the same time.

Sustainable harvest is the largest amount of a renewable resource that can be taken from the ecosystem without diminishing future use. The maximum sustainable harvest rate occurs, theoretically, at the midpoint between the population limit of a species and its extinction. Populations of animals and plants increase exponentially in a given ecosystem until either predation, disease, or starvation begin to control numbers.

Philosophies regarding the environment can be loosely grouped into four categories and placed on a continuum. The categories are exploitation, development, preservation, and conservation. Exploiters take resources directly from nature for human use, for instance oil and coal. Developers reshape nature to make it more useful to humans, for instance building a house or a highway. Preservationists attempt to protect natural resources for the sake of keeping them in a natural state, for instance protecting a rare bird from extinction. Conservationists believe in the wise use of natural resources to minimize ecological degradation.

Environmental science, on the other hand does not provide judgments about philosophical questions. When environmental scientists begin to advocate philosophical positions they move outside science and into philosophy or politics.

Management decisions on public domain resources are difficult enough. Often, short-term economic decisions result in environmental damage that has long-term impacts. But in the case of common properties, it is unusual for individuals to make wise long-term environmental management decisions. Essentially, the management of the environmental commons must remain in the hands of governments.

DISCUSSION QUESTIONS

1. What is a natural resource? Why do the things that we regard as natural resources change over time?

2. Why are natural resources classified in an anthropocentric manner?

3. What are inexhaustible, exhaustible, renewable, and nonrenewable natural resources?

4. How can an inexhaustible resource become in short supply? How could a renewable resource become nonrenewable? How could a nonrenewable resource become inexhaustible?

5. Explain the concept of multiple use. Of sustainable harvest.

6. What is the paradox of the commons and how is this concept related to environmental problems?

ADDITIONAL ACTIVITIES

1. Locate common properties in your community. Try to determine who is charged with the responsibility to manage them. Find out who, if anyone, actually maintains them.

2. Invite a soil conservationist or forester to speak to your class about multiple use and sustained harvests.

3. Contact your state's department of fisheries and wildlife. Find out how hunting seasons, bag limits, fishing seasons, and creel limits are set. Find out where the monies used for managing fish and wildlife populations come from.

4. Organize in your class a debate on hunters' rights vs animal rights. Strong cases can be made for both sides.

5. Have a class discussion to determine where your classmates would fit on the environmental philosophy continuum when it comes to cockroaches. A largemouth bass. A deer. An elephant. The spotted owl.

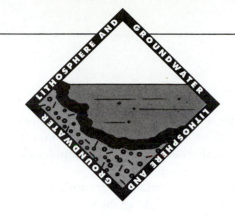

CHAPTER 4

Lithosphere and Groundwater

Courtesy United States Geological Survey.

Terms to Look for and Learn

Chemigation
Fertigation
Fungicides
Groundwater
Herbicides
Igneous Rock
Insecticides
Lithosphere
Metamorphic Rock

Methemoglobinemia
Nematicide
Nitrogen Fixation
Overdraft
Porosity (Soil)
Sedimentary Rock
Volatile
Water Table

Learning Objectives

After reading this chapter and participating in the activities, you should be able to:

- Discuss the implications of underground injection of chemical wastes in terms of the groundwater supply.

- Explain how chemicals such as nitrates and pesticides in the soil move into the groundwater.

- Discuss why some pesticides are relatively nonpolluting while others can be very damaging to the environment.

- Discuss the levels of nitrate nitrogens in well water in the United States.

- Outline steps that can be taken in agriculture to decrease groundwater pollution.

Overview

Our 4.5 billion-year-old Earth is in the shape of a ball, somewhat flattened at the north pole and south pole. Around the Earth at the equator it is about 25 thousand miles. The solid, rocky crust of the Earth is known as the **lithosphere**. Rocks on the surface of the Earth and to several hundred feet within the planet are classified as igneous, sedimentary, or metamorphic.

- Igneous—Cooled molten rock; examples: granite, basalt

- Sedimentary—Sediments deposited in water and later made solid; examples: limestone, sandstone, and shale (Figure 4-1)

- Metamorphic—Igneous or sedimentary rock changed by heat and/or pressure; examples: marble, slate, gneiss

The lithosphere is the rocky crust of the Earth and **groundwater** is water occurring beneath the surface. We see it and use it mostly as well water. It is logical to conclude that *most* groundwater occurs within the lithosphere. The principal exceptions are very shallow water tables that exist in earthy materials and water tables in deep sands.

The **water table** is the surface of groundwater in a well. Another way of describing the water table is that point in the ground that is completely saturated with water. Below this depth, usually for several hundred feet, water completely fills the spaces between soil or rock particles, known as pores or voids. **Porosity** can be defined as the proportion of the ground that consists of spaces or voids. The total porosity of the materials and their size and continuity determine the volume of groundwater that can be held available for use from a well. For example, a clay soil may have half of its total volume occupied by water but will release only two percent of this water for pumping in a well.

In contrast to clay, sand and fractured limestone may have only half of the total volume of voids that hold water but release for our use more than ninety

Figure 4-1. When bedrock is limestone, dolomite, or gypsum, solution caverns may develop, such as this one in Carlsbad, New Mexico. Under such conditions, polluted ground water may travel for many miles, and thus be difficult to trace the source of the pollution. (Note two persons at arrow.) (Courtesy United States Geological Survey.)

percent of the total water held in those voids. Granites holds almost no water and, therefore, would not yield enough to supply adequate groundwater in a well.

Groundwaters are important in the United States because about half of us get our drinking water from them—if we live in the country, about ninety-five percent of us depend on groundwater for drinking. On a hot day there is nothing as refreshing as a glass of cold water from a well. But is the water really safe to drink?

In 1988 the United States Environmental Protection Agency published a report on groundwater pollution. Their study showed that there were 46 pesticides found in groundwaters in 26 states. However, only in nine of these samples were the concentrations of pesticides greater than Health Advisory Levels; i.e., at a level hazardous to human health. (The World Health Organization estimates that each year one-half million persons in the world are poisoned by pesticides.)

Where could the pesticides in groundwaters have come from? **Herbicides** (weed killers) and **insecticides** (insect killers) are being used in "large quantities" by farmers, ranchers, and city dwellers. (The term "large quantities" means about one billion pounds of active ingredient per year.) Proper use of pesticides is guaranteed by following the directions on the package.

Nitrates are also a human health hazard in groundwater. The United States Environmental Protection Agency has set a human health standard of no more than ten milligrams per liter of nitrate nitrogen in drinking water. Recent studies in sixteen states of nitrate nitrogen in groundwater used for drinking resulted in the following information.

- More than twenty-one percent (over one in five) of the samples of drinking water contained more than 10 milligrams per liter of nitrate nitrogen in Kansas and Rhode Island.

- Ten to nineteen percent were in excess in California, Arizona, Oklahoma, and New York.

- Five to nine percent were in excess in Colorado, South Dakota, Nebraska, Texas, Minnesota, Iowa, Illinois, Pennsylvania, Maryland, and Delaware.

- In all other of the fifty states, less than five percent of the groundwater tested 10 or more milligrams per liter of nitrate nitrogen.

> Based upon pesticides and nitrates in drinking water, in which state would you prefer to drink the water?

Action to Control Pesticides and Nitrates in Groundwater

In 1990 the National Fertilizer and Environmental Research Center at Muscle Shoals, Alabama, confirmed that farmers and fertilizer/pesticide dealers are partly responsible for pesticides and nitrates in drinking water. Their action to control these pollution sources consisted of establishing twenty model demonstration sites to control such pollution on dealers' property where fertilizers and pesticides are mixed for delivery to farms. The first ten of such sites are located as follows:

- Agri-Liquids, Inc., Leighton, Alabama

- Shields Soil Service, Dewey, Illinois

- Agriform Farm Supply, Woodland, California

- Western Farm Service, Santa Maria, California

- John Pryor Co., Salinas, California

- Willard Chemical Co., Frederick, Maryland

- Alliance Fertilizer Corp., Mechanicsville, Virginia

- Ouachita Fertilizer Co., Monroe, Louisiana

- Cone Ag Service, Pierre, South Dakota

- Ranch Fertilizer Co., Okeechobee, Florida

In addition to establishing environmentally safe fertilizer and pesticide demonstration sites, the National Fertilizer and Environmental Research Center is offering "friendly" consultants to visit dealers' plants for the purpose of proposing improvements to reduce pollution of waters, including groundwaters. Problems of greatest environmental concern consist of:

- Metallic impurities in phosphoric acid fertilizer

- Gypsum pond leakage

- Treatment of contaminated soil

- Reducing pollution from liquids rinsed from application equipment

- Use of microbiological techniques for waste treatment

- Use of nitrate inhibitors for urea

- Techniques for sampling and analyzing groundwater

- Developing better containments for **chemigation** and **fertigation** (applying pesticides and fertilizers with irrigation water)

Technology and research results will be made available to dealers and farmers everywhere in the United States. Action to control pesticides and nitrates in groundwaters was taken in 1990 by establishing a seven-year, $6.8 million Iowa Big Spring Demonstration Project. The Project is in a farming area where deep, limestone-fractured soils favor deep and rapid percolation of applied pesticides and nitrates which pollute the groundwater.

Pesticides and Groundwater

The pesticides used in the 1990s are mostly synthetic organic compounds. The principal processes that influence their potential for loss from soil to groundwater are volatilization, decomposition, retention by the soil, and transport by percolating waters.

Substances used as fumigants must be relatively **volatile** (readily vaporized at a relatively low temperature) so they will vaporize and move in effective concentrations as gases throughout the soil. Pesticides that have a marked tendency to volatilize and a low solubility in water tend to be lost from the soil to the atmosphere; their residues are unlikely to reach the groundwater. Ethylene dibromide, a volatile liquid used for nematode control, is relatively soluble in water. This solubility in water, together with the typically large quantities applied and the low retention by soil solids, help to explain why ethylene dibromide has been found in groundwater under conditions favorable for downward movement.

Synthetic organic pesticides may be decomposed in different ways. Those applied to plant foliage or the soil surface may be broken down rapidly by sunlight.

Some pesticides react with water in soil to form new compounds. The **fungicide** (fungus killer) Captan, for example, reacts rapidly with water to form an innocuous product, so that downward movement of the parent fungicide to groundwater is not a matter for concern. The **nematicide** (nematode killer) dibromochloropropane also reacts with water to produce a new compound that lacks the pesticidal properties of the parent compound, but the rate at which this reaction occurs under the usual environmental conditions in soils is inconsequential. For this and other reasons, dibromochloropropane has been detected in groundwater.

Breakdown of organic pesticides in soils beyond that involving reaction with water is attributed to microorganisms and catalytic effects of soils. Pesticides such as dalapon, barban, 2,4-D, malathion, and parathion that break down rapidly (fifty percent decomposition in two weeks or less under favorable conditions) are not likely to be detected in groundwater. Some organic pesticides, such as chlordane, DDT, and dieldrin, decompose very slowly and may persist for years. These pesticides, however, are not of concern as groundwater contaminants from agricultural use because they are relatively insoluble in water and are retained strongly by adsorption by clay and humus. In other words, those chemicals stick tightly (adhere) to clay and humus particles.

The soil constituent of greatest importance in retaining pesticides is organic matter. Binding to humus in organic matter and clay decreases the potential for downward movements of many pesticides in soils. The capacity of the soil to hold pesticides such as paraquat and other pesticides is very important.

The principal mechanism by which pesticides are transported from soil to groundwater is downward percolation of water containing dissolved pesticides. The relative potentials for movement of different pesticides to groundwater in different soils can be estimated. This is done by applying known quantities of the pesticides to different soils, adding equal quantities of water, and measuring the content of the various pesticides in the drainage water or the distance to which the pesticides move in the soil. The potentials found in this way depend mostly upon the retention of the pesticides by the soil. They exceed the "worst case" situations for comparable thickness of soil in the field because they do not allow for the full effects of loss by volatilization and decomposition by microorganisms.

The U.S. Environmental Protection Agency has estimated that as many as 50 of the more than 1,000 registered pesticides possess the *potential* for detection in groundwater under conditions conducive to downward movement. According to a 1984 listing, 12 of the 50 *potential* pesticides have been detected in drinking water wells under certain conditions. Typical concentrations exceeded the health-advisory concentrations for four pesticides (bromacil, dibromochloropropane, dichloropro-pene-dichloropropane, and ethylene dibromide). Concentrations were below the health-advisory concentrations for seven pesticides (alachlor, atrazine, carbofuran, dacthal, dinoseb, oxamyl, and simazine). Concentrations approximately equaled the health-advisory concentration for one pesticide (aldicarb).

Despite the existence of residues of certain pesticides in some groundwaters in excess of the health-advisory concentrations, no verified adverse health effects are on record as a result of pesticide residues in groundwaters used for drinking. One reason may be the safety factors involved in the health-advisory concentrations.

To derive the health-advisory concentration for a pesticide, each of two animal species is treated with a range of doses of the pesticide in lifetime studies. The maximum dosage that produces no observable detrimental effect in the more sensitive of the two species then is divided by a safety factor, commonly 100. For a substance for which a safety factor of 100 has been used, a 22-pound child would have to drink 26 gallons of water with the health-advisory concentration per day every day to ingest an amount of the pesticide equivalent to the maximum daily intake that produced no observable detrimental effect in the test animal of the same average weight.

If a significant concentration of a pesticide is found in groundwater as a result of agricultural usage, further increases in the concentration may be limited by discontinuing the use of this pesticide in the affected area.

Integrated pest management may suggest ways to control the target pest using lower amounts of the chemical without loss in efficiency of pest control. Environmental warnings on pesticide labels should be followed. Techniques to avoid pesticides in potable (drinking) water include substituting alternative water sources, distilling the water, using ultraviolet light to decompose the pesticide residues, and passing the water though an activated carbon filter to remove the pesticide residues.

Nitrates and Groundwater

Among the dissolved inorganic substances occurring naturally in surface water and groundwater are the nutrients plants require for growth. Originally, agriculture depended almost entirely upon nutrients from the soil and from organic residues derived from agriculture. This system resulted in impoverished soils of low productivity. Eventually, agriculture was improved by involving **nitrogen fixation** by introducing legumes as an integral part of cropping systems. The legumes added available nitrogen as a result of fixation of atmospheric nitrogen by the bacteria that live in nodules on their roots. Now chemical fertilizers are used to supplement the natural supplies of nitrogen and other inorganic nutrients in soils that are required for plant growth. Consequently, agricultural productivity has been much increased.

Plants absorb only part of each of the nutrients present in soluble forms in the soil. As the concentration in the soil water increases (generally leading to greater uptake by plants and greater yields), the percentage of the total amount absorbed decreases, and greater residues are left in the soil. Most of the chemical ions added in fertilizers are retained by soils as a result of chemical interactions, but a few are not. Of those not retained, only nitrate is of environmental concern in groundwater used for drinking. Although loss of nitrate from soils to groundwater is a natural process, the potential for loss to groundwater is increased in local areas by high concentrations of livestock manures and in much cropland by applied nitrogen fertilizers.

Accumulation of nitrate in groundwater has been suggested as the cause of several human health problems, including birth defects, cancer, nervous system impairment, and **methemoglobinemia**. Only the last of these is well verified. Methemoglobinemia is caused by alteration of some of the nitrate to nitrite in the digestive tract and absorption of nitrite into the bloodstream. There the nitrite reacts with the hemoglobin in blood to produce methemoglobin, a form that does not carry oxygen to the body cells.

A number of infant deaths from methemoglobinemia occurred until the cause was discovered about forty years ago. Now the cause of and cure for methemoglobinemia are well known among medical doctors. Where water supplies are high in nitrate, bottled water of known low nitrate content is recommended for infants in their first year. Adult humans and livestock tolerate far higher concentrations of nitrate than do infants.

Various agricultural practices may be used to reduce the loss of nitrate to groundwater. These include:

- Reducing the amounts of nitrogen fertilizers applied at one time in current cropping systems

- Adjusting nitrogen fertilizer application on the basis of soil or plant-tissue tests

- Applying nitrogen fertilizer in small amounts as needed by plants during the growing season

- Using slow-release fertilizers

- Using chemical inhibitors to delay the formation of nitrate from the ammonium and other forms in which much of the fertilizer nitrogen is applied

- Avoiding fall applications of nitrogen fertilizers for crops to be planted in the following spring

- Spraying plants with mild solutions of urea in place of supplying nitrogen fertilizer to the soil

- Changing to cropping systems that derive their supplemental nitrogen from legumes

- Capturing animal wastes from dairy, feedlot, and hog operations in liquid manure pits for treatment and later use as organic fertilizer or safe disposal.

The effectiveness of the last two practices result in a lowering of the nitrates that enter the groundwater from farming operations. Cropping systems that rely on legumes for nitrogen mean a smaller addition of nitrogen from fertilizers and the slow release of the legume nitrogen. In the end, the legume nitrogen may be changed to the ammonium form of nitrogen which will not leach, or to nitrate which will move with waters of percolation. Both the ammonium and the nitrate forms of nitrogen are absorbed by plants. The capture of animal wastes in lagoons allows for their use as manures on a broadcast basis. That means lower concentration of nitrogen on the soil surface. Growing crops can use the nitrogen more effectively and less is leached into the groundwater as a result.

To date, the adoption or nonadoption of alternative agricultural practices that would reduce nitrate loss from soils to groundwater has been based upon economics and convenience. There has been no special incentive to override these practical considerations with health concerns for private wells. This is true because the use of bottled drinking water for infants makes it possible to adjust to the hazard of nitrate occurrence in groundwater at little cost and inconvenience. Some municipalities, however, have spent large sums of money to develop alternative sources of water to meet the public health standard of < 10 ppm of nitrate nitrogen (NO_3–N) and < 45 ppm of nitrate (NO_3)*. (Remember that 1 ppm is the same concentration as 1 milligram per liter.)

Groundwater Use

The states using the most groundwater are California, Texas, Nebraska, Idaho and Kansas, respectively, as shown in Table 4-1. Only surface water, and not groundwater, is used in Washington, D.C.

With so much groundwater use, there must be many overdrafts. **Overdrafts** occur when more groundwater is removed than is replaced by water from the surface. There are many overdrafts, as recorded in Figure 4-2 and Table 4-2. Of the 106 groundwater resource regions in the United States, 60 of them (57 percent) are using more water than is being replaced by recharge. In other words, in 57 percent of the water resource regions, the water table is falling. A falling water table means that the NOW generation is living on the edge of disaster.

The most serious overdraft of groundwater is from the largest of U.S. aquifers, the Ogallala Aquifer under parts of eight states and extending for 220,000 square miles, shown in Figure 4-3.

Most groundwater use is for irrigation, illustrated in Figure 4-4.

Underground Injection of Chemical Wastes

In 1988, 1.2 billion pounds of chemical wastes were disposed of by injecting them underground. Louisiana injected 423.3 million pounds underground, the most of any state, and more than one-third of the nation's total. Louisiana's total injections were dominated by two companies: American Cyanamid Co. (175.2 million pounds) and Shell Oil Co. (156.4 million pounds). Texas ranked second among the states in underground toxic releases (390.8 million pounds). Du Pont in Beaumont, Texas, injected 108.1 million pounds and Monsanto in Alvin, Texas, released 102.8 million pounds underground. See Table 4-3 for more statistics on underground injection of chemical wastes.

* *Note:* milligrams per liter (mg/1) is the same concentration as parts per million (ppm)

Table 4-1. Groundwater Use by State*

State	Total Groundwater Use (millions of gallons per day)	State	Total Groundwater Use (millions of gallons per day)
Alabama	343	Nevada	905
Alaska	72	New Hampshire	84
Arizona	3,090	New Jersey	667
Arkansas	3,810	New Mexico	1,510
California	14,800	New York	1,100
Colorado	2,310	North Carolina	435
Connecticut	144	North Dakota	127
Delaware	79	Ohio	730
Florida	4,050	Oklahoma	568
Georgia	1,000	Oregon	660
Hawaii	655	Pennsylvania	759
Idaho	4,800	Rhode Island	27
Illinois	930	South Carolina	214
Indiana	635	South Dakota	249
Iowa	471	Tennessee	444
Kansas	4,800	Texas	7,180
Kentucky	205	Utah	790
Louisiana	1,430	Vermont	37
Maine	66	Virginia	341
Maryland	219	Washington	1,220
Massachusetts	315	West Virginia	227
Michigan	596	Wisconsin	570
Minnesota	685	Wyoming	504
Mississippi	1,580	Puerto Rico	175
Missouri	640	Virgin Islands	1
Montana	203		
Nebraska	5,590	Total: United States and Territories	72,702

* **Source**: U.S. Geological Survey

Figure 4-2. Ground waters over a large part of the nation and surface waters in semiarid and arid regions are being depleted. (Source: National Assocaition of Conservation Districts, "Soil Degradation: Effects on Agricultural Productivity," National Agricultural Lands Study, Washington, D.C., 1980.)

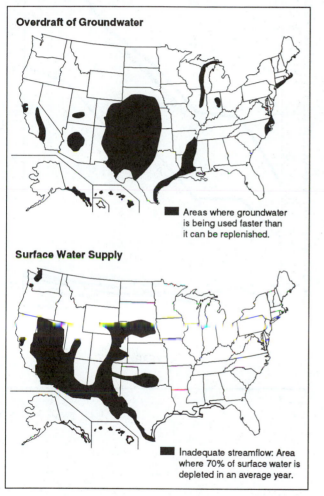

Overdraft of Groundwater

■ Areas where groundwater is being used faster than it can be replenished.

Surface Water Supply

■ Inadequate streamflow: Area where 70% of surface water is depleted in an average year.

Table 4-2. Groundwater Overdraft by U.S. Water Resource Regions*

Water Resources Region	Overdraft In Region (%)
New England	0
Mid-Atlantic	1.2
South Atlantic Gulf	6.2
Great Lakes	2.2
Ohio	0
Tennessee	0
Upper Mississippi	0
Lower Mississippi	8.5
Souris-Red-Rainy	0
Missouri	24.6
Arkansas-White-Red	61.7
Texas-Gulf	77.2
Rio Grande	28.1
Upper Colorado	0
Lower Colorado	48.2
Great Basin	41.5
Pacific Northwest	8.5
California	11.5
Alaska	0
Hawaii	0
Caribbean (Puerto Rico)	5.1

* **SOURCE**: United States Water Resources Council (1978). *The nation's water resources*, Vol., 1: Summary. Washington, D.C.

Figure 4-3. This is the Ogallala Aquifer, which lies beneath parts of eight states and covers 220,000 square miles. (Source: United States Department of Agriculture.)

Figure 4-4. United States trends in ground water use, 1950–80. (Source: United States Geological Survey.)

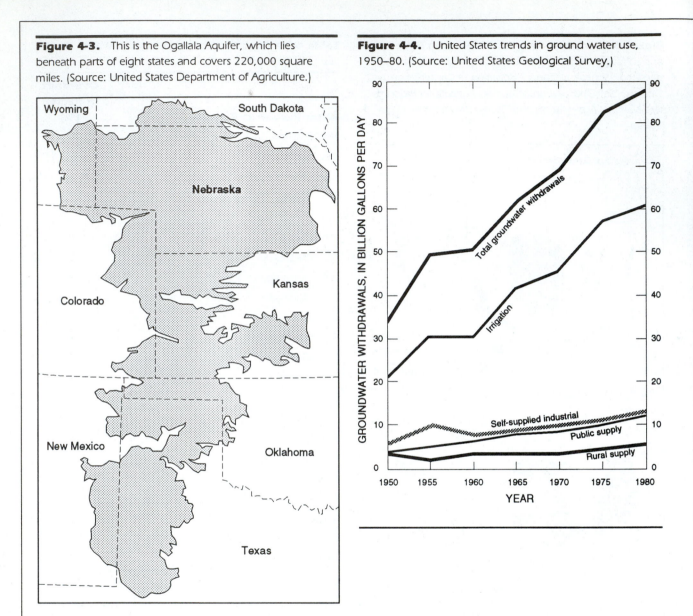

Table 4-3. Underground Injection of Chemical Wastes in 1988 by State*

State	Chemical Waste Injected Underground (pounds)	State	Chemical Waste Injected Underground (pounds)
Alabama	1,634.717	Nevada	0
Alaska	1,018	New Hampshire	0
Arizona	0	New Jersey	2,750
Arkansas	7,036.201	New Mexico	0
California	946,853	New York	251
Colorado	0	North Carolina	250
Connecticut	0	North Dakota	0
Delaware	0	Ohio	30,202,293
Florida	34,651,596	Oklahoma	6,353,464
Georgia	52,800	Oregon	1
Hawaii	1,051,509	Pennsylvania	750
Idaho	1,400	Puerto Rico	0
Illinois	7,340,184	Rhode Island	0
Indiana	34,820,650	South Carolina	0
Iowa	0	South Dakota	0
Kansas	90,766,710	Tennessee	49,906,110
Kentucky	30,000,250	Texas	390,826,922
Louisiana	423,320,002	Utah	0
Maine	0	Vermont	0
Maryland	2	Virginia	1,373
Massachusetts	4,000	Washington	0
Michigan	5,617,060	West Virginia	97,712
Minnesota	0	Wisconsin	250
Mississippi	46,806,563	Wyoming	27,113,559
Missouri	500		
Montana	0	Total: United States	1,215,343,908
Nebraska	68,208		

* **Source**: United States Environmental Protection Agency (1988). *Toxics in the community: National and local perspectives.* Washington, D.C.

Summary

The lithosphere is the rocky crust of the Earth. In addition to providing a surface on which we live and the soil in which we grow our food, the lithosphere also holds our groundwater. In the United States, about half of us get our drinking water from groundwater.

Not all water in the soil is considered groundwater. The free water that fills the voids between soil particles and rocks is groundwater. Groundwater is held in the soil at and below the water table. That is the level in the ground below which those voids are filled with water. If a hole is drilled into the ground (a well for instance), it will fill with water at the water table. When water is removed from the well, more water will move into the well from the ground. The rate at which the water being removed can be replaced is dependent of the porosity of the ground around the well. That is dependent largely on the kind of materials making up the ground at that point. Clay releases water very slowly. Sand and fractured limestone release water rapidly.

Pesticides and nitrates make up the primary contaminants in groundwater. In earlier years, pesticides were developed with little concern for how long they lasted in the environment. Persistent pesticides were often considered desirable because they did not break down quickly—they were long-lasting. It is now understood that such pesticides remain active in the environment, sometimes for many years. They can contaminate our water supplies.

As a result of a growing understanding of this problem, pesticides that are nonpersistent have replaced persistent pesticides. Nonpersistent pesticides break down into harmless compounds when exposed to air, sunshine, or water. Some nonpersistent pesticides are volatile so that they enter the atmosphere quickly. There, exposure to sunlight breaks them down. Other pesticides bind tightly to the organic matter in surface soil and so do not percolate into the groundwater supply.

Nitrates in the drinking water supply are a major problem in some areas. Nitrogen compounds are naturally occurring in the soil. They are essential for plant growth. The problem occurs when more nitrogen than is being used by crops is added to the soil. Part of the excess nitrogen salts can move into the water supply where they can become a health hazard in concentrations that are too high.

DISCUSSION QUESTIONS

1. How are health-advisory limits determined for pesticide residues?

2. What agricultural practices can lead to reduced nitrates in groundwater?

3. What states allow the most underground injection of chemical wastes?

4. Why are some pesticides more polluting than others?

5. How do chemicals such as nitrates and pesticides get into the groundwater?

6. What are some steps that can be taken in agriculture to decrease groundwater pollution?

ADDITIONAL ACTIVITIES

1. Locate a well in your community. If you live in a large city with no wells, perhaps you have a relative who owns a well. Find out how far below the surface of the land the surface of the well water is. That depth is the water table where the well is located.

2. Find out where the water for your community comes from. Does it come from groundwater or from surface water? If it is groundwater, find out whether the water table has been rising, falling, or stable for the past decade?

3. Invite a speaker to your class from your community's water treatment facility. Another activity would be to conduct a field trip to your local water treatment facility.

CHAPTER 5
Soils and Environment

Courtesy United States Department of Agriculture, Soil Conservation Service.

Terms to Look for and Learn

Clods	Permafrost
Ecological Erosion Control	Regolith
Geologic Materials	Salinity
Hermaphrodite	Sanitary Landfill
Infiltration	Sediment
Mechanical Erosion Control	Soil
Peds	Soil Conservation District
Percolation	Soil Survey

Learning Objectives

After reading this chapter and participating in the activities, you should be able to:

- Explain the relationship of the vegetative cover and soil erosion.

- Describe major mechanical and ecological means of erosion control.

- Discuss the effects of soil organisms, such as earthworms, on soil erosion.

- Discuss the effects of soil pollution on soil erosion.

- Explain how salt concentrations in water can affect the soil.

- Discuss the effects of chemical fertilizers on the Earth's soil.

- Discuss appropriate soils and geologic materials for landfills.

Overview

Soil is at the beginning of the food chain that sustains all plants, animals, and human beings. Soils are the filters that absorb potentially toxic elements such as lead, copper, and mercury. However, soils are also the greatest polluters of the environment—our environment.

In agriculture, the surface of the Earth in which plant roots are growing is usually designated as **soil**, and the underlying substratum as parent material. Soil is also considered as the surface six feet or the depth to bedrock or other layers that restrict root extension.

The terms soil and **regolith** often have the same meaning to a geologist and include all unconsolidated materials that rest on bedrock. Thus, the word **soil** to a geologist may apply to all materials on the surface of the Earth overlying bedrock, including weathered bedrock, windblown sand, loessial (wind blown) silt, glacial materials, alluvium, volcanic ash, and organic accumulations known as peats and mucks.

An engineer considers **soil** to be natural aggregates of mineral grains, with or without organic matter, that can be separated by gentle mechanical means such as agitation in water. By contrast, rock is a natural aggregate of mineral grains usually bound by strong and cohesive forces.

For the purposes of this book, soil is defined as the natural medium for the growth of land plants and extends to the depth of their roots.

The Earth could be compared to an apple. As the peeling protects the apple, so soil and the plants it supports are the Earth's protective covering. Peel an apple and within minutes deterioration sets in. Remove the soil and its vegetation and the Earth is immediately degraded. It is no longer as productive for living things or as effective in pollution control. In fact, once its vegetative cover is removed, the soil becomes the number one polluter of the environment.

Soil originated from rocks in place or from weathered rock materials moved and deposited by water, wind, ice, or gravity. Bacteria, fungi, actino-mycetes, lichens, mosses, and higher plants such as grasses and trees add the living component to soil formation, soil stabilization, soil production, and environment enhancement.

Soils are such an intimate part of the environment that the quality of enhancement or degradation is interdependent. Soils may pollute, be polluted, or enhance the chemical, physical, biological, and aesthetic environment. Soils pollute by supplying erosion sediments when the soil has been abused.

Soils are sometimes polluted by heavy metals, excess salinity, or permafrost. Soils may be used to control pollution and to enhance the environment by supporting luxuriant vegetation, making carbon monoxide harmless, neutralizing excess nitrates and phosphates, providing a safe and productive medium for the disposal of sewage sludges and wastewaters, and serving as a sanitary landfill (see Figures 5-1 and 5-2).

Soils and Polluters

In 1939, H.H. Bennett, who established the Soil Conservation Service, estimated annual losses of soil in the United States at 3 billion tons. Twenty-eight years later, Donald A. Williams, then Director of the Soil Conservation Service, estimated that four billion tons were lost each year. Fifteen years later, the erosion estimate was increased to 6.4 billion tons per year. Either different criteria were used for determining erosion losses in 1939, 1967, and 1982, or erosion had accelerated during the forty-three years the Soil Conservation Service had made estimates of erosion.

The facts are that soil erosion sediments in 1982 were the major source of pollution of the land and water environment. Assuming that 6.4 billion tons a year is the current sediment loss, half of this is estimated to come to rest at the base of slopes and the other half reaches streams and lakes to pollute them.

Figure 5-1. The soil—the Earth's protective cover—has eroded and is no longer productive or effective for control of pollution. In fact, the eroded sediment is now a major pollutant (see Figure 5-2). (Courtesy United States Department of Agriculture, Soil Conservation Service.)

Figure 5-2. The foot of sediment that has smothered all vegetation here could have come from the hillside gullies in Figure 5-1. There must be a better way to treat the soil and the environment! (Courtesy United States Department of Agriculture, Soil Conservation Service.)

Table 5-1 presents the latest data on erosion rates under cropland, pastureland, rangeland, and forestland. The relative erosion rates by water are cropland (cultivated), cropland (all), forestland (grazed), pastureland (all), rangeland (all), and forestland (not grazed), respectively. Relative erosion rates by wind are cropland (cultivated), cropland (all), rangeland (all), and forestland (grazed), respectively.

Table 5-1. Erosion Rates and Acreages of Cropland, Pastureland, and Rangeland in the United States in 1982*

Land Use	Acres (millions)	Erosion (tons/acre/year **) By Water	By Wind
Cropland (all)	421.4	4.4	3.0
Cropland (cultivated)	323.0	4.8	3.3
Pastureland (all)	133.3	1.4	0.0
Rangeland (all)	405.9	1.4	1.5
Forestland (all)	393.8	***	***
Forestland (grazed)	***	2.3	0.1
Forestland (not grazed)	***	0.7	0.0

*National Research Council (1986). *Soil conservation: Assessing the national resouces inventory.* Vol. 1, pp. 7, 8. Washington, D.C.: National Academy Press.

Data Source: United States Soil Conservation Service (1984). *1982 national resources inventory*. Washington, D.C.: United States Department of Agriculture.

***Data not available.

Soil erosion can reduce crop yields by *reducing* plant nutrients, soil organic matter, clay content, water-holding capacity, and the proliferation and depth of rooting of plants. Modern technology and management have included better liming and fertilization, plant breeding for higher yields, more timely and better use of pesticides, and more timely harvesting. So far, plant breeding, management, and nutritional technologies have been sufficient to overcome loss of original productivity of soil by erosion. How long can this continue, and at what cost?

A six-year study on three soils in Indiana concluded that past severe soil erosion reduced corn yields 15 percent and soybean yields 24 percent. In high rainfall areas, erosion and sedimentation rates from construction sites and surface mining sites have been measured and estimated at rates as high as 50,000 tons per square mile per year, as illustrated in Figure 5-3. When this sediment is deposited to a six-inch depth in areas such as a black walnut plantation in Illinois, almost all of the trees die. Surface mining in the United States for coal, phosphate ore, oil shale, and uranium is displayed in Figure 5-4.

Control of Soil Erosion and Sedimentation

Particles of soil normally tend to form aggregates, or clumps. Aggregates that form from natural forces are called **peds**. Peds result from the actions of decaying organic matter, earthworms, plant roots, and sticky soil particles (clay). Aggregates that form when the soil is manipulated (like plowing) when it is too wet, are called **clods**. These peds and clods are important in a healthy soil because they are separated by pores that allow water to enter the surface (**infiltration**) quickly and move downward readily (**percolation**).

Surface soil erosion starts when falling raindrops strike bare soil and shatter the peds and clods. The mud so formed fills the pores in the soil and slows water movement downward. When rainwater cannot infiltrate, it moves over the surface of the soil along with soupy mud. This is sediment. The muddy sediment moves across the soil surface, scouring other soil. The net result is rill, sheet, and gully erosion. The control of this erosion and sedimentation must start with establishing a soil cover of vegetation. When raindrops strike living or dead plants, their velocity is reduced and the water moves down plant stems to soak into the soil. The net result is greatly reduced erosion (much less sediment), and water which infiltrates the soil for greater plant growth.

Thirty-two grassed soils in the thirteen North Central States had water infiltration rates an average of 47 percent greater than the same soil series when cultivated. Fifteen of these soils had the same infiltration rate, whether grassed or cultivated; eleven had a rate on grassed soils at least twice that of the same soil series when cultivated; and one grassed soil series had

Figure 5-3. Strip mining for coal in North Dakota disturbs both soil and water drastically. Bare and barren slopes erode before they can be regraded and revegetated. (Courtesy United States Department of Agriculture, Agricultural Research Service.)

an infiltration rate more than five times the rate on the same soil series when cultivated.

Greater infiltration means less runoff, less erosion, and less sedimentation. It also means more water stored in the root zone (rhizosphere) for plant growth.

Surface crusts form when raindrops, at a speed of twenty miles per hour, strike bare soil. The crusts reduce infiltration and force more water to move over the soil surface and cause more erosion and sedimentation, as shown in Figure 5-5. A dense vegetative cover, living or dead, reduces surface soil crust formation and almost eliminates erosion and sedimentation, as shown in Figure 5-6.

Although the ecological control of soil erosion in principle is to establish a cover of living plants or plant residues, in practice this is feasible in many places but not in all places. **Ecological erosion control** can include planting trees in some areas, pasture grasses in others, and cultivated crops can be planted in narrow rows and scientifically fertilized or planted in last year's residues. The use of mulches in urban developments and in highway construction projects is becoming very popular as a means of stabilizing the soil until seeded perennial vegetation grows sufficiently to protect the soil against the violent splash of the beating raindrops.

When ecological methods of erosion control are not adequate, **mechanical erosion control** or a combination of the two are used. Mechanical erosion controls are measures taken to reduce erosion, beyond simple vegetative cover of the soil. Mechanical methods of erosion control consist of:

- Establishing contour strips of cultivated crops alternating with sodlike crops, as shown in Figure 5-7

- Constructing terraces on sloping cropland to slow the rate of surface runoff

- Practicing no-till culture. This means planting a crop directly into the residue of the previous crops. Weeds are controlled by an herbicide. In a North Carolina experiment with no-till corn vs. conventional corn culture, infiltration and corn yields were up to fifty percent greater on no-till, as shown in Figure 5-8.

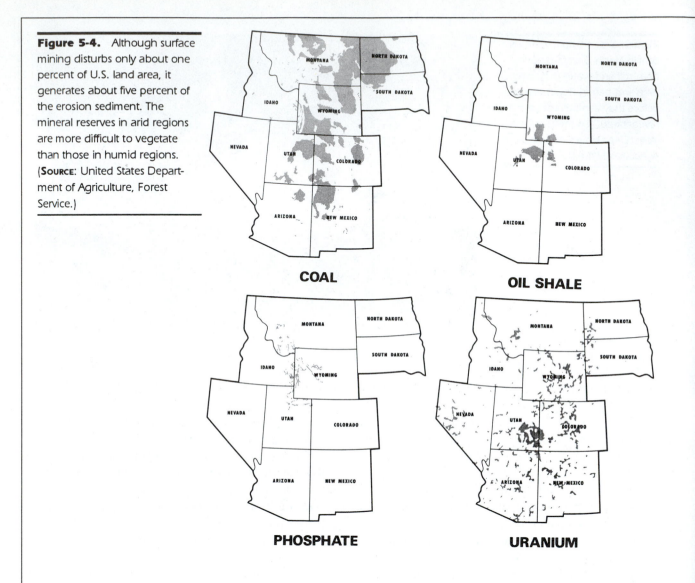

Figure 5-4. Although surface mining disturbs only about one percent of U.S. land area, it generates about five percent of the erosion sediment. The mineral reserves in arid regions are more difficult to vegetate than those in humid regions. (**Source**: United States Department of Agriculture, Forest Service.)

COAL

OIL SHALE

PHOSPHATE

URANIUM

■ Covering exposed soil with mechanical coverings such as mulch, gravel, rocks, or even concrete.

The April, 1992 "Missouri Conservationist" reported that the Missouri Soil Conservation Districts in 1991 saved an estimated ten million tons of soil on 136,000 acres of farm land. The soil was conserved by farmers using no-till farming, seeding soil-stabilizing crops, establishing terraces, and building ponds. The cost was estimated at $1.82 per ton of soil saved.

Earthworms and Environment

No-till farming leaves crop residues on the soil surface. These are food for earthworms. Earthworms in turn make burrows in the soil and in reality till the soil into a desirable seed bed. Earthworm casts are also more fertile than the native soil because the nutrients are partly digested. Compared with the soil in which

Figure 5-5. Where crop residues or living vegetation do not break the velocity of raindrops, soil crusts form (at arrows) that decrease plant productivity by restricting plant emergence and reducing soil oxygen and water to plant roots. More sediment is also produced to pollute the environment. (Courtesy United States Department of Agriculture, Soil Conservation Service.)

Figure 5-6. A dense growth of crimson clover intercepts raindrops and breaks their velocity; erosion and sedimentation are thereby reduced nearly to zero. (Courtesy United States Department of Agriculture, Soil Conservation Service.)

Figure 5-7.   Contour strips of corn alternating with strips of alfalfa in Wisconsin. Soil washed from the cultivated corn strip will come to rest on the adjoining lower strip of alfalfa. (Courtesy United States Department of Agriculture, Soil Conservation Service.)

Figure 5-8.   Corn planted no-till in last year's wheat residue is shown here. (Courtesy United States Department of Agriculture, Soil Conservation Service.)

earthworms live, earthworm casts are much richer in potassium, phosphorus, nitrates, magnesium, and calcium (see Figure 5-9).

Earthworms occur throughout the world, primarily in the surface of the Earth. Any environment supplying adequate water, oxygen, and food may be a habitat for one or more of the estimated hundreds of species around the world.

Manure piles, sewage beds, and septic tank drain fields are ideal sites for certain species of earthworms. Important sites in humid tropical climates include rotten trees and logs and axils of banana leaves. Earthworm species also live in caves, mines, deep moss on high mountains, and in nonfunctional gutters where leaves accumulate.

Longevity of earthworms varies with the species and the environment. When protected, a life span of two to ten years is normal for most species. However, this period is shortened by pesticides, fertilizers, clean cultivation, excess heat or cold or dryness, and predators.

Earthworms are about 85 percent water and cannot withstand a dry environment for long. In fact, earthworms have the capability of surviving under aerated water (as in an aerated fish tank) for as long as a month or more. (Fishermen are happy about this.) It is also a fact that earthworm capsules (eggs) will hatch under water.

Food for earthworms consists of organic matter such as selected forest tree leaves, most crop residues,

Figure 5-9. (A) Clean tillage; (B) No-till culture. Note mound of earthworm casts; (C) An earthworm. (Courtesies: (A) and (B), Texas A&M University; (C) Julian P. Donahue.)

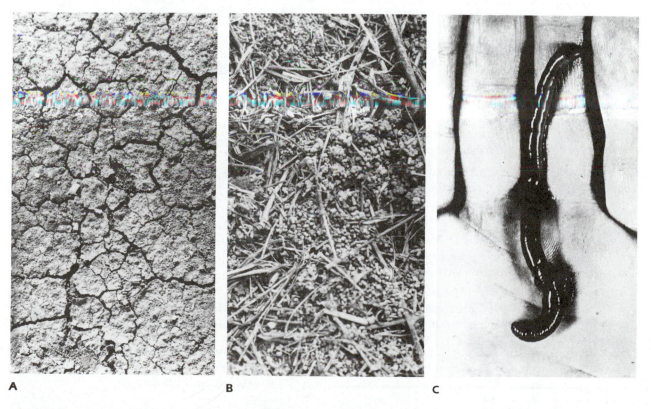

A   B   C

almost all grasses, animal manures, animal carcasses, sewage sludge, feed grains, and milk.

Predators of earthworms include robins, moles, badgers, shrews, centipedes, and a few species of slugs, beetles, and leeches. The six most common species of earthworms in the United States are listed in Table 5-2, together with comments on their adaptation and use.

Each individual earthworm is an **hermaphrodite**; i.e., each has functioning male and female sex organs. Technically, this permits each earthworm to fertilize itself, known as parthenogenesis. Seldom, however, has this sex act been studied by scientists.

Research workers throughout the world have shown that a soil rich in earthworm casts contains a large percentage of water-stable aggregates. The net result is that beating rains and flowing irrigation waters will not destroy this soil structure which is so desirable for plant growth.

Soils with many earthworm burrows not only are less apt to puddle, they also contain perhaps five percent of the total soil volume in holes made by the earthworms. These holes permit water to move into the soil more rapidly, thus reducing surface runoff and erosion. Oxygen exchange from the atmosphere to plant roots is also hastened. In addition, soils with earthworms are capable of holding more available (capillary) water. Plant roots actually grow down the pores made by the earthworms. Thus, we can say that earthworms are "plant-root friendly."

Earthworms have some drawbacks. They can carry these diseases: gapeworm of poultry and lungworm of swine. This swine lungworm can carry type A influenza, a human disease. Also, earthworms may be carriers of several fungi spores, a potato nematode, and the foot-and-mouth disease virus.

Table 5-2. Earthworms Most Common in the United States*

Common Names	Comments
Fieldworm	Adapted to warm humid South; not popular for commercial production.
Greenworm	Common in fields; good for fishing; not adapted to commercial production.
Manureworm	Abundant in well-manured soils; sold as "red wiggler." Well adapted to commercial production.
Nightcrawler	Largest of U.S. species; common in wet, productive soils; excellent for fishing; not adapted to confined commercial production.
Redworm	Abundant in manure piles and other organic refuse; well adapted to commercial production.
Slimworm	Exists in cool, humid regions; too thin for baiting fishhooks; not adapted to commercial production.

*Adapted from Edwards, C.A., & Lofty, J.R. (1972). *Biology of earthworms*. Ontario, CA: Bookworm Publishing Co.

Polluted Soils

Soils may be so polluted that no vegetation will grow. This is the situation with the Copper Basin in Tennessee. In southeastern Tennessee is a ten-thousand-acre watershed that has a long and sad history of toxicity. Copper was smelted from local ore in four plants between 1854 and 1879. Ore was roasted over fires from charcoal made from local hardwoods. Smelting resumed in 1891 and by 1907 sulfur dioxide fumes had destroyed so much vegetation that the companies were forced to control these fumes. The net results of the sulfur fumes are as follow:

■ Six thousand acres around the smelters remain mostly bare and barren in 1990

■ Twenty thousand additional acres support only grass and weeds, the trees having been killed by the SO_2 (sulfur dioxide) fumes

- From bare eroded soil, sediment washed into Ocale Reservoir 3, and eliminated all of its water storage capacity

- The sediment contains toxic SO_2 (strongly acid), Cu (copper), Fe (iron), Zn (zinc), and Pb (lead)

Some of the barren areas have been reclaimed by deep tillage and seeding and planting of adapted herbaceous and woody species, as shown in Figure 5-10. This illustrates that some forms of contamination can be very long lasting. Even though this area was polluted long ago, the results are still with us and are predicted to continue long into the future.

Salinity and Environment

Waters available for irrigation in the arid and semiarid West are usually salty, and more salts remain in the soil when additional irrigation water is evaporated or transpired. Salts thus accumulate to degrade soil productivity and result in **salinity**. Approximately half of the irrigated acres in the West are affected by salt.

Rate of salty degradation of soils varies as a result of the original saltiness of the soil, internal drainage of the soil, saltiness of the water, the amount of water applied, evaporation, transpiration, and the amount and intensity of rainfall. When a soil has accumulated enough salt to be injurious to certain crops, perhaps a more tolerant crop can be grown successfully. Table 5-3 lists the relative soil salt tolerance of selected field crops; forage crops; vegetable crops; fruit, nut, and vine crops; and ornamental shrubs.

Permafrost

The word "**permafrost**" was coined by S.M. Muller in 1947 as a short way of saying permanently frozen soil. Seasonal frost is a surface phenomenon which means frozen soil at the surface during winter and thawing during summer. Permafrost means permanently fro-

zen soil under the seasonal frost layer. The permanently frozen soil, however, may be horizontally or vertically continuous or discontinuous, depending upon latitude, snow cover, vegetation, topography, and soil texture.

Permafrost, both continuous and discontinuous, occurs on the highest of the Rocky Mountains in the United States as far south as 36° north latitude, and in Canada and Alaska at lower elevations north of 60° north latitude. Continuous permafrost in general occurs north of the Arctic Circle (66-1/2° north latitude) and south of the Antarctic Circle (66-1/2° south latitude). In Alaska it also occurs in the western part of the Yukon areas adjacent to the Bering Sea. Permafrost occurs under one-fifth of the world's land surface at a depth of one to three thousand feet.

Moving groundwaters tend to melt permafrost. For that reason the soils along drainage areas have less permafrost than those on top of river divides. A stand of birch or spruce trees shades the soil and permits the formation of more and thicker permafrost. A heavy mat of sphagnum moss or other living or dead ground cover also is conducive to more and thicker permafrost. Most unusual is the fact that heavy snow cover reduces the formation of permafrost. Areas of solid rock, gravel, or sand contain very little permafrost. Fine textured soils such as those high in silt or clay, or organic matter, encourage the formation of permafrost.

Permafrost is extremely fragile when used as foundations for buildings, subgrades for roads, septic tank sewage disposal fields, lagoon sewage disposal, sanitary land fills, or oil pipelines as shown in Figure 5-11. Heat from a furnace room or stove in a building will melt the permafrost and the foundation will fracture and sink differentially into the soil. Subgrades for roads will receive more heat from the sun after trees are cut and the road is established. Permafrost melts and the road collapses.

Septic tank disposal fields require a porous soil for about six feet in depth. Any water from a septic tank drain field that moves into the soil does so through melting caverns in the permafrost and quickly

Figure 5-10. Bringing dead soil back to life is shown here. A copper smelting plant at Cooperhill, Tennessee spewed sulfur dioxide (SO_2) fumes and traces of copper and other toxic metals into the air from 1854 to 1907. Nearby, all vegetation was killed over several thousand acres. (A) Deep tillage plus planting of adapted species; (B) one year later; (C) two years later. (Courtesy Tennessee Valley Authority.)

A

B

C

Table 5-3. Relative Soil Salt Tolerance of Various Crops*

Sensitive	Moderately Tolerant	Tolerant	Highly Tolerant
		Field Crops	
Field bean	Soybean	Wheat (grain)	Barley (grain)
	Castorbean	Oats (grain)	Rye (grain)
	Sesbania	Safflower	Sugar beet
	Rice	Cotton	
	Flax	Sunflower	
	Guar	Triticale	
	Sorghum (grain)		
	Corn (field)		
		Forage Crops	
White clover	Reed canarygrass	Hardinggrass	Bermudagrass
Dutch clover	Oats (hay)	Kleingrass	Crested wheatgrass
Alsike clover	Orchardgrass	Alfalfa	Barley (hay)
Red clover	Bromegrass	Birdsfoot trefoil	Rye (hay)
Ladino clover	Big trefoil	Hubam clover	Panicgrass
Crimson clover	Grama grasses	Dallisgrass	Alkali sacaton
Meadow Foxtail	Sour clover	Tall fescuegrass	Rhodesgrass
Kentucky bluegrass	Milkvetch	White sweetclover	Saltgrass
	Timothy	Yellow sweetclover	Western wheatgrass
	Sudan-sorghum hybrids	Perennial ryegrass	
	Sorghum (forage)	Wheat (hay)	
	Corn (forage)	Johnsongrass	
		Vegetable Crops	
Carrot	Lettuce	Tomato	Asparagus
English pea	Corn (sweet)	Beet	
Radish	Potato	Kale	
Celery	Squash	Spinach	
Green bean	Onion	Broccoli	
Lima bean	Sweet potato	Cabbage	
Kidney bean	Bell pepper	Cauliflower	
Cucumber	Blackeyed pea	Watermelon	
Rhubarb	Muskmelon		

Cont'd

Table 5-3 Continued. Relative Soil Salt Tolerance of Various Crops*

Sensitive	Moderately Tolerant	Tolerant	Highly Tolerant
Fruit, Nut, and Vine Crops			
Grapefruit	Pecan	Pomegranate	Date palm
Orange	Peach	Fig	
Lemon	Apricot	Olive	
Avocado	Grape		
Pear	Quince		
Apple			
Cherry			
Plum			
Walnut			
Blackberry			
Raspberry			
Strawberry			
Boysenberry			
Ornamental Shrubs			
Viburnum	Spreading juniper	Oleander	Purple sage
	Arborvitae	Bottlebrush	Saltcedar
	Lantana		
	Pyracantha		
	Privet		
	Japonica		

*PRIMARY SOURCE: Publications of the U.S. Salinity Laboratory, Riverside, California.

pollutes wells and streams. A lagoon built for sewage disposal will contain waters warmed by livestock; the warm waters may melt the permafrost differentially and result in seepage losses and pollution of the groundwaters.

An oil pipeline built underground in permafrost, because of the higher temperature of the oil, will melt its way through the permafrost and the pipeline will collapse. Plans for the Alaska Pipeline Service Company to transport oil from the Prudhoe Bay Oil Field south across Alaska to a port at Valdez, a distance of 789 miles, included construction of the pipeline through the permafrost area on supports several feet *above* the soil surface.

Chemical Soil Testing

Poorly managed soils are the source of sediment, which is the greatest polluter of the environment. Well-managed soils are necessary for the support of

Figure 5-11. This house in Alaska was built over permafrost. Because of differential melting from heat in the house, the house settled and could not be lived in. A soil survey could have identified a suitable building site. (Courtesy United States Department of Agriculture, Soil Conservation Service.)

Table 5-4. Major Nutrients, Secondary Nutrients, and Micronutrients Essential for Plant Growth and Reproduction

Element	Symbol
From Air and Water	
Carbon	(C)
Hydrogen*	(H)
Oxygen	(O)
Nitrogen**	(N)
From Soil and Fertilizer	
(Major Nutrients)	
Nitrogen**	(N)
Phosphorus	(P)
Potassium	(K)
(Secondary Nutrients)	
Calcium	(Ca)
Magnesium	(Mg)
Sulfur	(S)
(Micronutrients)	
Iron	(Fe)
Boron	(B)
Manganese	(Mn)
Copper	(Cu)
Zinc	(Zn)
Molybdenum	(Mo)
Chlorine	(Cl)
Nickel	(Ni)

* Hydrogen for plant growth comes from water.

** Nitrogen comes from the air but only legumes and a few nonlegumes with the aid of bacteria can use the nitrogen from the air.

SOURCE: U.S. Department of Agriculture.

luxuriant vegetation to stabilize the soil and control sediment pollution. Where rainfall or irrigation water are adequate, some soils are fertile enough to support adequate vegetation to hold the soil in place. A chemical test is necessary to determine the adequacy of plant nutrients in most soils (see Table 5-4).

For many years chemical fertilizers have been one of the "best buys" for use in increasing protective vegetative cover and crop yields. Costs of land, labor, machinery, and marketing have increased at a faster rate than the cost of fertilizers.

The logic of many farmers has been that they should apply as much fertilizer per acre as is recommended *plus a little more for insurance.* For many, the net result has been not only a waste of money but a pollution hazard to the environment, especially to the groundwater.

Nitrogen and phosphorus fertilizers sometimes are a pollution hazard. Potassium, although slightly leachable when used on sandy soils and on organic soils, does not seem to pollute the environment—even when used in excess.

The only logical way to reduce pollution of the environment by nitrogen and phosphorus fertilizers is by periodic soil tests and by following the recommendations. To obtain a soil test, contact the local County Extension Director or the land-grant university in the state where you live. If the fertilizer recommendation based upon the soil test is followed (and never exceeded), the hazard of pollution of the water environment by fertilizer will be reduced to almost zero.

Each of the fifty states in the United States offers a scientific soil testing service for farmers, golf greens keepers, vegetable producers, backyard gardeners, and homeowners who are interested in luxuriant plant growth produced efficiently. In addition, many private soil testing laboratories exist to serve special-interest groups.

Until about 1970, soil testing was performed solely to serve as a basis for making a scientific fertilizer recommendation for the economic production of plants. With the recent awareness of fertilizers as possible pollution hazards to the surface water and groundwater environments, soil testing now serves the dual purpose of efficient plant production and a reduction in the hazard of polluting the nation's waters.

Suggestions on how to take a soil sample for chemical testing are presented in Figure 5-12.

Soils for Sewage Wastewaters

One of the best ways to recycle sewage wastewaters is to treat them and then use them to irrigate crops. Hazards to the environment inherent in this method of recycling include the spreading of pathogens and the degradation of the soil. With the use of treated effluent, the disease-spreading hazard is nearly zero; but possible injury to the soil is more difficult to ascertain.

A fourteen-year study of the influence of wastewater effluent on soil properties was conducted by the

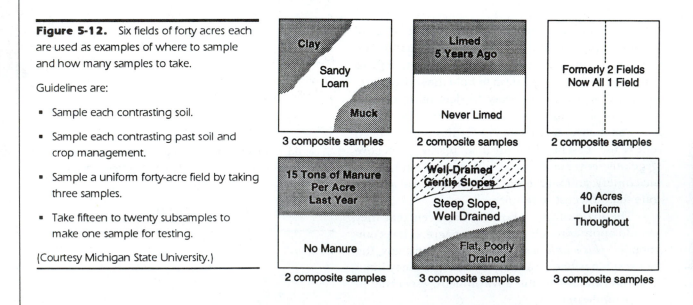

Figure 5-12. Six fields of forty acres each are used as examples of where to sample and how many samples to take.

Guidelines are:

- Sample each contrasting soil.
- Sample each contrasting past soil and crop management.
- Sample a uniform forty-acre field by taking three samples.
- Take fifteen to twenty subsamples to make one sample for testing.

(Courtesy Michigan State University.)

staff of the Agricultural Experiment Station at the University of Arizona. A comparison was made of soil characteristics of the soil resulting from irrigating with well water vs. sewage wastewater effluent from the city of Tucson, Arizona. After fourteen years, the soil receiving wastewater treatment, as compared with well water, had the same pH, more total soluble salts, more phosphorus, more nitrates, more organic matter, but a *lower infiltration rate.* Crop yields were about the same on both treatments.

Until someone proves a better method, it appears that the best method of recycling sewage effluent wastewater has been demonstrated in Arizona, especially when the soil is suitable.

A soil suitable for accepting large and continuous amounts of sewage effluent wastewaters may be characterized as follows:

- Drainage (internal): must be good (at least six feet to the water table)

- Slope: must be nearly level

- Texture: medium textured (loams), avoid deep sands and fine-textured clays

- Structure: stable, coarse

- Permeability: surface horizon at least 0.5 inch of water infiltration per hour and subsurface horizons at least 0.15 inch per hour

- Depth: at least six feet to bedrock, fragipan, or other very slowly permeable layer. If soil is a coarse sand, depth must be at least twenty-five feet to a water table.

Note that in Figure 5-13, three inches of water will wet a sand soil to four feet compared with one foot in a clay soil.

Soils for Sanitary Landfills

A **sanitary landfill** is a technique for disposal of solid wastes in soils without causing excess pollution to the water, air, or aesthetic environment. The selection of the most suitable soil site has often been neglected because of the communication gap among the public health sanitarian, the engineer, the geologist, and the soil scientist/soil surveyor.

Soils are usually described as the surface mineral and organic materials that support plant growth. Other attributes of soils are that they support plant roots, extend to the depth of bedrock; or in the absence of roots or bedrock, extend to a depth of six feet.

Geologic materials always touch soils at the soils' lower extremities and extend to the center of the Earth. These geologic materials may range in compo-

Figure 5-13. Three inches of water will wet each soil texture approximately to the depth shown. **Note**: Rain and/ or irrigation water on sands in excess of three inches may contaminate ground waters. (**Source**: United States Department of Agriculture.)

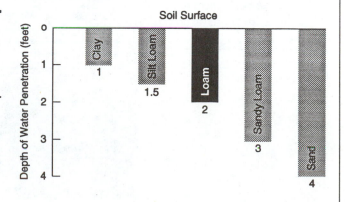

sition from loose sands and gravels to compact clays to sedimentary and metamorphic rocks to igneous rocks.

Fluctuating water tables often exist within depths that are considered as soil; permanent water tables usually occur below the soil in the geologic materials.

In planning for the use of the soil as a sanitary landfill, the underlying geologic materials should also be examined and evaluated. For example, in areas of limestone bedrock, vertical openings in the bedrock made by the waters of percolation dissolving the lime over geologic time, serve as channels for rapid pollution of the permanent water table. A sanitary landfill located over solution fissures, for example, represents faulty and hazardous planning.

A geologic deposit of coarse sand and gravel from the soil to the water table is almost as dangerous because rapid leaching can contaminate the water table. Abandoned limestone rock quarries can be used as sanitary landfills only if a water-tight substance such as fine clay or heavy plastic is used to seal the fractures. Granite rock quarries no longer in use are usually naturally waterproof. Deep sands can be used if the depth to the water table is more than twenty-five

feet. The ideal soil and geologic materials for use in sanitary landfills are loams, silt loams, sandy clay loams, and clays.

Soil Surveys to Enhance the Environment

Soil surveys are the systematic mapping, classification, and description of the soils and soil characteristics in a given geographic area. Soil surveys originated in the United States in 1899, and in 1952 the National Cooperative Soil Survey was created. These soil surveys are published on a county basis in a comprehensive report that includes a county soil map on a photographic base at a scale of approximately three to four inches to the mile. A soil survey is being performed in Figure 5-14.

They were originally made for agricultural planning and management, but newer uses of the county soil maps and the descriptive and interpretive reports include serving as a guide to the best location of

Figure 5-14. Converting farm and ranch land to urban use can be done environmentally correctly by first making a detailed soil survey, as is being done here in Kentucky. (Courtesy United States Department of Agriculture, Soil Conservation Service.)

urban developments, roads, foundations of buildings, airfields, and for sanitation. A soil survey can also determine suitability of a home site, which could have been helpful to the flood victims shown in Figure 5-15. Environmental uses of these soil maps and reports include the prediction of the suitability of each mapping unit for septic tank sewage disposal, lagoon sewage disposal, and sanitary landfills. Such interpretive information on three soil mapping units in the State of Maine is displayed in Table 5-5. Similar information is available for many counties of the United States.

Information can be obtained on the availability and use of the soil maps and reports from the local County Extension Director, Soil Conservation Service, Soil Conservation District, or State Agricultural University.

Each soil mapping unit can be accurately examined for many uses. For example, each mapping unit has a specific drainage class. Soil drainage classes are important for predicting crop adaptation, need for

Table 5-5. Soil Mapping Units in Maine and Their Relative Suitability for Enhancing the Environment

| Mapping Unit | Soil Suitability for | | |
	Septic Sewage Disposal	Lagoon Sewage Disposal	Sanitary Landfills
Adams fine sandy loam	Poor	Fair	Very Poor
Berkshire fine sandy loam	Good	Poor	Good
Melrose fine sandy loam	Poor	Good	Good

Note: Only level to gently sloping fields were used in making this table.

Source: U.S.D.A. Soil Conservation Service.

Figure 5-15. A soil survey of this area could have warned this family to not build a home on flood-prone soils. (Courtesy United States Department of Agriculture, Soil Conservation Service.)

artificial drainage, suitability of soil for use as a septic tank drainage field or lagoon, or use for receiving sewage treatment plant effluent. Determining the depth to a water table throughout the year on each prospective building site is very expensive and time consuming. Drainage classes, as mapped on standard soil surveys in the United States, could be used to substitute for infiltration capacity (percolation) tests or periodic but year-round readings on the depth of the water table. By using drainage classes the economy in time and money should make people happy by saving them millions of dollars. This hypothesis was tested and the results follow.

For a period of fifty-six months, water table readings were made at thirty-five locations in Montgomery County in Southeastern Pennsylvania near Philadelphia. The primary purpose of the study was to correlate drainage classes of the soil mapping units, as recognized in making soil survey maps, with actual field measurements of depth to water tables.

Field soil mapping units are classified into drainage classes—meaning *internal* drainage, not surface draining as recognized by engineers. The four usually recognized drainage classes are listed below, and shown in Figure 5-16.

A—Well drained

B—Moderately well drained

C—Somewhat poorly drained

D—Poorly drained

In field mapping, the drainage classes are differentiated on the basis of color of the horizons in the soil profile. A uniform bright color (often red) throughout the soil profile is indicative of a soil that is well drained. A poorly drained soil will be a dull gray from lower depths almost to the soil surface. The two intermediate soil drainage classes will be partly dull gray to partly mottled in color. Each county, state, and national soil legend will specify if present the depth to mottling and to a gray colored horizon for each soil mapping unit.

Table 5-6 summarizes for the four drainage classes the percentage of time the measured water table was at a depth of 6, 12, 18, 24, 30, or 36 inches. A brief study of this table will reveal that the criterion of using soil colors to predict soil drainage classes and depth to water tables in standard soil surveys is accurate.

Soil Conservation Districts

Soil Conservation Districts are administrative (and sometimes political) organizations of people in spe-

Figure 5-16. The relationship between the four soil drainage classes used in soil mapping, plant root depths, and generalized soil colors used in differentiating the drainage classes: A—Well drained; B—Moderately well drained; C—Somewhat poorly drained; D—Poorly drained. (Courtesy United States Department of Agriculture, Soil Conservation Service.)

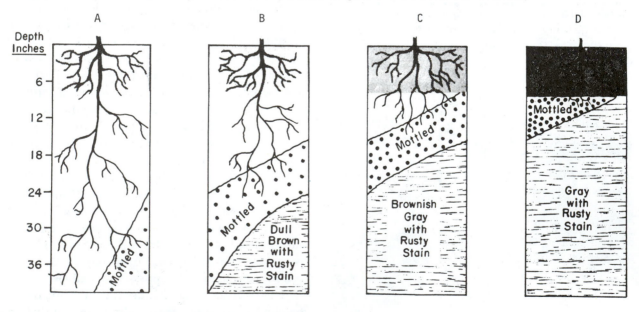

Table 5-6. The Four Soil Drainage Classes Recognized in Soil Mapping and the Corresponding Measured Water Table Depths

Soil Drainage Class as Mapped	Percentage of Time Water Table is at Specified Depth (inches)					
	6	12	18	24	30	36
Well drained	0	0	0	0	0	0
Moderately well drained	0	0	2	9	21	34
Somewhat poorly drained	0	11	24	34	47	58
Poorly drained	24	39	53	63	69	76

SOURCE: U.S.D.A. Soil Conservation Service.

cific areas, with interests in planning and implementing soil and water management programs. The first Soil Conservation District in the United States was established as the Brown Creek Soil Conservation District in North Carolina on August 4, 1937. In ten years all states, plus Puerto Rico and the Virgin Islands, had enacted enabling legislation to establish Soil Conservation Districts. Unlike the Soil Conservation Service, which is a U.S. government agency, soil conservation districts are state agencies of private citizens who receive no salary. They are comparable to a school district and its governing school board.

By 1972, there were 3,027 soil conservation districts, established mostly on county lines and comprising almost ninety-nine percent of the privately-owned land in the United States. They are supervised by a state agency which is usually the soil and water commission.

All soil conservation districts started by law on a voluntary basis. Since the acute awareness of pollution by the public in 1970, some states have enacted new legislation to make soil erosion control mandatory and to include water conservation.

By 1987 all fifty states, the District of Columbia, and the Virgin Islands had passed laws regulating soil disturbance, soil erosion, and soil sedimentation. Major provisions of these laws include prior approval for all major soil-moving activities and compliance with established soil loss tolerance limits (T-values).

Summary

Soil can be defined in many different ways. In agriculture the definition emphasizes growing plant roots. In engineering, the emphasis is on the mechanical make-up of the soil.

Soil starts with inorganic materials, such as rocks and minerals, which are broken down into particles of different sizes, mostly by weathering. At some point, organic matter becomes incorporated into the mixture, and the result is soil. It is the thin layer of soil (up to a depth of six feet) that provides the growing medium for the production of human and animal food.

But soils also contribute to our environmental problems, particularly when they are not used carefully and properly. Soil erosion is by far the biggest source of water pollution in the environment. Soil particles that are knocked loose by falling rain and torn free by running water become suspended in runoff as sediment. When that happens, the water becomes muddy and the soil particles are deposited elsewhere when the runoff slows. The results are clogged sewers, filled-in lakes, river channels filled with sediment, and the list goes on.

Soil particles can also be dislodged by wind. In many parts of the world, wind-caused erosion is a major problem. Current estimates of wind and water erosion indicate that as much as 6.4 billion tons of soil per year enter this cycle of erosion in the United States alone. The most serious erosion problems occur on land being cultivated for crop production. Land being graded for construction or surface mining may erode more severely than cropland, but there is much more cropland than construction or mining sites in this country as well as worldwide.

There are basically two main ways of addressing erosion control: ecological and mechanical. Ecological control measures involve land use while maintaining as much vegetative cover as possible. In agriculture, that may mean more no-till farming in the future. Mechanical erosion control measures include terracing, contour farming, and the use of constructed water runoff controls.

Living soil organisms make up an important component of soil. Most plants grow with their root systems in the soil. Soil bacteria, fungi, and other microorganisms are a vital part of the soil complex. When they die, the organisms decompose into organic matter which is also an integral part of the soil. One very important soil organism is the earthworm. Earthworms digest complex organic matter converting much of it into compounds more usable by other plants and animals. They increase the amount of airspaces in the soil and loosen it by burrowing.

Soil pollution does not receive as much publicity as water pollution, but both forms of pollution are major environmental concerns. Soil pollution occurs when compounds are introduced into or onto the soil which make it more difficult for desirable plants to grow. The biggest soil pollution problem involves increasing salinity resulting from the buildup of salts from irrigation water. A second major source of soil pollution is the injection of waste materials into the ground.

Soil Surveys and Soil Conservation Districts have been used in this country to help address the soil erosion problem for most of the twentieth century. Soil surveys are documents that describe the kinds of soils in a region. They also examine climate, topography, and other factors that affect how the soil can be safely used. Soil Conservation Districts represent an organizational and management system that helps land-use planners and landowners to plan for the safe use of the land.

DISCUSSION QUESTIONS

1. What are some differing definitions of soil?
2. What relationship does soil have to water pollution and environmental damage?
3. How does the vegetative cover of the soil affect erosion rates?
4. How do salinity levels increase in agricultural areas?
5. How do earthworms affect the soil erosion rate?
6. What is permafrost?
7. Why are oil pipelines, paved roads, septic systems, and agricultural drainage particularly problematic for groundwater in a permafrost region?
8. How can farmers reduce pollution of the environment from chemical fertilizers and pesticides?
9. What are the environmental concerns from the use of sewage wastewaters for fertilization?
10. What are soil surveys? What are their uses?

ADDITIONAL ACTIVITIES

1. Fabricate some artificial soil. It should include inorganic particles of all sizes (sand, silt, and clay). It should also include some organic matter, such as the partially decomposed layer of leaves on the forest floor or "organic soil" that you could purchase at a garden center. Add a small amount of complete fertilizer. Try growing some plants in the mixture. Incidentally, in horticulture, this would be called "growing medium."

2. Find a road bank or dig a pit in the soil that will allow you to look at the profile of the top three to four feet of soil. What do you see? Are there differences between the soil near the surface and deeper into the ground? What causes the differences? How deep do the plant roots grow?

3. Visit a site where erosion has occurred. What appears to be causing the erosion? Analyze the site to determine what ecological or mechanical erosion countermeasures might be helpful.

4. Take a series of soil samples around your school. Send them to the appropriate state agency for analysis. Before the results come back, make projections of what you think the recommendations will be. After the results are returned, compare the actual recommendations to what you predicted. You can contact your local agricultural extension agent for help in this activity if necessary.

5. Repeat activity 4 for the soil around your own home.

CHAPTER 6

Forests

Courtesy United States Department of Agriculture, Forest Service.

Terms to Look for and Learn

Biochemical Oxygen Demand

Biomass

Chlorosis

Clearcutting

Environmental Diseases

Fertilizer

Forest Type

Hardwood

Mycorrhiza

Organic Sediments

Prescription Burn

Selection Cutting

Shelterwood Cutting

Shifting Cultivation

Slash and Burn

Softwood

Learning Objectives

After reading this chapter and participating in the activities, you should be able to:

- Discuss the role of trees in human thinking.

- Outline some of the major contributions of trees to the environment.

- Discuss the negative impacts on the environment from some forestry activities.

- Outline the forest resource base in the United States.

- Outline the uses of trees and shrubs in controlling and improving the human environment.

- Discuss fires in the forest—both wildfires and prescribed burning.

- Describe and explain the major kinds of tree diseases.

Overview

Trees add to human pleasure by their beauty, shade, coolness, and soundproofing. Trees also reduce air pollution by absorbing carbon dioxide and acid rain and by physically trapping dust particles. Of special value to people and animals is the release of oxygen into ambient atmosphere (see Figure 6-1). Green belts of trees and shrubs have unique value to screen ugly sights and tame violent winds. And whoever heard of beautiful birds and friendly squirrels without trees for them to nest in? Forestland, however, is decreasing; and when forests are harvested improperly, serious pollution can occur.

Land in forests in the United States in 1987 totaled 652 million acres. This is about twenty-nine percent of the country's total land surface. From 1959-1987, the acreage of forestland was reduced by ten percent while cropland increased twenty-nine percent. Yet today, forests in the United States produce more new wood each year than was produced by the larger area three centuries ago.

Even though the total forested area in this country has declined, the annual growth of wood from those forests has increased. This has been made possible by a number of factors. The most important factor is the difference in net annual biomass produced by a young forest as opposed to a mature forest. In a mature forest, the trees are mostly at or near their full size. Although trees never completely stop growing as long as they are alive, their rate of growth slows as they reach maturity. A mature forest has no net increase in biomass each year—it is near equilibrium in terms of total biomass. In other words, younger tree growth may equal old tree death and decomposition in a mature forest. For the purposes of this book, biomass refers to the amount (weight) of materials that are currently, or have recently been, living or parts of living organisms.

Ironically, what that means is that to produce the most net increase in biomass (growth) from a forest, it must be harvested before it reaches full maturity. A mature forest is more eye-appealing than an area that

Figure 6-1. Multiple use of our national forests is shown here with pulpwood coming **out** while an outboard motor is going **in** (Cherokee National Forest in Tennessee). (Courtesy United States Department of Agriculture, Forest Service.)

has been harvested, but the harvested forest may quickly renew itself (either by natural reforestation or by being planted), producing more new biomass, including wood, each year than the more mature forest would have.

Trees Are Something Special

According to traditional beliefs, for a century or more before Christ, the Druids were high-caste philosophers, politicians, and magicians who met in the oak forests of the British Isles to practice human sacrifice. As stated by one angry man, the Druids believed in worshipping trees and sacrificing people (Donahue 1973).

Trees are such long-lived and beautiful perennials that many people become emotional and possessive about them—even when the trees are owned by another person.

Starting with the Book of Genesis, trees were created on the third day; trees were books to Shakespeare; trees were to be spared, according to Morris; and trees were more lovely than poems to Kilmer.

"The Earth brought forth vegetation, plants yielding seed according to their own kind, and trees bearing fruit in which is their seed, each according to its kind. And God saw that it was good. And there was evening and there was morning, a third day." (Genesis 1:12, Revised Standard Version).

"These trees shall be my books." William Shakespeare, 1564–1616, in Act III, "*As You Like It.*"

"Woodman, spare that tree! Touch not a single bough! In youth it sheltered me, and I'll protect it now." (George Pope Morris, 1802–1867).

"I think that I shall never see a poem lovely as a tree . . . Poems are made by fools like me, but only God can make a tree." (Joyce Kilmer, 1888-1918).

Trees have also been memorialized by being named as "state trees." Certain trees have special historic, religious, aesthetic, or sentimental associations, partly because they are the oldest living things and partly because they are often associated with historical events or childhood memories.

Some giant sequoias are three thousand–four thousand years old. The oldest living tree is 4,700 years old—a **bristlecone pine** on the U.S. West Coast. There is a Washington Friendship Tree; a horse chestnut at Bath, Pennsylvania; a Washington Elm at Cambridge, Massachusetts; a John Quincy Adams Elm on the White House grounds; a Grant Elm in front of Woodstock Academy in Connecticut; a Lincoln Oak at Hodgenville, Kentucky; a Charter Oak in Hartford, Connecticut; and scores of others to honor the memory of famous persons or events.

A more personal example of emotionalism and possessiveness about someone else's trees occurred when Roy Donahue cut four trees in his own yard in the Ozarks, and two of his neighbors told him that he should not have done this.

Trees and Water Environment

Water *yields* from forested lands are higher than from any other type of land use, primarily because forests naturally occur in the most humid regions. Water *quality* from forests is higher than from any other type of land use or vegetative cover. This is due to the interception of rain by tree foliage and branches, a dense and protective forest floor, and large root cavities down through which water moves rapidly. Water quality is high also because of the relatively low concentration of sediment, fertilizers, and pesticides used in a forest (see Figure 6-2).

If water quality is all that good from forestland, where is the environmental problem?

There is often water quality degradation by sediment when forests are cut, especially clearcut, and where skidding and yarding are done by a tractor. Additional pollution of air and water results when slash is burned in preparation for natural or artificial seeding or planting or when used to reduce plant competition of less valuable tree species.

Rarely, pesticides, fertilizers, and fire retardants are additional sources of water deterioration, espe-

Figure 6-2. Water pollution from forested lands occurs only during and soon after logging operations. Clear stream on left came from a forested area logged according to the best state-of-the-science. Water from the right is raging and muddy because the watershed was logged with no regard for the environment. (Courtesy Coweeta Hydrologic Laboratory, Franklin, North Carolina and United States Department of Agriculture Forest Service.)

Figure 6-3. The Iguazú Falls on the Parana River in Paraguay, South America are the largest falls in the western hemisphere. The watershed is protected by forest trees. (FAO photo courtesy Peyton Johnson.)

cially when applied by fixed-wing aircraft. Application by helicopter can easily avoid water bodies if these are clearly marked and the helicopter operator is careful. Helicopters are, however, more expensive to operate. Water is further degraded when trees along streams are cut and direct sunlight strikes the water surface to increase its temperature. Oxygen absorption by water is less at elevated temperatures and aquatic life is thereby reduced.

Although water pollution from forestry activities is a fact and can often be seen, it is extremely difficult to quantify. Forested areas are often very large, so there is an apparent dilution over a large geographic expanse. Quantification means monitoring, and monitoring a 500-million-acre area is a formidable task (see Figure 6-3).

Research Findings on Pollution

Water pollution from forestry activities comes mostly from mineral soil sediments, organic sediments, and occasionally from pesticides and fertilizers. **Mineral soil sediments** are the principal water pollutants. Their primary sources are erosion from skid trails, haul roads, and clearcutting.

Where tractors are the source of power in logging, soil compaction and surface soil disturbance further add to water quality degradation by mineral soil sediments. When the soils are fine-textured (clay) and wet, the sediment pollution hazard is greater. Forest fires are an additional cause of bare soil and more erosion and sediment pollution of water courses.

Even **organic sediments** such as leaves, twigs, and bark can sometimes become a serious local pollutant. Forests always produce large quantities of organic matter. But, logging operations produce massive amounts of such organic matter in a short amount of time. When organic matter decays, it causes an increase in **biochemical oxygen demand**. When much of the organic matter gets into surface water, its decay results in a consequent decrease of dissolved oxygen in the water, causing many forms of aquatic life to die. Forest fires contribute more organic matter, such as charcoal, that moves more readily than soil sediment into streams.

Pesticides may be water pollutants, especially when applied in or near streams or lakes. Insecticides, fungicides, silvicides, and rodenticides are the principal pesticides that are used in woodland management activities. Research in Oregon has confirmed the potential hazard of the use of pesticides in a forest, primarily because only fifty percent or less ever reaches the target tree species. The other fifty percent or more presumably drifts away into the unknown environment—our environment. Control measures of water pollution from pesticides include the selection of rapidly biodegradable formulations that are applied uniformly and in recommended quantities.

Urea [$CO(NH_2)_2$], 46% nitrogen (N) in the amine form (NH_2), is the principal fertilizer used to enhance forest tree growth. Ammonium nitrate (33% N) is also used. In the South, poorly drained soils are often low in available phosphorus. A chemical soil test should be used to indicate forest soils where trees are likely to respond to the application of phosphorus fertilizers. The most common forms are triple superphosphate (45% P_2O_5) and a combination of N and P as diammonium phosphate (21% N plus 53% P_2O_5). Urea, ammonium nitrate, and diammonium phosphate are readily converted by bacteria to ammonia—and ammonia, even in relatively low concentrations, is toxic to many species of fish.

Forests of the World

The top of a tree where the branches and leaves grow is called the crown. The area under the crown is said to be under or below crown cover. A forest with the tree crowns touching each other has a very high percentage of the ground under crown cover.

There are an estimated 7.5 billion acres of forest land with twenty percent or more tree crown cover in the world. Most of the **hardwood** forests (such as oak) are in Latin America and the tropical regions of Africa and Southeast Asia. The **softwood** acreage (such as pine and spruce) is concentrated in the countries of the former Soviet Union and in North America (see Table 6-1).

Technically, hardwood trees are broad-leafed trees with true vessels in their wood. Some hardwoods are very soft. Softwood trees have seeds that develop in cones or conelike structures and have wood blacking

Table 6-1. Forest Growing Stock in the World, by Area and Species Group*

Area	(billions of cubic feet)		
	Total	Hardwood	Softwood
North America	1,288	335	953
Latin America	3,260	3,168	92
Europe	526	191	335
Africa	2,134	2,130	4
Asia-Pacific (except Japan)	1,330	1,129	201
Japan	71	32	39
Soviet Union **	2,790	424	2,366
World	11,399	7,409	3,990

* **Source**: U.S.D.A. Forest Service.

** **Note**: As of September, 1991, all fifteen of the republics of the former Soviet Union became independent. Many joined the newly organized "Commonwealth of Independent States."

true vessels. Some softwoods have very hard wood. In general, softwoods tend to be less dense and softer than hardwoods.

Hardwood forests in many regions of the world, including the United States, could support higher levels of harvest in the next several decades. Most of this apparent potential is in the tropical hardwood forests of Latin America, Southeast Asia, and Africa.

The tropical hardwood forests are extensive and have a large capacity for timber growing. Yet serious problems exist which off-set the capability of these forests to continue to supply high-quality timber products to world markets.

Much of the tropical forest area is relatively inaccessible. Only twenty-two percent of the Brazilian humid, tropical forest is considered accessible. Hence, development of timber resources is slow and expensive.

Utilization of timber is complicated by the great numbers of widely different species with unknown genetic characteristics. Such problems of biological diversity occur in all regions but are particularly acute in Latin America. The future of tropical forests in all regions is further complicated by the expanding need for agricultural land for food and fodder to accommodate rapidly growing populations.

Improved utilization can have a more immediate effect on supplies. The largest part of the expected increase in world demands for industrial timber products is for pulp and particleboard. This should enhance the possibility of expanded management and utilization since smaller trees, lower quality logs, and manufacturing byproducts can be more easily used for such products than for lumber manufactured directly from solid woods.

Conservation of wood fiber through expanded recycling of paper and paperboard in the industrialized countries of the world offers another possibility for meeting a significant portion of growing world demands for pulp products. In the United States about nineteen percent of paper and paperboard is recycled and in Japan, forty percent is recycled.

Possibilities for expanded output of softwood lumber and plywood outside the United States in the years immediately ahead seem to be limited to currently undeveloped resources in the northern parts of Canada and Siberia. Both Canada and countries in the former Soviet Union have indicated a desire to further develop their forests. Unused timber in both countries is under government control. Hence, government policies—as well as trends in prices, markets, and availability of investment capital—will be significant factors in determining how rapidly expansion of softwood timber output takes place.

The softwood timber resources of Canada are of special significance to the United States. Both geographic and economic ties make Canada a primary timber supply region for this country. Canada is the leading softwood timber-exporting nation in the world, with three-fourths of its exports going to the United States.

For the period 1963 to 1983, the use of wood throughout the world was slightly less than the human population, as shown in Figure 6-4.

Forestland in the United States

The percentage of land in each state that is forestland is displayed in Figure 6-5. The amount of forest varies from one percent in North Dakota to ninety percent in Maine. States with fifty percent or more land in forest include most of New England, Middle Atlantic, and Southern states plus Michigan and Washington. Forestland acreage by regions is shown in Figure 6-6.

Sixty-two percent of forestland in the United States is privately owned by individuals or corporations. The United States government owns thirty-eight percent, nineteen percent of which is managed by the United States Department of Agriculture (U.S.D.A.) Forest Service and another nineteen percent by other Federal agencies, mostly the Bureau of Land Management in the Department of the Interior, as shown in Figure 6-7.

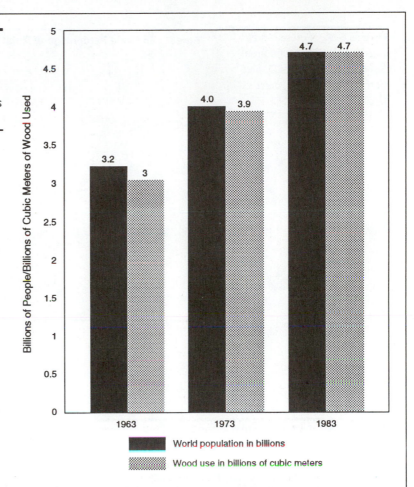

Figure 6-4. The use of wood in the world rises slower than human population, 1963 to 1983. **Note:** To convert cubic meters to cubic feet, multiply cubic meters by 35.314. (**Source:** World Resources 1986 and United States Census Bureau.)

Major Forest Types of the United States

A **forest type** is a kind, class, or group of forest tree species that occurs in a large area to be worthy of separate management. Figure 6-8 shows the generalized forest types in the United States. For example, Forest Type 2, "Central Forests of Oak-Hickory," is named after the most abundant forest tree species. "Oak" refers to many species of oak and there are several species of hickory. The implication of the name is that in this forest type, there are more oak trees than hickory trees.

Trees and Shrubs to Control the Human Environment

Scientifically selected tree and shrub species planted at proper locations around the home add to the pleasure of the human environment. For instance, in areas with alkaline soils, the iron deficiency disease of **chlorosis** can be a problem for some trees. In such areas, homeowners should consider that pin oak and silver maple are especially at risk and should probably select other species (see Figure 6-9).

Figure 6-5. Forestland as a percentage of total land area. (**Source**: United States Department of Agriculture, Forest Service.)

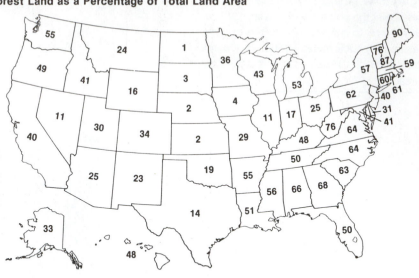

Forest Land as a Percentage of Total Land Area

Figure 6-6. Forestland areas of the United States. Forests occupy approximately one-third of the nation's total land area. Two-thirds of the forest area is timberland, or productive forests, owned by the public, forest industries, farmers, and other individuals and firms. In total, U.S. timberland contains 830 billion cubic feet of timber; fifty-seven percent of softwood species and forty-three percent hardwoods. (**Source**: United States Department of Agriculture. 1987 data. Hawaii and Alaska included in Pacific Coast.)

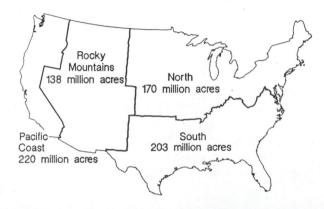

Figure 6-7. Ownership of forestland. (**Source**: United States Department of Agriculture, Forest Service.)

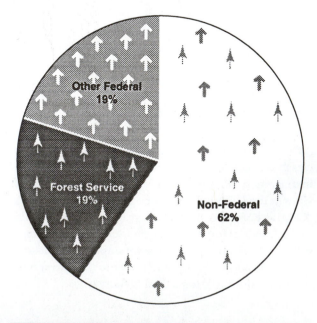

Figure 6-8. Major forest types of the United States. (**Source**: The 1983 Yearbook of Agriculture, p. 131.)

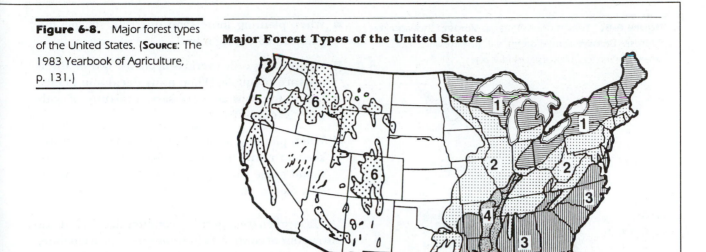

Major Forest Types of the United States

1. Northern forests of white-red-jack pine, spruce-fir, aspen-birch, and maple-beech-birch groups.
2. Central forests of oak-hickory.
3. Southern forests of: Oak-pine, loblolly-shortleaf pine, and longleaf-slash pine groups.
4. Bottom land forests of oak-gum-cypress.
5. West coast forests of: Douglas-fir, hemlock-sitka spruce, redwood, and some western hardwood groups.
6. Western interior forests of: Ponderosa pine, lodgepole pine, Douglas-fir, white pine, western larch, fir-spruce, and some western hardwood groups.

Desirable functions of trees and shrubs in *summer* are to:

- Move cooling breezes into areas where you want them

- Add moisture to dry air

- Reduce glare away from buildings and from the direct rays of the sun

- Add oxygen to the ambient air

- Reduce carbon dioxide levels in the ambient air

- Make more favorable wildlife habitats

- Reduce wind speeds

Desirable functions of trees and shrubs in *winter* are to:

- Reduce speed of wintry winds and, therefore, save on fuel bills

- Allow warming of home walls by the direct rays of the sun

- Control drifting snow

- Make domestic animal and wildlife habitats more favorable

To grow best, evergreen species should be planted on the northwest side of a building and deciduous trees and shrubs on the southwest side (see Figures 6-10 and 6-11). Other scientific suggestions include:

- When planting several rows of trees or shrubs, the ultimate heights of the rows should vary to give an uneven rather than an even upper edge to "confuse" the wind.

- Windbreaks should allow some wind penetration. Impenetrable windbreaks create a partial vacuum on the protected side, which reduces their effectiveness.

- Trees are most effective when they branch to ground level.

- The wider the planting, the more effective the windbreak.

Figure 6-9. Leaves of silver maple. **Above:** showing iron chlorosis; **Below:** normal green leaf. (Courtesy U.S.D.A. Extension Service, University of Idaho.)

- When planting more than one row, stagger the plants in adjoining rows.

- When using only evergreen plants, two or three rows are adequate. When using deciduous plants, four to five rows are necessary. A mixture of both types is most effective.

- When planting a windbreak to deflect the snow, keep in mind that a twenty-foot tall windbreak would cause snow to accumulate from twenty to sixty feet to the leeward side.

Evergreen tree species recommended for human environment control include junipers, Austrian pine, mugho pine, white spruce, and yews. Deciduous trees recommended are flame maple, green ash, Siberian peashrub, black locust, and white oak.

Environmental Tree Diseases

Environmental tree diseases are those caused by nonliving agents such as lack of water or high or low temperature. This type of disease is not transmitted from one plant to another. Other causes of environmental diseases are chemical substances in the soil such as common salts or fertilizer, too much or too little water, transplant shock, and mechanical injuries.

These nonliving disease agents are a major cause of loss in transplanting forest and landscape trees.

Figure 6-10. Evergreens adjacent to the northwest sides of a house reduce wind speed and create dead air space for insulation.

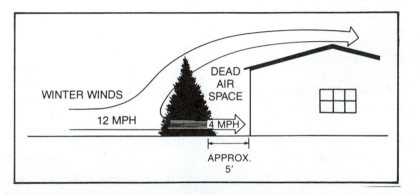

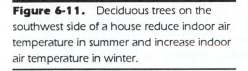

Figure 6-11. Deciduous trees on the southwest side of a house reduce indoor air temperature in summer and increase indoor air temperature in winter.

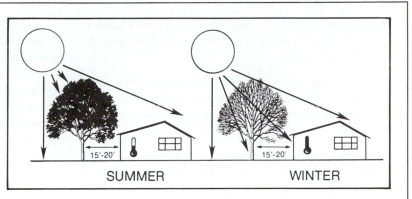

SUMMER   WINTER

Often they weaken trees, enabling living agents such as fungi, bacteria, viruses, nematodes, and insects to attack and further injure or kill the trees. Numbers of these living agents may build up on trees weakened by noninfectious agents and threaten healthy nearby trees.

The symptoms of noninfectious diseases may resemble those produced by insects or fungi. If no signs of these organisms are present, the tree may have been affected by one of the agents described here. Even if signs of fungi or insects are present, a nonliving agent may be the underlying cause. In most cases, prevention is the key to minimizing injury.

Listed below are selected tree species that *tolerate* low temperatures; high temperatures; poor, droughty, acid soils; wet soils; city smog; partial shade; and salt.

- Low temperature tolerance—white birch, wild black cherry, Douglas fir, hemlock, sugar maple, red pine, Norway spruce, yellow birch, and basswood.

- High temperature tolerance—acacia, Norfolk-island pine, silk-cotton-tree, coconut, mango, weeping willow, and tamarind.

- Tolerance of poor, droughty, acid soils such as those around gullies and in mine-spoils— European birch, dogwood, bristly locust (rose acacia), autumn olive, Osage orange, most pine species, and European birch (see Figure 6-12).

Figure 6-12. European birch (**Betula pendula**, family Betulaceae) has been used successfully to revegetate this acid mine spoil in Pennsylvania. (Courtesy United States Department of Agriculture, Forest Service.)

- Wet soils tolerance—white birch, arborvitae, black gum, cypress, white pine, tamarack, black spruce, and most willows.

- City smog tolerance—tree-of-heaven, silky dogwood, hemlock, little-leaf linden, honey locust, London plane-tree, Norway maple, pin oak, red oak, Russian olive, Austrian pine, and white pine.

- Partial shade tolerance—most dogwoods, hemlock, sugar maple, white pine, Norway spruce, redbud, and tamarack.

- Salt tolerance—saltcedar, date palm, live oak, slash pine, loblolly pine, redcedar, sweetgum, water oak, all hickories, red maple, red oak, and dogwood.

Forest Insects and Diseases

A substantial part of the heavy losses of timber to insects and diseases occurs in old-growth stands in the national forests. Consequently, insect and disease control are important in national forest management.

The insect and disease control programs and detection and evaluation surveys are conducted to locate and appraise the potential damage of insect and disease outbreaks. Surveys cover all classes of forest ownership. Detection and evaluation surveys are carried out annually on about 213 million acres of federal land and on an additional 300 million acres of private and state lands in cooperation with the state and federal agencies. They involve all of the following:

- Both remote sensing and ground observations to verify damage and kind of insects or diseases involved.

- Evaluation of population trends of pests.

- Determination of control alternatives.

- Appraisal of the feasibility and environmental consequences of suppression projects such as the aerial use of pesticides.

Control projects to suppress outbreaks of destructive insects and diseases may include aerial spraying of infested areas with pesticides, release of insect predators, or the felling (cutting) and removal of host trees. Recent control projects have been directed particularly against the spruce budworm in the Pacific northwest and in Maine, southern and mountain pine beetles, and Gypsy moth, shown in Figure 6-13. Disease suppression

Figure 6-13. Gypsy moths are a very serious pest. Their caterpillars eat leaves of oak trees, conifer needles, field crops, vines, and grasses. (Dark colored moth (top) is male and light colored (bottom) is female, shown here in a mating position.) Caterpillars may be controlled by suitable chemicals or by biological control using the bacteria **Bacillus thuringiensis.** (Courtesy Tennessee Valley Authority.)

projects have been directed mainly toward reducing growth losses caused by dwarf mistletoe in the West.

Forest Fires

There are 188 million acres of National Forest under the control of the U.S.D.A. Forest Service. Each year about 230,000 acres of these forests are burned by wildfires. Of the wildfires, about seven thousand are people-caused and about six thousand are lightning-caused (see Figure 6-14).

Controlled (Prescription) Burning

Especially in the South, forest fires often are set on purpose. The reasons for **controlled (prescribed) burning** are:

■ To kill less valuable hardwoods in mixture with more valuable pines (see Figure 6-15).

■ To reduce the accumulation of slash, pine needles, and leaves so that accidental fires will be more manageable.

■ To make a better seed bed for natural reseeding of the more desirable tree species such as pines.

Figure 6-14. Are forest fires beneficial to the forest?

(A)   No, in New York State (Results of an uncontrolled wildfire)

(B)   Yes, in Florida in pine stands (A prescribed burn for environmental management)

(Courtesies: (A) United States Department of Agriculture, Soil Conservation Service and (B) Florida Forest Service.)

A

B

Figure 6-15. **Left**—typical stand of southern pines and hardwoods. **Right**—after the forest had been control-burned. The more valuable pines are more fire-resistant than the hardwoods. However, burning pollutes the air. This is an example of environmental trade-offs. (Courtesy United States Department of Agriculture, Forest Service.)

A

B

- To control the brown spot needle disease of long-leaf pine.

- To improve the environment for wildlife.

Forest Fires and Longleaf and Loblolly Pines

In the virgin forest, **longleaf pines** existed in pure stands or in association with other species under a wide variety of environmental conditions in the South. In managed ecosystems, however, longleaf has often failed to perform as well as other southern pines, even on sites where it once existed in pure stands. At best, foresters' past attempts to manage the species have produced erratic results. As a consequence, the area of longleaf pine type in the South has declined from thirty to sixty million acres in colonial times to less than four million today.

The longleaf pine type became established and persisted on sites with frequent wildfires. The fires

reduced exposure of seedlings to the brown-spot needle blight fungus disease and held back the competing vegetation. The longleaf pine also needed areas where flooding was infrequent or of short duration. In a managed ecosystem within its natural range, longleaf pine has the potential to compete favorably in yields with other major southern pines on a wide range of site conditions. But, without the frequent wildfires, humans must take appropriate measures to assure prompt emergence of well-stocked stands from the grass (seedling) stage.

Growth of loblolly pine seedlings in the South was enhanced by years of annual winter burning. Seedling heights, biomass, and nutrient accumulations were greater on soil from burned plots. This suggests that nitrogen and phosphorus were more available to the seedlings. Phosphorus amendments greatly increased loblolly pine seedling height, biomass, and total uptake of all nutrients. These results suggest that the systematic use of annual prescribed fires in the management of loblolly pine stands in the South can be beneficial to seedling establishment. This may be true because of increased nutrient availability and faster seedling growth, as well as removal of competing species of plants. Fires also help to control a needle disease of longleaf pine.

Forest and Agricultural Fires in Oregon

In Oregon, as in many other western states, controlled burning by the agricultural and forestry industries is a common practice. Such burning is the largest source of smoke emissions in the state, exceeding industrial emissions by a factor of seven. Approximately 367,000 acres are burned each year in Oregon, generating about ninety-seven thousand tons of smoke.

Forestry burning accounts for eighty-four percent of these emissions. The remainder is from agricultural field burning. Brush and other unusable wood remaining after clearcutting, called "slash," are burned routinely to reduce fire hazards, permit reforestation, and reduce infestations of insects harmful to trees, such as the bark beetle. This controlled-burning method is called **slash and burn.**

Most field burning occurs in the Willamette Valley where approximately half of American grass seed is grown. Fields are burned in the late summer after the seed harvest to control insects and to prepare for replanting. Wheat fields in the Umatilla Plateau region of eastern Oregon also are burned after harvest. After rangelands are burned, the regrowth of grasses and leaves controls weeds and provides more green forage for cattle and sheep. Approximately ten to fifteen thousand acres of rangelands in eastern Oregon are burned each year.

The potential human health effects of smoke and odor and reduction in visibility have been controversial environmental issues in Oregon for many years. Smoke from forest burning, regulated by the U.S. Forest Service and the Oregon Department of Forestry, has been reduced by about thirty percent since the late 1970s. A recently adopted visibility strategy is designed to further reduce it by twenty-two percent by 1995. The Oregon Department of Environmental Quality has been successful in limiting the effects of field burning on populated areas through its smoke management program.

As in many other parts of the country, environmental and economic values are in conflict. Concerns about public health and welfare must be balanced with the economic importance of burning as a land management practice. Logging and grass seed growing are two of Oregon's key industries. To make further progress in pollution control, Oregon must come to terms with competing economic and environmental values.

Woodlands: To Graze or Not to Graze?

When cattle, sheep, or goats are turned into the woods to graze, they usually have very little grass. They do, however, eat many hardwood leaves, buds, and twigs. Goats especially, eat the tender parts of hardwood tree seedlings and strip off and eat bark. Livestock do very little damage to pine forests. What effect does grazing have on the hardwood forests in the United States?

- Water erosion is increased by grazing by seventy percent, from 0.7 tons to 2.3 tons per acre per year (see Figure 6-16).

- Wind erosion increases from zero in forests not grazed to 0.1 ton per grazed acre per year (National Research Council 1986).

- Most tree seedlings will be destroyed. This means that when the large trees are harvested there are few young trees to replace them.

Tree Harvest Methods

Forest trees are the best protectors of the air, water, and soil environment. Some careless tree harvest techniques, however, can cause severe pollution for a

Figure 6-16. Erosion is kept to nearly zero by not grazing the woods in this hardwood forest in northern Mississippi where annual rainfall is about 50 inches. The thick layer of fresh and rotting leaves breaks the velocity of the raindrops and water percolates gently through the soil. Note earthworm burrow at arrow. (Courtesy United States Department of Agriculture, Forest Service.)

few years. The principal tree harvest methods and their effect on the environment are clearcutting, seed tree harvesting, shelterwood harvesting, and selection cutting.

Clearcutting removes all of the trees from a particular tract. This harvest technique is selected when the tree species requires full sunlight for regeneration of the next crop or when replanting of seedlings is in the master forest plan. Clearcutting is required for Douglas fir on the West Coast, for the southern pines, and for the wild black cherry in the Allegheny mountains. This method of harvest is strongly opposed by many people because it is ugly and soil erosion hazard is great for a brief time after the harvest.

Seed tree harvest method involves nearly clearcutting an area but leaving a few trees of the desired species that are of seed-bearing age. These seed trees are left standing for a few years until the seeds they produce sprout and the next crop of tree seedlings are started. This technique is also adapted to the southern pines.

Shelterwood cutting removes all mature trees in a series of harvests over a period of years. This technique is well adapted to tree species such as the oaks and tuliptree because their seeds are heavy and are dropped mostly beneath the canopy of the seed tree. Oak seedlings tolerate partial shade as they grow from acorns into mature trees. Almost no soil erosion occurs under the shelterwood system of forest harvest (see Figure 6-17).

Selection cutting means selecting and cutting only the individual mature trees as they mature. Erosion is at a minimum.

Trees and Shrubs Get Help from a Fungus

Most fungi are diseases on plants but one fungus species helps at least seventy-five species of trees and shrubs in thirty-eight states in the United States and in thirty-three countries of the world. This fungal spe-

Figure 6-17. Tuliptree logs are used for making veneer. Tulip tree forests are adapted to the shelterwood system of harvest. (Courtesy United States Department of Agriculture, Forest Service.)

cies is known by the single common name of **mycorrhiza** (meaning fungus root).

This fungus attaches itself to the tree root hairs and the fungal rootlike growths (hyphae) function as extensions of the tree root hairs. The hyphae absorb water and nutrients from the soil and pass them on to the tree roots to increase their growth. Strangely, mycorrhizae are more plentiful in soils of low fertility. In a sense, it is an example of welfare among plants.

Tropical Rainforest

About 5.75 million square miles, seven percent of the Earth's land surface, support trees in what is known as tropical rainforest. Most of these forests are in South America and equatorial Africa. Soils supporting this large forest area are classified in the soil orders of Oxisols and Ultisols—the most highly weathered of all soils. Both soil orders are low in organic matter, highly acidic, low in calcium and phosphorus, and high in toxic aluminum and manganese.

The numbers of species of trees, shrubs, birds, mammals, and snakes in the tropical rainforest are guessed to include half of those of the world. Scien-

tists give this a name—"biological diversity." As an example, the Amazon rainforest supports thirty thousand species of plants and 30 million species of insects. (*Note:* Dr. Donahue once worked in the Amazon rainforest exploring for wild rubber trees.) For more on biological diversity, see Chapter 19.

With an increase in human population in the tropics of about two percent per year, people become desperate to clear the forest to raise vegetable and animal food crops. As a consequence, an estimated fifty-eight thousand square miles of tropical rainforest are destroyed each year.

Logging cuts another seventeen thousand square miles, fuelwood—ten thousand square miles, and cattle raising—eight thousand square miles. At this rate ninety-three thousand square miles of trees disappear each year. If these estimates are accurate, slightly less than two percent of the tropical forest is disappearing each year. This is nearly three times faster than that of U.S. forests.

The word, "disappearing" may not be accurate, for the reasons listed below. In most of these cases, the forest is not actually destroyed. More accurately, the mature trees are removed, but the forest remains forest and rapidly regenerates itself.

■ When trees are cut to plant food crops, the land is left in stumps and food crops are planted, tended, and harvested by hand. When food crop yields decline, another patch of trees is cut and planted to food crops while the former clearing has sprouted to start a new forest. This process is repeated for five to ten years, at which time the first-cleared area is a forest and is ready to clear and plant again. This system of clearing the tropical rainforest for raising food crops is known as **shifting cultivation**.

■ Logging removes only trees that can be harvested and sold at a profit. Under the high rainfall and warm temperatures, trees sprout and grow rapidly. In no sense will the forest disappear under this management system (see Figure 6-18).

■ Fuelwood use no doubt cuts and removes the trees easiest to cut that will burn readily, but in a scientific sense the forest remains and does not disappear, except in an emotional sense.

■ Cattle raising requires the trees to be cut and adapted grasses and legumes to be planted. However, stumps left in the pasture will sprout and tree seeds will blow in from the surrounding trees to start a new forest in the pasture. This is true even in the United States in forested areas.

■ The most rapid growth in terms of biomass production as well as in terms of the animal life that is supported occurs in a young forest. The trees are growing quickly in height and total weight. Undergrowth is very lush. Animals have access to easy food sources. In a mature forest, the net production of new biomass is zero because old vegetation must die and decompose at the same rate that new biomass is generated.

A more valid criticism of the wholesale cutting of these tropical rainforests lies in its side effects on the environment. Soil erosion is increased until the forest regenerates (reproduces itself). Increased runoff means less water entering the groundwater supply (usually not a major problem in the tropics where rainfall is plentiful). The tropical soil is shallow and low in organic matter, so disturbing the surface by heavy machines is very harmful to the soil structure. And finally, as you will learn in Chapter 19 (Species Diversity), many tropical plants and animals have very small ranges. Clearcutting a large tropical rainforest can cause the extinction of such species.

Figure 6-18. Logging in the tropical rainforest of South India with elephants. (Courtesy Roy Donahue.)

Summary

The total forested area in the United States has declined substantially over the centuries since European settlers arrived, yet the net amount of wood grown by those forests each year has never been higher than today. Mature forests produce new biomass at approximately the same rate that old biomass dies and decomposes. Only in immature (young) forests is there a net increase in biomass.

Trees contribute to the environment in many ways. They improve the ability of the soil to absorb water. That means both more and purer groundwater. They reduce erosion from both water and wind.

They reduce flooding. Trees provide a cooling effect in hot weather and a warming effect in cold weather.

Trees have always held a special place in the minds of humans. They have been praised in mythology, poetry, and religious writings for thousands of years. Some trees grow to massive sizes—the largest living things on the planet are trees. Some trees survive for thousands of years and continue to produce new growth throughout the entire period.

Exploitation of the products of trees has been absolutely essential to the development of human society as we know it. Yet, certain practices of forest management can produce serious environmental degradation. Clearcutting is by far the most economical way to harvest trees, but it temporarily removes the tree cover from the harvested area. The result can be serious erosion and sedimentation problems downstream. Careless use of chemical fertilizers and pesticides can mean contamination of the water supply, both surface water and groundwater.

In reality, the forests of the world could be harvested even more heavily than they are now. The problem is not the total amount of forests in the world. The problem is that the vast majority of the forests are inaccessible. That means the accessible forests tend to be overused. The result can be unproductive mature forests in many regions and overused forests that do not have time to recover from harvesting and regenerate in other regions.

Trees are affected by many diseases—some are environmental and others are caused by pathogens. Environmental diseases result when something in the environment damages the tree—temperatures too hot or too cold, drought, excessive soil water, city smog, deicing salt, and others. Diseases caused by pathogens include fungi, viruses, and bacteria. Insect damage is another serious problem in forests.

Most of us think of forest fires as always bad. In truth, fires have always been a natural part of the forest environment. Over the years, natural fires promoted growth and renewal in the forests. The real objection to forest fires is not the damage they do to nature, but the damage they do to humans and to their property.

Planned use of fire has been a management technique of humans for thousands of years. Prehistoric people used fires to clear the forests, to promote new growth, and to provide feeding areas for wild game animals. Today, foresters use fire to manage pests, insects, and diseases. Planned fires are also used to decrease the likelihood and severity of wildfires. Such planned forest fires are called prescription burning or prescribed fires.

The use of forested areas for grazing livestock, particularly where the trees are interspersed with grasses, has been controversial. Grazing the forests means a cheap source of food for herbivores such as cattle, sheep, and goats. At the same time, overgrazing reduces the vegetative cover and increases erosion problems. It can mean decreased water absorption into the soil, increased flooding, and increased water pollution.

Trees can be harvested in a number of ways. The most economical in the short run is clearcutting, but clearcutting is also the most ecologically damaging harvesting method. Other methods such as shelterwood and selection harvesting are less economically desirable, but cause less environmental damage.

DISCUSSION QUESTIONS

1. Many people feel strongly about trees. Why do you think this is so?

2. What are the direct benefits to the environment from trees?

3. What harmful effects on the environment can result from harvesting trees?

4. How can the impacts of pesticide and fertilizer contamination of water supplies be minimized?

5. The total amount of hardwood being produced by the world's forests is more than adequate for present demands. If that is true, then why is there still a problem in supplying the world's hardwood demand?

6. How much forestland is there in the United States and how is it distributed? Who owns it?

7. How are trees and shrubs used to control and improve the human environment?

8. What is prescribed burning? How is it used as a management technique?

9. What are environmental tree diseases? What causes them? What are some of their symptoms?

10. Why is "destroyed" probably not an accurate term for what is happening to much of the tropical rainforest that is being "destroyed" each year?

ADDITIONAL ACTIVITIES

1. Visit a National Forest Service or state forestry office. Find out the kinds of forestry management practices being used in the forests they manage.

2. Prepare a report on the history of forestry in the United States.

3. Find an article in a magazine or newspaper about the rainforests. Does the article mention the beneficial aspects of clearing the forests (economic return to people, food production for those living there, increased biomass production that will result) or does it deal only with emotional issues?

4. Examine your home to find as many wood and other forestry products as you can. How many of these products would you be willing to give up to "save a tree"?

CHAPTER 7

Environmental Crop Production

Courtesy DMI, Inc., Goodfield, Illinois.

Terms to Look for and Learn

Corn	LISA
Cotton	Maize
Crownvetch	No-till
Cultivar	Organic Gardening
Disease Resistance	Photoperiodicity
Endophyte	Salt Tolerance
Fescue Foot	Sorghum
Grass Tetany	Soybeans
Hybrid	Sweetclover
Insect Resistance	Wheat
Legume	White Clover

Learning Objectives

After reading this chapter and participating in the activities, you should be able to:

■ List, describe, and compare the major crops grown in the United States and worldwide.

■ Discuss the concept of Low Input Sustainable Agriculture (LISA).

■ Discuss biological pest and disease control procedures that can be used in crop production.

■ Discuss the implications of organic gardening and organic farming on human society.

■ Explain the techniques involved in no-till crop production.

99

Overview

Humans live on all continents, in all kinds of climates, and in all kinds of conditions. Yet, for all of our differences, most of the food that is consumed by our species comes from just a few crops. The United States is in a particularly advantageous position in the world community, because all of the world's major food and feed crops can be successfully grown either in the contiguous forty-eight states, Alaska, Hawaii, or Puerto Rico. In this chapter we will examine the major world food and feed crops.

The United States is a relatively large country, but only a small part of the total land area is suited to crop production. The country has a bounteous food supply, but to produce as much food as they do, American farmers must use much land that is not considered prime agricultural land to grow crops. As a result, soil erosion from water runoff and wind and salt buildup on irrigated land has become a serious problem in this country. Federal legislation passed in 1985 aims to reduce the use of marginal farm land, reduce erosion, and increase farm income.

All plants, including agricultural plants, require certain nutrients for growth and reproduction. They must get those nutrients from air, water, soil, lime, or fertilizers. Lime is a material containing calcium oxide, which is applied to soil to raise the soil pH (i.e., to neutralize acid soil). Fertilizers are natural or synthetic materials applied to soil to provide plant nutrients. There are no other important sources for plant nutrients. In the last half of this century, fertilizers and lime have become major sources for those nutrients, irrigation has become an increasingly important source of water for farm production, and pesticides have become a critical part of this country's farm production system. Because in the past scientists did not understand the effects of overuse of such agricultural inputs, farmers have sometimes applied too much fertilizer, lime, pesticides, and irrigation water. In some cases, the results have been nitrate and pesticide buildup in our water supply and salt buildup in surface soils from irrigation.

We cannot simply stop using fertilizer, lime, pesticides, and irrigation—unless all of us in the industrialized world are willing to greatly increase what we pay for our food and drastically reduce the world's human population. Even if we were willing to pay more, abandoning chemical agriculture is not a workable alternative given the present state of technology and world population. In this chapter, we will examine some realistic, partial solutions to the problems that have resulted from excess fertilization, liming, pesticide use, and irrigation in farm production.

Food Crops of the World

The ten major food crops of the world are wheat, rice, corn, Irish potato, barley, sweet potato, cassava, soybeans, sorghum, and sugarcane. Botanically, six of these ten major food crops are members of the grass family. They are wheat, rice, corn, barley, sorghum, and sugarcane.

Five of the crops produce most of the world's food and feed grains (wheat, rice, corn, barley, and sorghum). Sugarcane is the only grass family crop not grown for grain. Of the four remaining major food crops, three are staple root crops—Irish potatoes, cassavas, and sweet potatoes—and one is an oilseed protein crop, soybeans. When grown under dryland or rain-fed cropping practices, all of these crops have some rather specific climatic requirements.

Eight out of ten of these world food and feed crops are successfully grown within the continental limits of the forty-eight contiguous states. Crops requiring true tropical climates, such as sugarcane and cassava, are not grown extensively within the contiguous states. But true tropical climates exist in Hawaii and Puerto Rico.

As a result, climates of the fifty states and Puerto Rico are diverse enough to provide all crop-producing climates in the world. These extensive and diverse crop-producing climates are the main reason why the

United States leads the world in food and feed production.

Table 7-1 indicates the climatic type and range in latitude in which these ten crops are adapted. Note that rice, corn, cassava, sorghum, and sugarcane can be grown from the equator (0° latitude) to at least 30° north or south latitude.

Cropland Acreage in the United States

In 1987 there were 464 million acres of cropland in the United States, only half of which are classified as *prime agricultural land*. Prime agricultural land is "The most productive lands for raising the common food and fiber crops . . ." (Herren & Donahue, 1991).

This cropland acreage was on 2,087,759 farms with a total of 964 million acres. Diverted from production in 1987 was 76.2 million acres that were to be planted to perennial grasses or trees to control erosion and sedimentation. Cultivated cropland in 1987 was being eroded by water at the average rate of 4.1 tons per acre per year and by wind at the average rate of 3.6 tons per acre per year.

Food Security Act of 1985

The Food Security Act of 1985 established these goals for the Conservation Reserve Program:

- Reduction of sediment, the number one pollutant in the United States

- Improvement of water quality

- Creation of a better environment for fish and other wildlife

- Curbing of overproduction of surplus farm products

- Adding to farmers' income

The Conservation Reserve Program is voluntary for farmers. It is designed to take forty million acres of the most highly erodible soils out of cultivation and plant them to perennial grasses or trees. This action is predicted to reduce soil erosion on these set-aside acres from about twenty tons to less than two tons per acre per year.

Nutrient Requirement of Crops

Crops require seventeen elements for growth and reproduction. The sources of these elements—from air, water, soil, lime, and fertilizers—as follow:

Table 7-1. The Ten Major World Food Crops, Their Climatic Adaptations, and Ranges in Latitude *

Crop	Climatic Type (dryland or rainfed)	Latitude Range (north/south)
Wheat	temperate humid, subhumid to semiarid	25° to 55°
Rice	tropical-subtropical humid	0° to 45°
Corn	tropical-warm temperate humid	0° to 45°
Irish Potato	cool temperate	30° to 60°
Barley	cool temperate	35° to 65°
Sweet Potato	subtropical to warm temperate	25° to 40°
Cassava	humid tropical	0° to 30°
Soybeans	subtropical to warm temperate	25° to 45°
Sorghum	semiarid tropics to semiarid warm temperate	0° to 40°
Sugarcane	humid tropics	0° to 30°

* **Source**: Food and Agriculture Organization of the United Nations.

■ From air—carbon as carbon dioxide, oxygen as ionic and molecular forms, and nitrogen (indirectly via bacteria on legume roots and manufactured fertilizers such as urea).

■ From water—hydrogen.

■ From soil and lime—calcium and magnesium.

■ From soil and fertilizers—phosphorus, potassium, nitrogen, sulfur, iron, boron, manganese, copper, zinc, molybdenum, nickel and chlorine.

Note: Neither "low input" nor "sustainable" has been adequately defined in the references used, but "low" must mean "less than usual" and "sustainable" has to mean "capable of continuous maintenance." In blunt words, as far as their critics are concerned, whatever farmers have been doing with chemical inputs must have been all wrong.

There are very few soils that can supply all fourteen of the seventeen essential elements to crops for satisfactory yields. Chemical soil tests are available in all states for determining what essential nutrients are needed to obtain economic crop yields. This soil testing service is available through the various state agricultural extension services with representatives serving most counties in the United States.

Low Chemical and Irrigation Inputs for Agriculture

Fertilizers, herbicides, insecticides, and irrigation water have been used in excess in some locations, resulting in pollution of groundwater. This is especially true on deep sandy soils with shallow groundwater. The problem of pollution has become so serious in so many places in the United States that a national program has been established, known as **Low-Input Sustainable Agriculture** (LISA). Field monitoring of low input of chemical fertilizers, herbicides, and irrigation water are being researched and demonstrated more intensively on farmers' fields, starting with more soil tests and more scouting for insects and diseases.

Soil tests can be used more often to determine which fields need lime and fertilizers and at what levels. Scouting for threshold levels of pests is the monitoring device used in **integrated pest manage-**ment. Rather than spraying as a preventive measure or upon first observation of a particular pest, chemicals should be used only when the potential damage of that pest exceeds the cost of the pesticide and its application.

Benefits from irrigation efficiency extend far beyond conserving scarce water supplies. Surplus irrigation runoff can contaminate wetlands and waterways with damaging concentrations of salts, pesticides, and various toxic elements, such as cadmium and selenium that occur naturally in many desert soils. Such pollution problems have increased significantly with the doubling of farmland under irrigation since 1945.

A three-year demonstration project tested the potential of one method to conserve irrigation water in California. Small electronic devices, called Bouyoucos gypsum blocks, were used to determine when to start irrigating and when to stop. The costs for gypsum blocks and labor amounted to about two dollars per acre. Resulting benefits from lower water and energy costs and improved crop yields were fifty dollars or more per acre (Council on Environmental Quality, 1990).

Integrated pest management programs in the U.S. Department of Agriculture have a much longer history than LISA. The Extension Service received its first half-million dollar appropriation for LISA in 1972, largely in response to the environmental concerns raised in Rachel Carson's book, *Silent Spring.* Funding gradually increased to a peak of $7.5 million in fiscal year 1981, and it has since continued at about

$7.1 million per year. The success of integrated pest management in the 1980s is attributed largely to the proven economic benefits accruing to many farmers.

States also are undertaking initiatives to explore and demonstrate the potential for low-input agriculture. A prime example is found in Iowa, where legislation enacted in 1989 authorized at least five model farm demonstration projects across the state. While all projects are intended to implement integrated efficient farm management, each will be tailored to address special problems or characteristics of its region. For example, a project in southeastern Iowa will focus on conservation tillage. Another project in the south-central part of the state will focus on forage management. In total, nearly two hundred farms will participate in the demonstration program, and an aggressive information program of their experience is intended to demonstrate to 140,000 farmers more economic but less polluting practices on the use of fertilizers, pesticides, and irrigation water.

Gardening without Pesticides

Pesticides are chemicals, both natural and synthetic, used to kill pests. Gardening without pesticides starts during winter months when garden catalogs are first available. Many seed companies now offer **cultivars** (cultivated varieties) that are **insect-resistant,** as well as resistant to drought damage. If your garden soil is too salty for some vegetables, see Table 7-2 to determine crop salt tolerence.

Many cultivars are more **disease-resistant** than are resistant to insect damage. Plants that are disease-resistant are less susceptible to specific diseases than other similar plants. Insect-resistant plants tend to suffer less insect damage than other similar plants. For example, when a tomato cultivar is designated "V, F, N" this means it is proclaimed to be resistant to the diseases verticillium, fusarium wilts, and nematodes. Very few cultivars are promoted as being resistant to specific insect species. Avoiding heavy insect infesta-

Table 7-2. Relative Salt Tolerance of Selected Crops*

Low Tolerance	Moderate Tolerance	High Tolerance
Banana	Broccoli	Asparagus
Beans	Cabbage	Beets
Celery	Carrot	Bermudagrass
Grapefruit	Cauliflower	Coconut
Orange	Corn	Date palm
Radish	Cucumber	Spinach
Soybeans	Fig	Zoysiagrass
Sugarcane	Grape	
	Lettuce	
	Onion	
	Peas	
	Pepper	
	Potato, Irish	
	Potato, sweet	
	Rice	
	Sorghum	
	Squash	
	Sunflower	
	Tomato	

*SOURCE: Soil Salinity Laboratory, USDA, Riverside, California.

tions is the most common practice. This can be accomplished by buying early-maturing varieties and planting them as early as possible. Other techniques of gardening without pesticides are as follow.

■ Rotate crops in the garden to avoid diseases and insects that live in the soil. As an example, one year plant cabbage in a certain location, the next year plant tomatoes there.

■ Insects often find food by odor. You can scatter the plantings of a specific crop throughout the garden, rather than putting it all in the same part of the garden. This can help to confuse the insects.

■ Use as many biological control techniques as are available. Special mention should be made of the bacterium, *Bacillus thuringiensis*, which can be sprayed on plants. It kills larvae of the butterfly/moth order of *lepidoptera* insects. Many seed houses and garden supply centers now sell products containing this biological control bacterium.

Salt Tolerance of Selected Crops

Salt injury to crops is a major problem in the irrigated arid west and an occasional problem on soils adjacent to salt water and in areas adjacent to where salt was spread on highways for ice control. The most easily applied technique of avoiding salt injury to crops is to select cultivars with high **salt tolerance**, as listed in Table 7-2. Some plants cannot live in a soil with heavy concentrations of salts. Others are less susceptible to damage from salt build-up. They are said to be salt tolerant. The number of garden vegetables with high salt tolerance is limited to asparagus, beets, and spinach; however, there are many vegetables with moderate salt tolerance. Since chemical fertilizers are salts, the use of fertilizers on salty soils should be limited. All plants can withstand a higher concentration of salt in the soil when the organic matter level is maintained at high levels.

Organic Gardening

Organic gardening and farming refers to the production of plants without the use of synthetic fertilizers or pesticides. Organic gardening and farming are very ancient practices. In early history, animal manures were the *only* crop-yield-increasing fertilizers available. There is an old proverb from central India which has been translated to: "If there be enough animal manure, even an idiot can be a successful farmer."

In about 1910, Sir Albert Howard of Great Britain taught that only lime and raw rock phosphate were to be used on the farm from off-the-farm sources; all other soil applications such as compost and animal manures were to originate on the farm. Sir Albert took this concept to India and to the United States.

Sir Albert's preachings started before synthetic fertilizers were readily available. Even though available, manufactured chemicals have been largely avoided by followers of the organic gardening cult of, "All synthetic chemicals are harmful."

The closed cycle of "nothing-foreign-except rock phosphate, and lime" cannot possibly produce crops for long periods of time because all crops and livestock sold from the farm contain plant nutrients and organic matter which must be replaced if crop yields are to be maintained at economic levels.

The organic gardening and organic farming movement has been extended to include alternative agricultural systems based upon less science, less technology, fewer chemical inputs, use of more animal and plant residues, and fewer corporate farms.

Organic gardeners and farmers are now able to receive higher prices in many markets for products labeled "organically grown" or "naturally grown." Some states and the United States Department of Agriculture are trying to *define* and *police* organically grown products.

No-till Crop Production

The no-till crop residue management system actually started in the 1930s, a time known as the "dirty thirties." The soils of the Great Plains were being blown to deposit on the desks of Congress in Washington, D.C., one thousand miles east.

Research scientists F.L. Duley and J.D. Russell left grain stubble unplowed. At the time of planting the next crop, flat-bladed, wide sweeps were developed to cut stubble and weeds (Duley & Russell, 1942). A modification of this stubble mulch technique is now being used and is known as **no-till**. Implements have been developed to plant the seeds through last year's

Figure 7-1. Planting, fertilizing, and pesticiding into last year's corn stalks with no prior land preparation. This is known as no-till culture. *Saved* are fuel costs, labor costs, and soil. (Courtesy DMI, Inc., Goodfield, Illinois.)

crop residue and spread fertilizer and herbicides all in one pass over the field as shown in Figure 7-1. Figure 7-2 shows no-till corn coming up through last year's wheat stubble.

Much research has been conducted on no-till farming. The effects of no-till culture on yields and water infiltration in a study on corn in North Carolina and a study on cotton in Mississippi can be summarized briefly.

On sandy loams or fine loamy sands under corn in North Carolina, infiltration was twenty-five to fifty percent greater with no-till than with conventional culture. When subsoiling was done between the corn rows, infiltration was increased only in no-till plots. Corn grain yields were sixteen to fifty percent greater with no-till culture (Naderman & Wagger, 1990).

There was no difference in yield of cotton lint between conventional (turning plow and harrow plus winter cover crop) and no-till (planting in last year's residues without tillage and without a winter cover crop of hairy vetch or wheat). Relative water erosion was estimated for the two cultural systems. Soil erosion loss for conventional tillage was estimated to be 1.56 tons per acre per year. For no-till culture, the estimated soil erosion loss was 0.36 tons

Figure 7-2. No-till corn coming up through wheat stubble in Maryland. (Courtesy United States Department of Agriculture, Soil Conservation Service.)

per acre per year. Water erosion on no-till cotton culture was, therefore, estimated to be only twenty-three percent of that of conventional tillage for cotton in Mississippi. Use of no-till also resulted in a saving of the costs of winter cover crop seed and from less cost of tillage.

No-till culture has one major ecological hazard—with no cultivation to control weeds, herbicides are needed. Most herbicides can eventually cause weeds to develop resistance to them. When this happens, the solution is to:

■ Change the herbicide or rotate herbicides, or

■ Rotate crops to include one crop that can be cultivated at least one year in four.

Atmospheric Nitrogen Fixation

About eighty percent of the lower atmosphere of the Earth is nitrogen gas (N_2); however, plants cannot use this nitrogen until it is in the nitrate or ammonium form. Fortunately for people, there are several groups of bacteria capable of utilizing gaseous nitrogen and eventually producing nitrate and ammonium nitrogen for use by plants.

Best known of the crops that work together with bacteria to fix atmospheric nitrogen are the **legumes**. Less well known are free-living bacteria that can also fix atmospheric nitrogen. In addition, some algae, and one species of an actinomycete (fungus) have this ability. Least known perhaps is the fact that many seed-bearing (higher) plants and nonseed-bearing plants of the tropics have this near-miraculous capability.

Symbiotic legume bacteria and the common crops with which they have mutual, specific, nitrogen-fixing relationships are as follow:

■ Alfalfa and sweetclover—*Rhizobium meliloti*

■ True clovers—*Rhizobium trifolii*

■ Peas—*Rhizobium leguminosarum*

■ Beans—*Rhizobium phaseoli*

■ Lupines—*Rhizobium lupini*

■ Soybeans—*Rhizobium japonicum*

Note: It should be emphasized that the species of *Rhizobium meliloti*, compatible for alfafa and sweetclover, will not inoculate any other plants listed here; i.e., no *Rhizobium* is promiscuous.

Characteristics of Some Important Legumes

Table 7-3 presents characteristics of some environmentally important legumes. It includes alfalfa, three clovers, crownvetch, two lespedezas, and soybeans. The table indicates for each legume the growth habit, life span, and principal uses.

Alfalfa

Alfalfa has been traced to southwestern Asia as far back as 490 BC where it was grown to feed horses used by defending and invading armies. Now alfalfa is grown in most countries of the world principally as a livestock forage. It is made into hay, haylage (hay silage), and is grazed by all classes of livestock.

Alfalfa cultivars are of five groups: common, variegated, Turkistan, nonhardy, and rhizomatous (new plants come from specialized roots). Plant breeders have done remarkable things with alfalfa. They have developed cultivars that are high yielding, disease-resistant, cold-resistant, drought-resistant, weevil-tolerant, and aphid-tolerant.

An Ohio study confirms that alfalfa contains organic compounds that prevent or inhibit germination of alfalfa seed. This explains why efforts to interseed old

Table 7-3. Characteristics of some legumes that enhance the environment by helping bacteria fix atmospheric nitrogen, providing food and cover for wildlife and forage for domestic animals, and by reducing soil erosion and sedimentation.

Legume	Growth Habit	Life Span	Principal Uses
Alfalfa	Taproot	Perennial	Hay, Pasture
Clover,			
red	Taproot	Perennial	Hay, Pasture
white/ladino	Taproot	Biennial	Hay, Pasture
Crownvetch	Rhizomatous	Perennial	Erosion Control, Pasture
Lespedeza,			
sericea	Bunch	Perennial	Wildlife, Hay, Pasture
annual	Fibrous	Annual	Wildlife, Hay, Pasture
Soybeans	Fibrous	Annual	Wildlife, Food, Feed, Oil
Sweetclover	Taproot	Biennial/ Annual	Hay, Pasture for livestock and bees

SOURCE: U.S.D.A. Soil Conservation Service.

alfalfa fields have had little success. This phenomenon, known as autotoxicity, also occurs when alfalfa is seeded where an old alfalfa stand has been plowed up recently. Studies are underway to characterize the compounds that are already known to be water soluble, and seem to be more prevalent in leaves and stems than in crowns and roots.

White Clover and Ladino Clover

White clover and ladino clover are botanically similar; they vary mostly in size. They are generally known as small (wild white), medium (cultivars), and large (ladino). The wild white is the least productive and ladino the most productive.

Ladino clover is preferred as a pasture legume, especially when planted with bromegrass (see Figure 7-3). The amount of atmospheric nitrogen fixed by ladino clover varies from about eighty to 100 pounds of nitrogen (N) per acre per year.

Sweetclover

There are annual white-flowered sweetclovers, biennial white-flowered sweetclovers, and biennial yellow-flowered sweetclovers. The annual cultivars are used primarily in the South as a winter legume. It has special use in the Texas Blacklands as a **biological control** of the cotton root rot fungus (see Figure 7-4).

Sweetclover grows best on well-drained, neutral to alkaline soils. It is more drought-resistant than alfalfa. Sweetclover makes high-quality hay but must not be cut as low as alfalfa. Whereas, alfalfa forms new shoots at its crown at ground line, sweetclover forms new shoots from the stems at a height of ten to twelve inches above ground level. When cut as low as alfalfa, sweetclover dies.

Figure 7-3. Holstein cows grazing a bromegrass-ladino clover pasture in the White Mountain Region of New Hampshire. (Courtesy University of New Hampshire.)

Figure 7-4.   Now it is called **alternative agriculture**, but in the **fall** of 1947, Roy Donahue, the County Agent, L.M. Hendley, and farmer Dale Stockton arranged this demonstration: **Left**—46 lbs P_2O_5 fertilizer on annual white (Hubam) sweetclover, followed by cotton in spring, 1948. Yield was doubled. **Right**—No P_2O_5 and no sweetclover (cotton after cotton). Yield was half of that on left. (Courtesy Roy Donahue.)

Livestock will graze sweetclover but it may take a few days before they learn to like its flavor. Good hay is also made with sweetclover, but spoiled hay may cause the blood of animals to not clot properly. Other uses of sweetclover are as an excellent bee pasture and as a very good soil building crop, especially on fine-textured clay soils.

The following ten characteristics are the principal environmental benefits and limitations of sweetclover.

- Sweetclover is an excellent legume to provide atmospheric nitrogen in symbiotic relationship with *Rhizobium* bacteria and improve soil tilth.

- It must have a high soil pH, but phosphorus and potassium levels are seldom critical for survival.

- Because of the way it regrows from buds along its main stems, during its first year of growth sweetclover must not be grazed down to less than ten to twelve inches in height.

- It is not as palatable to livestock as most other legumes.

- Sweetclover is not as consistent as red clover in producing high-quality forage over the summer period.

- Sweetclover is very tolerant of droughty soils, heat, and grasshoppers.

- The biennials are true biennial (two growing seasons) and will survive only one winter. Annual sweetclover survives only one growing season.

- It is susceptible to attack by the sweetclover weevil.

- It is not a very good hay crop, but is a good bee crop for making honey.

- It is better adapted to short rotation pastures than to long-term permanent pastures.

Crownvetch

Crownvetch (*Coronilla varia*, family Fabaceae) is a native of the Eastern Hemisphere but is well adapted to humid central and eastern United States. It is a very hardy, long-lived, perennial legume that spreads by rhizomes (underground roots). It becomes established very slowly but, once established, it is very persistent. Crownvetch provides only fair grazing because of its high tannin content. It does not belong in a garden or a flower bed because it is too aggressive and difficult to control or eradicate.

Crownvetch has been planted for its attractive white and rose-to-violet flowers; however, it must have room to spread, such as in wild areas. Its best use is for erosion control on cut and fill slopes, other construction sites, and on surface mine spoils (see Figure 7-5).

Crownvetch has two very specialized uses in field crop production:

Figure 7-5. Crownvetch is an excellent legume for stabilizing soils disturbed by construction and mining activities. (Courtesy United States Department of Agriculture, Soil Conservation Service.)

■ On fine clay, high lime soils (Vertisols) where cotton rootrot (*Phymatotrichum omnivorum*) occurs, crownvetch makes a good green-manure crop because it is the only warm-season legume resistant to this disease.

■ In the Corn Belt and in the higher elevations of the South, crownvetch is being established as a perennial crop and corn is planted in rows in the crownvetch stand. Herbicides are used to control the growth of crownvetch at corn planting time. Such a companion-crop system is especially suited to sloping lands to control erosion.

The name crownvetch is derived from its "vetchlike" leaves and the shape of the pink and white flowers arranged to resemble a crown. Reproduction is by seed and adventitious buds on the roots which produce rhizomes. Its potential as a perennial legume cover crop for no-tillage crop production is just being discovered.

A cover crop of crownvetch improves the nitrogen fertility of hillsides and also holds more of the water, allowing it to soak into the soil. The soil water can then be used later by the crownvetch and no-tillage crops planted into it. Surface water runoff is reduced; this reduces the loss of nutrients and pesticides which may be carried with it. Compared with conventional tillage, corn planted in a stand of crownvetch reduced water loss by 98.5 percent, soil loss by 99.9 percent, and herbicide loss by 95.4 percent.

Soybeans

Soybeans are a legume crop native to the Orient and used in the United States mainly for high-protein feed and oil. Soybeans rank third in total U.S. cash receipts with an annual farm value of over $13 billion, after cattle and calves, and dairy products. Corn ranks fourth and wheat fifth.

States ranking highest in cash receipts from soybeans are: Illinois, Iowa, Indiana, Missouri, Ohio, Arkansas, Louisiana, Mississippi, and Tennessee, respectively. Note that the first five are the Corn Belt states and the others are Cotton Belt states. Illinois,

Iowa, and Indiana also rank one, two, and three, respectively, in corn.

In the Corn Belt, soybeans are frequently rotated with corn. Often, the rotation is corn-corn-soybeans-wheat. Again, a wheat-soybean system is used as a winter-summer crop, respectively. In the Cotton Belt, the rotation may be cotton-cotton-soybeans or rice-rice-soybeans. Continuous soybeans (monoculture) is not recommended because of an increase in soil-borne diseases and weeds. Insect pests are usually a minor problem.

Soybeans are a warm-season annual legume that require a growing season of about 120 days and soil temperatures between 60° F and 90° F. In general, soybeans are more sensitive than any other major field crop to the length of day. This characteristic is known as **photoperiodicity**. Yields of soybeans are commonly reduced by tall weeds (see Figure 7-6).

Grass Crops to Enhance the Environment

Grass crops that are adapted to a wide range of environmental conditions include tall fescuegrass,

weeping lovegrass, orchardgrass, redtopgrass, annual ryegrass, timothy, switchgrass, sorghum, and foxtail millet. Their growth habits, life spans, and principal uses are presented in Table 7-4.

Although tall fescuegrass is readily established and has seedling vigor, it can transmit a disease of cattle known as **fescue foot**. More than ninety percent of all tall fescuegrass pastures are estimated to be infected with a fungus that is toxic to mammals. Eating grass infected with the fungus can cause poor growth, lameness, and hooves to be shed.

The fungus grows inside the grass and, therefore, is classified as an **endophyte** (Greek *endon* [within]). However, the fungus (*Acremonium coneophialum*) repels certain insects and nematodes and makes fescuegrass grow faster and be more tolerant of drought. The fungus is transmitted through the seed. As of 1990, there appears to be no practical treatment to control this fungal disease except to breed a cultivar of fescuegrass resistant to fungal attack.

Fescuegrass is an excellent crop for stabilizing soils on steep slopes, as shown in Figure 7-7. It also has special characteristics to recommend it for planting on construction sites and mine spoils to stabilize them.

Figure 7-6. Stages in the germination and early growth of a soybean seed. (Courtesy Texas A&M Instructional Services.)

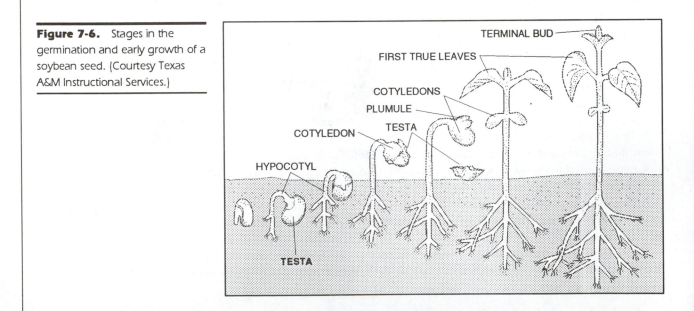

Table 7-4. These plants of the grass family (Gramineae) help to enhance the environment by their ease of establishment, strong seedling vigor, and use for wildlife and domestic animals.

Name of Grass	Growth Habit	Life Span	Principal Use
Fescuegrass, tall	Bunch	Perennial	Erosion Control, Hay, Pasture
Lovegrass, weeping	Bunch	Perennial	Erosion Control, Wildlife
Orchardgrass	Bunch	Perennial	Hay, Pasture, Wildlife
Redtopgrass	Rhizomatous	Perennial	Hay, Pasture
Ryegrass, annual	Fibrous	Annual	Cover Crop, Erosion Control
Switchgrass	Rhizomatous	Perennial	Pasture, Wildlife
Timothy	Bunch	Perennial	Hay, Pasture
Sorghum	Fibrous	Annual	Wildlife
Foxtail Millet	Fibrous	Annual	Wildlife

SOURCE: U.S.D.A. Soil Conservation Service.

Grass Tetany

Grass tetany (Latin—Tetanus, *tetanicus*—[a spasm, staggers]) is a disease of cattle caused by a deficiency of magnesium in the blood of domestic animals. It results from a magnesium deficiency in forage grasses. Low levels of magnesium in forage can be caused by:

- Low magnesium in soils, resulting in a low level of magnesium in forage eaten by cattle.

- Excess potassium fertilizer, which reduces magnesium uptake by forage crops.

- Excess ammonium fertilizer, resulting in low uptake of magnesium by plants.

- Excess application of poultry manure which reduces plant uptake of magnesium. Poultry manure is high in potassium and ammonium.

Grass tetany can be avoided by applying dolomitic (high magnesium) limestone, calcined magnesite, or magnesium sulfate to the soil.

Figure 7-7. This steep slope in Tennessee has been successfully seeded on the contour to fescuegrass. (Courtesy United States Department of Agriculture, Soil Conservation Service.)

Sorghum

Sorghum is used for many purposes and is also known by many common names. It is not surprising that confusion exists when referring to the commodity. Sorghum is erroneously called "milo," "maize," "milo maize," "feed," "grain," and "Egyptian corn." "Milo" is only one type of sorghum as is "kafir," "hegari," "shallu," and "zerazera."

Sorghum for grain is more drought-resistant than corn but most kinds of sorghum have lower palatability and lower feeding value than corn. In semiarid areas, sorghum can serve as a field windbreak that aids also in catching snow to add to the limited soil moisture supply, as shown in Figure 7-8.

Other names associated with sorghum arise from association only. Until about ten years ago, "grain sorghum" was the official U.S. terminology for the crop. But now, the USDA has deleted grain and the crop is now sorghum, officially. Other types or descriptions of sorghum include:

- Yellow sorghum
- Brown sorghum
- White sorghum
- Bronze sorghum
- Waxy sorghum
- Yellow-Endosperm sorghum
- Twin-seeded sorghum
- 3-Dwarf sorghum
- Food-type sorghum
- Conversion program sorghum
- Bird-resistant sorghum
- Hybrid sorghum

Sorghum appears to have been domesticated in Ethiopia about five thousand years ago. Sorghum has a number of features which make it a drought-resistant crop. It is extensively grown under rainfed conditions for grain and forage production. In dry areas with low or erratic rainfall the crop can respond very favorably to supplemental irrigation. However, considerable differences exist among cultivars in their response to irrigation. Those that are considered very drought-resistant respond slightly while others produce high yields under irrigation but are poor yielding when water is limiting.

Figure 7-8. Rows of grain sorghum have been used in semiarid Colorado as a windbreak and a barrier to catch snow in order to add to soil moisture. **NOTE:** The prevailing winds in this photo are from right to left. (Courtesy Colorado State University.)

Temperature is an important factor in variety selection. Optimum temperatures for high-producing varieties are over 90° F but some varieties are adapted to lower temperatures and produce acceptable yields. Most sorghums are short-day plants, which means they flower in the early spring while the days are relatively short and the nights are long. However, day-neutral varieties do exist. The crop does well on most soils but grows better in sandy loam soils. The soil should preferably be well-aerated and well-drained. Sorghum is relatively tolerant to short periods of waterlogging. Sorghum is moderately tolerant to soil salinity.

Sugar Crops

The principal sugar crop throughout the world is sugarcane, and second-ranked is sugarbeets. In the United States, however, both crops are grown but since 1965 the amount of refined sugar derived from sugarbeets has been slightly higher than from sugarcane. Third-ranked is maple sugar from the sap of sugar maple trees. Honey from honeybees ranks fourth as a source of sweeteners. Corn sugar production is a minor source but its use is on the increase (see Figure 7-9).

An interesting interaction has been demonstrated in Colorado between the amount of fertilizer nitrogen applied per acre and the yield of sugar (sucrose) per acre. Excess nitrogen, as determined by field research plots, results in less sugar per acre than optimum nitrogen. Less applied nitrogen fertilizer also means less chance of contaminating surface waters and groundwaters with hazardous nitrates (see Figure 7-10).

Wheat

People in the United States eat more wheat products than all other grain products combined. The average

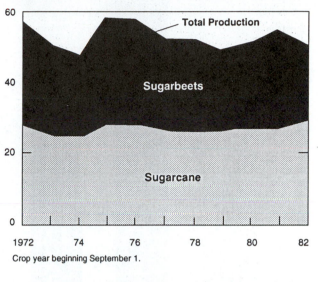

Figure 7-9. The production of sugarbeets varies from year to year more than does sugarcane. About half of U.S. production comes from each source. (**Source:** United States Department of Agriculture.)

Million Tons

Crop year beginning September 1.

consumption of wheat per person per year is 120 pounds. This compares with ten pounds of corn, nine of rice, three of oats, one of barley, and 0.7 pound of rye.

Wheat ranks second after corn in the United States but comprises seventy-five percent of all *small grain* production. Among small grains, oats rank second in acreage, followed by barley, rice, and rye. However, worldwide, wheat is by far the dominant cereal grown. States ranking highest in wheat production are, in order, Kansas, North Dakota, Oklahoma, Washington, and Texas.

Wheat is an annual herb of the grass family which is one of the most important crop plants of the world. It furnishes grain, one of the principal sources of human food, animal feed, and manufactured prod-

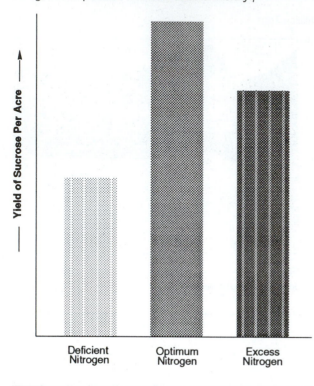

Figure 7-10. Sugarbeets produce the most sugar (sucrose) per acre at "optimum" levels of nitrogen fertilizer. Field plot research is needed to determine the "optimum" nitrogen rate. (**Source**: Colorado State University.)

ucts, such as alcohol and glucose. The wheat grain is not commonly fed to livestock in the U.S. but the byproducts of milling are extensively used. Place of origin of wheat is unknown but presumably originates from western Asia.

Wheat is grouped into five principal grain classes:

- Hard Red Spring—Hard red spring wheat is grown principally in the northern Great Plains States where there exist a fairly deep, dark, black soil and hot, dry summers. These environmental characteristics are believed important for the production of a high grade of spring wheat suitable for milling into bread flour.

- Durum—The center of durum wheat production has continued to be in the Dakotas, Montana, and Minnesota.

- Hard Red Winter—The center of production of hard red winter wheat is in Kansas and the adjoining states of Nebraska, Oklahoma, Colorado, and the Panhandle of Texas. The annual precipitation for this area averages less than twenty-five inches, with frequent dry periods and subzero winter temperatures. Hard red winter wheat is the most important class of wheat in the United States on the basis of the number of acres grown.

- Soft Red Winter—The production of soft red winter wheat centers in Ohio, Indiana, and Illinois.

- White—White wheat may be either spring or winter. It is produced mostly in the Pacific Coast States, with some acreage in southern Michigan and in western New York.

Corn

Corn is known as **maize** in most countries outside of the United States. Corn is a member of the grass family of Gramineae. It is one of the most important crops in the United States, grown for human and livestock food, feed, and for oil. Worldwide, corn is grown on about 290 million acres as compared with about eighty million acres in the United States (see Figure 7-11).

Corn is grown in climates ranging from temperate to tropical during the period when mean daily temperatures are above 59° F. Adaptability of cultivars in different climates varies widely. Successful cultivation depends largely on the right choice of cultivars. It is critical that the length of growing period of the crop matches the length of the growing season and the purpose for which the crop is to be grown. Corn is moderately sensitive to soil salinity.

Corn tassels are the male parts. The ears and silks are the female parts. The tassels produce pollen, which blows in the wind and may fertilize the same

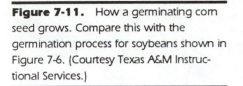

Figure 7-11. How a germinating corn seed grows. Compare this with the germination process for soybeans shown in Figure 7-6. (Courtesy Texas A&M Instructional Services.)

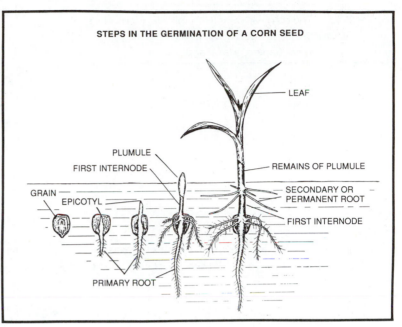

STEPS IN THE GERMINATION OF A CORN SEED

LEAF

PLUMULE

FIRST INTERNODE

GRAIN

EPICOTYL

REMAINS OF PLUMULE

SECONDARY OR PERMANENT ROOT

FIRST INTERNODE

PRIMARY ROOT

plant or others. Initially the silks are usually a light green, almost white. Pollen grains land on the silks, and one grows down each silk as a "pollen tube" to its base, which is the ovary. Eventually each fertilized ovary grows into a kernel.

The first step in **hybrid** corn production is inbreeding one or more open-pollinated varieties. In open-pollination, the pollen from any plant may fertilize the silks on any plant. The many kinds of corn that contribute to the genetic makeup of these varieties must be separated into what are called inbred lines. Inbreeding is done by fertilizing the silks of a selected plant with pollen from the same plant. The process is repeated in successive generations until the resulting corn plants do not change from the parent.

After the inbred lines are obtained, they are evaluated by crossing them. That involves fertilizing the silks of the ears from one inbred line with the pollen from the tassels of another. The crosses are then tested by growing the seed that results. The objective is to find crosses that have the highest yields, and the most desirable agronomic qualities.

After many inbred lines have been developed and their crosses tested under various conditions to assure that they are good, they are ready for commercial production. In a seed corn field, four rows of the inbred line to be used as the female parents are planted for each two rows of the inbred line to be used as the male parent. As the plants grow, the tassels are removed from the plants to serve as females, and are allowed to remain on the plants to serve as males. The wind then blows the pollen from the male plants to the female plants, and the seeds produced are hybrid seeds that are used in commercial production the next year.

The corn earworm (*Heliothis zea*) is also known as cotton bollworm, tomato fruitworm, vetchworm, and false tobacco budworm. Other plants attacked include alfalfa and peanut. Widely distributed in both North America and South America, the corn earworm is most destructive in southern United States (see Figure 7-12).

Natural enemies of the corn earworm include several species of birds, toads, spiders, larvae of lace-

Figure 7-12. The corn earworm (Heliothis zea) is one of the most destructive insects during the larvae stage on corn, cotton, tobacco, tomato, alfalfa, and peanut.

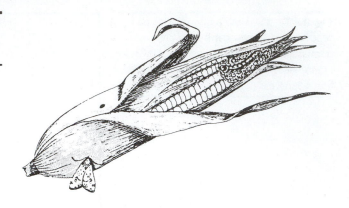

wing flies, ladybugs, and several species of wasps. *Trichogramma evanessens,* a small wasp, is an important egg parasite.

Cotton

Cotton is any plant of the genus *Gossypium,* family Malvaceae. These species are the most important fiber plants in the world. Some are native to Asia, others to Central and South America. Its seed yields an edible oil and cottonseed meal is an important stock feed.

Cotton is grown in areas of the United States, mostly south of the 36th parallel. The states with greatest acreage, ranked from high to low are California, Texas, Mississippi, Arizona, Louisiana, Arkansas, Alabama, Oklahoma, Tennessee, and Missouri, respectively (see Figure 7-13).

Cotton requires a frost-free season of at least 180 days. Summer temperatures must average 77° F or more, and the average annual rainfall must be at least twenty inches unless irrigation is practiced. Wet weather seriously interferes with the harvesting of cotton and reduces its grade and selling price. Therefore, autumn rainfall should not be more than ten inches.

Cotton is a warm-season crop and the seed will not germinate until the soil temperature at an eight-

Figure 7-13. **Above**—An open cotton boll ready for harvest. **Below**—An unopened cotton boll. (Courtesy The National Cotton Council.)

inch depth reaches 60° F. This soil temperature can be expected between twenty and thirty days after the average last freeze in the spring.

Most cotton in the United States is harvested in October and November, except in the Plains of Texas. In that area, cotton strippers are used to harvest the crop after the first freeze which causes the leaves to drop and, therefore, not interfere with the harvest. Most cotton grown in the United States is upland cotton with a staple length of 3/4" – 1 1/4" or longer and the fibers are of medium coarseness.

The principal cotton diseases are cotton anthracnose (boll rot), cotton wilt, and cotton root rot. In addition, there are many insects that reduce cotton yields. Of major importance are the cotton boll weevil, shown in Figure 7-14, cotton aphid, cotton bollworm, cotton leafworm, cotton pink bollworm, cotton fleahopper, cotton square borer, and cotton stainer.

Most cotton farmers use public or private cotton insect scouting programs or cotton insect integrated pest management programs. Scouting can help determine if or when to spray or dust and the target species to spray or dust for. No longer is it legal or even wise to spray or dust as a preventive measure on a regular schedule with the cotton planter's favorite pesticide. The environmental reasons are as follow.

- The favorite pesticide may be illegal or ineffective against the target insect pest.

- Applying a pesticide on a regular schedule such as "every Tuesday" will kill beneficial as well as target insects—and target insects may reproduce faster than the beneficial ones.

Nitrogen has been recognized as the first limiting nutrient in the production of most plants. Unlike phosphorus, potassium, calcium, and magnesium, soil tests for nitrogen are not routinely performed. Soils are sometimes analyzed for nitrates and ammonium but interpretations into nitrogen fertilizer recommendation are difficult. As you will learn in several other chapters in this book, excess nitrogen is a serious groundwater pollutant.

The University of Arkansas has solved this difficulty in this way:

- Apply a preplant application of nitrogen fertilizer based on yield goal.

- Monitor the nitrogen status of the leaf petioles each seven days for eight weeks starting a week before the first bloom. Vary the nitrogen applied on cotton leaves or to the soil based on the level of nitrogen in the petioles (leaf stems).

Rice

After wheat, rice is the second greatest source of human food. In Asia, more rice is eaten than any other food grains. An estimated forty percent of the world's population eat rice as their major source of food. Asia and all tropical, subtropical, and a few temperate countries produce rice—a total of 111 countries.

Rice breeders have developed compact rice plants that will not lodge (fall over) when setting a heavy

Figure 7-14. Cotton boll weevil.

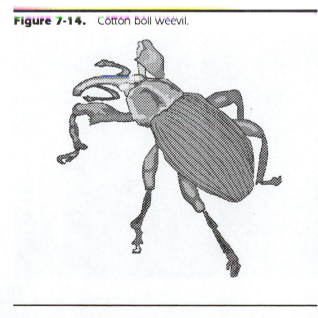

crop. Also available are improved rice cultivars that are insensitive to length of day and night. These are called day-neutral plants (see Figure 7-15).

A large part of the world's rice is grown in flooded soils, known as paddy rice. This means rice is classified as a hydrophyte (water-loving plant). Depth of water maintained over paddy rice plants is usually three to six inches. However, there are cultivars of rice capable of growing in deeper water, as deep as four feet. Rice is usually grown in flooded soil because yields are higher, even though rice can be grown on upland soils as is corn.

When rice soils are flooded, they become less acid as well as less alkaline, phosphorus, and iron are more readily available, fewer weeds will grow, more algae fix atmospheric nitrogen, fewer soil-borne diseases occur, and rice yields are higher.

Summary

Of all the food produced for human use, most comes from just a few crops. Just five grasses produce most of the world's food and feed grains. Eight of the ten major world food crops can be grown successfully in the forty-eight contiguous states. The remaining two are grown extensively in Hawaii and Puerto Rico. Thus, within the United States, all major world food and feed crops are grown. This diverse climate is part of the reason that the United States leads the world in food and feed production.

Agricultural crops, like all kinds of plants, require seventeen elements for growth and reproduction. Those elements are supplied from the air, water, and the soil. When a plant grows, it takes all of the elements it needs from those sources.

Figure 7-15. The first high yielding, stiff stemmed, rice variety (IR-8) developed by the International Rice Research Institute in the Philippines. It is not sensitive to the length of day, meaning it will produce over a wide geographic range where any rice will grow. (Courtesy International Rice Research Institute.)

In a natural ecosystem, the elements return to the soil when the plants die and decompose. But when a crop is harvested, the chemicals go with it. That means the chemicals that are not taken from air (carbon, oxygen, and nitrogen) or from water (hydrogen) are removed from the soil in the form of plant material. Those elements removed from the soil eventually must be replaced if the soil is to continue to produce crops for long. That simple truth gave rise to the use of animal manures in farming thousands of years ago. As the human population grew and as technology advanced, it gave rise to the use of lime and synthetic chemical fertilizers.

One important source of plant nutrients is atmospheric nitrogen. Bacteria that live on the roots of certain plants, take nitrogen directly from air. They "fix" the nitrogen gas as proteins. When these proteins decompose, nitrate and ammonium nitrogen are produced. The nitrogen compounds thus produced can then be used by the plants for growth and reproduction. The major group of plants that harbor nitrogen-fixing bacteria are the legumes.

In recent years, the concept of low-input sustainable agriculture (LISA) has been promoted as a means of minimizing the environmental impact of intensive farming practices. LISA is a fairly new term and all of its meanings and implications are not clear yet. But, clearly, the levels of fertilizer and pesticide use during the recent past bear close examination.

Groundwater contamination from excess nitrates and pesticides is an important environmental problem. Maintaining application rates of such chemicals at their minimum effective levels makes both economic and environmental sense. Pest control systems that do not rely totally on pesticides can be effective. By the same token, cultivation practices that have come to be known as no-till or low-till hold promise for helping to reduce the soil erosion problem.

The use of such farming practices means more work for the farmers. Close monitoring of field conditions is necessary for integrated pest management to work. No-till cultivation requires more frequent crop rotations. Low input of fertilizers means more frequent soil testing and more frequent applications of smaller amounts of fertilizer. Lower rates of irrigation are possible, but that may mean the use of Bouyoucos gypsum blocks and putting out complicated networks of drip lines rather than large irrigation pipe and boom systems or flood irrigation. Low input refers to chemical inputs not necessarily labor or management inputs. With food prices at the farm at current levels, it will be difficult to justify many of the practices that would be needed to go to a true sustainable agriculture; yet, eventually such systems must be put in place.

Continued irrigation, use of salt compounds in highway ice control programs, and natural salt deposits produce saline soil conditions. In very salty soils, most crops cannot be grown successfully. Salt injury is a growing problem in farm regions which rely heavily on irrigation. Salt tolerant crops can be successful, but in the long run, different farming practices will be needed to reduce the salinity of the topsoils in such regions.

Organic gardening has been very much talked about in the past few years. But for thousands of years, organic farming was the only kind of farm production. In fact, heavy reliance on chemical fertilizers and pesticides has only developed since about World War II. In many developing countries, organic farming practices are still the norm.

DISCUSSION QUESTIONS

1. What are the ten major world food crops? Of them, which are grasses? Which are root crops?

2. What is the Food Security Act of 1985? What are some of its major goals for United States agriculture?

3. What is LISA? Why has it arisen? What would be the results for humanity, if all of agriculture were to change to a LISA approach?

4. What are some techniques that have been used to help control insects and disease in crops, without relying on pesticides?

5. What are the real advantages of organic gardening over the use of fertilizers? What are the disadvantages?

6. What is no-till crop production? What are some of its effects on yields? On the environment?

7. What is nitrogen fixation? How is it accomplished?

8. What dangers to livestock are associated with fescuegrass?

9. What is grass tetany? What causes it?

10. Why are sorghums important as an agricultural crop?

11. What role does rice play in human food production? What parts of the world are most reliant on rice for food? Compared to most other farm crops, what is unusual about rice production?

ADDITIONAL ACTIVITIES

1. Select ten items of prepared foods like frozen dinners or canned soups. Make a list of all of the grain, root, and sugar crops that are listed in the ingredients. Which of the major world crops are and which are not included?

2. Invite your local agricultural extension agent or soil conservation agent to visit your class to discuss agricultural practices and how they affect the environment.

3. Organize a class discussion to contrast the results of organic farming to the results of farming with pesticides and chemical fertilizers.

4. Discuss the paradox that all farmers are really organic farmers and that all organic farmers are really chemical farmers.

5. Conduct a survey of the farms in your area to find out what crops and animals are being produced. You can do this as a class project. Get a copy of your state's agricultural census (available from the U.S. Bureau of the Census at your state agricultural college). Compare the crops and animals listed there to what you found in your own survey.

CHAPTER 8

Water Resources, Uses, and Major Sources of Pollution

Photo courtesy Bill Camp.

Terms to Look for and Learn

Aquifer
Evaporation
Evapotranspiration
Flux
Hydrologic Cycle
Latent Heat of Evaporation
Latent Heat of Melting
Leaching
Meteoric Precipitation

Percolation
Permeability
Reservoir
Residence Time
Specific Gravity
Sublimation
Transpiration
Water Budget

Learning Objectives

After reading this chapter and participating in the activities, you should be able to:

- Explain some of the characteristics of water that make it so important in the ecosystem.

- Outline the hydrologic cycle.

- Outline the world's water supply in various reservoirs.

- Explain the concepts of flux and residence time as they relate to the hydrologic cycle.

- Discuss the effects of major sources of water pollution on water supply.

121

Overview

This chapter deals specifically with fresh water resources, but in addition, it will provide some basic scientific concepts about water that are important in discussing marine water, brackish water, or fresh water (see Figure 8-1).

> The Earth is a water world. From outer space the oceans are visible as great swirls of blue, green, and gray covering over seventy-one percent of the surface. The blue-white of ice and snow cover another three and one-half percent. Below the patches and linear streaks of clouds are lakes and rivers, ponds and streams: unseen beneath the surface are vast reservoirs of groundwater.   (Speidel, Ruedisli, & Agnew, 1988, preface)

Figure 8-1.   Oceans are by far the major reservoir of the world's water supply. (Photo courtesy Bill Camp.)

Water—A Most Unusual Substance

Water is a most remarkable substance. As much as warmth from the sun and carbon from the atmosphere and soils of the Earth's crust, water made possible the development of life. It is the most plentiful compound on the planet's surface. It occurs in the form of a solid, liquid, and gas — all at the same time and in a temperature range of only a few degrees. Yet at the same time, it requires so much heat energy to move from one state to another that it is the planet's most effective heat storage "battery."

Water is so plentiful that the quantity available can only be approximated. It has so many uses that the amount consumed by humans can be only roughly estimated. The vast majority of our daily bodily intake is in the form of water — even for people who seldom drink just plain water. Our very bodies are mostly (about sixty-five percent) water. We use water for bathing, cooling, heating, manufacturing, creating energy, moving goods, recreation, decorating, producing aquatic food, growing agricultural crops, and for thousands of other things.

Water has a very small molecule, made up of only three atoms, two of hydrogen and one of oxygen (H_2O). Its liquid structure is such that it has a high viscosity (it flows easily). Think of water in comparison to honey or vegetable oil to get an idea of its viscosity.

Water is often referred to as the universal solvent. It will dissolve many other substances — a characteristic that makes it useful in living organisms. Plants and animals rely on water (in blood and sap) to transport food and other materials. Water conducts electrical current, particularly when substances such as salts are dissolved in it. It is that characteristic which allows neural (nerve) impulses to travel along our nerves to make our bodies function.

Water at sea level remains a solid (ice) up to 32° F (0° C) and begins to melt at that point. To raise the temperature of an ounce of ice (about 28.3 grams) from -100° F (-73° C) to its melting point 32° F (0° C) at sea level requires about 1,916 calories of heat. Yet to change the same amount of ice at 32° F (0° C) to water *at the same temperature* requires an additional 2,257 calories. The heat necessary to change the solid form of a substance to its liquid form at the same temperature is called the **latent heat of melting**.

Pure water at sea level boils at 212° F (100° C). To raise 1 ounce (28.4 g) of water from 32° F (freezing at sea level) to 212° F (boiling at sea level) requires 2,835 calories. But to convert it to water vapor without even changing its temperature, requires an additional 15,319 calories of heat (see Figure 8-2). That additional heat is called the **latent heat of evaporation.** (Speidel, et al., 1988).

The specific gravity of water compared to that of ice is another important characteristic. **Specific gravity** is a measure of relative weight. Water is at its greatest density at 39° F (4° C). At that point its specific gravity equals 1. Specific gravity is defined as the mass of a substance in grams per cubic centimeter. That is, one cubic centimeter of water at 39° F weighs 1 gram. In terms of American Standard measurements, 1 cubic inch of water at that temperature would weigh 0.6 ounce. As water freezes, it takes on a more open structure which causes it to expand by about 11%. That causes it to have a lower specific gravity (about 0.92) than that of liquid water, which has a specific gravity of 1.0. Because it is lighter, ice floats on water.

Since ice floats in water, standing water freezes from the top down. The latent heat of melting has a direct counterpart in the process of freezing. As the surface of a body of water freezes, the same amount of heat required for melting is released by the freezing water (about 2,257 gram calories per ounce or 80 gram calories per gram). In addition, the layer of ice is an excellent insulator. Both the heat released by freezing and the insulation provided by the ice act to keep deeper water liquid even under the coldest conditions. Thus, as a river or lake freezes, the underlying water continues in a liquid state and the fish and other aquatic animals are able to survive. As an extreme example, much of the polar ice cap in the Arctic Ocean is composed of floating ice over deep ocean water, through which fish can swim and submarines can operate.

Figure 8-2. Latent heats of melting and vaporizing of H_2O. Notice how many calories are needed to change ice to water without changing its temperature. Even more are required to change water to water vapor.

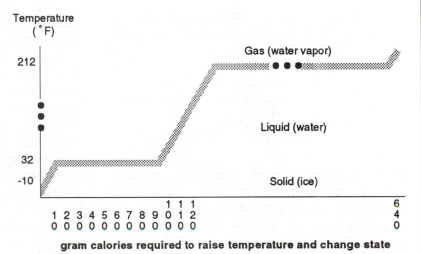

Water has six basic characteristics (Speidel, et al., 1988). Water is:

- *Ubiquitous.* It is everywhere. Anyplace on the surface of the planet where living organisms can survive, there is water. Even in the driest sand dunes of the Sahara Desert, there is some water clinging very tightly (adsorbed) to the surfaces of the individual grains of sand.

- *Heterogeneous.* Ice, water, and water vapor exist simultaneously in the environment, all within a range of temperatures which will support life.

- *Renewable.* Water is an extremely stable compound. Very little water enters into chemical reactions that permanently convert it from water into some other compound. The vast majority of the water that is used returns almost immediately to the hydrologic cycle. Practically all water that enters into other compounds eventually returns to water again.

- *Common Property.* Water is not a stationary resource. It is in constant motion. Water either flows, freezes, melts, evaporates, or condenses. Its borders are not fixed. Traditionally, water belongs to no individual but is available to everyone on a first-come, first-served basis. It is a common property — a common good.

- *Used in Vast Quantities.* Humans use far more water than any other resource. The worldwide production of all minerals, oil, coal, and metals has been estimated at almost 9 billion (8,816,000,000) tons per year. The annual water usage by humans in all activities was closer to 3 trillion (3,086,000,000,000) tons per year in 1975. That worked out to about 826 tons of water per year for every human on Earth at that time.

- *Very Inexpensive.* The fact that water is generally a common property means that it costs essentially nothing in most cases. The cost of water is a result of the expenses involved in its capture, treatment, transportation, storage, and dispensing. The actual cost of municipal water at the point of use averages about $0.03 per ton (three cents per two thousand pounds). That compares to the wholesale price of sand or gravel, probably the next cheapest resource, of about 100 times greater cost ($3.31 per ton).

The Hydrologic Cycle

The total quantity of H_2O on the planet is fairly constant. The availability of water is not so constant, however. Water is continually moving from place to place by means of the hydrologic cycle. A more common name for the hydrologic cycle is the water cycle. It can be defined as the process by which water moves from water bodies into the atmosphere, then back to the land and the water bodies.

The **hydrologic cycle**, like so many other things in the planetary ecosystem, is powered by solar energy. Energy supplied by the sun provides warmth to surface ice and water causing it to convert to its gaseous state (water vapor) and move into the atmosphere. The process of conversion from liquid water to water vapor is called **evaporation**. The amount of water moving through a particular system is referred to as the system's **water budget**. The planetary ecosystem's annual water budget can be viewed as the total amount of water that evaporates in one year.

A special form of evaporation is called **transpiration**. When plants absorb water from the soil, they conduct the water to the green leaf surfaces for use in photosynthesis. Water is also used as a medium for conducting suspended solids and compounds in solution within the plant. In the leaves of plants are small openings called stomata. Much of the water in the plants evaporates from the leaves by passing through the stomata. That process is called transpiration.

Transpiration is important in conducting plant nutrients upward in the plant. As water is transpired from the leaves, more water can then be absorbed by the roots along with the nutrients from the soil. In that way, water moves upward in the plant much like a cargo elevator.

A third process, similar to evaporation, is **sublimation**. Sublimation occurs when a solid, such as snow or ice, moves directly into a gaseous state, like water vapor. It really never becomes water. You can recognize sublimation has occurred in your refrigerator freezer. Ice cubes left alone for a long time become smaller and smaller. They have not melted, but the ice is smaller anyway. The same process takes place in the environment constantly. Ice cover in the Arctic and Antarctic sublimate constantly.

The total process of water entering the atmosphere (evaporation, transpiration, and sublimation) is referred to as **evapotranspiration**. As it changes from liquid water to water vapor, it gives up the latent heat of evaporation (15,308 calories per ounce). Part of it freezes and releases the latent heat of melting (2,257 calories per ounce). The resulting water or ice (sleet, snow, or hail) falls to the surface in the form of a liquid or solid. Water, ice, or snow falling from the sky is known as **meteoric precipitation**.

Once it reaches the surface, the water either evaporates again, enters the soil, or becomes runoff. Runoff may stop in a lake, wetland, or inland sea, or it may eventually make its way to the ocean. Surface water that does not enter the groundwater supply eventually returns to the atmosphere by evapotrans-

piration. This endless, tireless process is the hydrologic cycle, shown in Figure 8-3.

Water enters the surface soil by infiltration. The rate of infiltration is determined by the **permeability** of the surface soil. Permeability refers to the rate of water movement through a substance. Sand is a highly permeable soil material. Clay has a very low permeability.

The permeability of the surface layer of soil depends on a number of factors. The relative amounts of sand, silt, and clay is one important factor. The sandier the soil, the faster water can infiltrate. A second factor is the amount of organic matter in the soil. In general, the more organic matter the higher a soil's permeability. A third factor is the compaction of the soil, as is done by heavy machinery. In general, the looser the soil, the more rapidly water can infiltrate.

Another factor on the amount of water that can infiltrate into the soil, is the length of time available. A one-inch rainfall that arrives in a slow drizzle may infiltrate almost completely into the soil. The same amount of rainfall in a violent fifteen-minute downpour will produce very little soil moisture and much runoff and probably cause erosion sediment.

Figure 8-3. The hydrologic cycle moves water from the surface into the atmosphere and redistributes it to the surface.

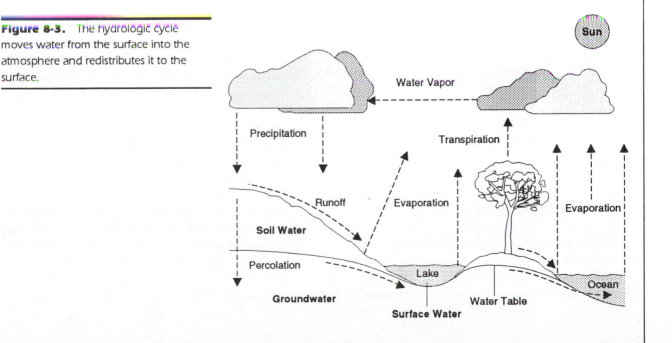

Once it enters the surface layer of soil, water continues to move downward. The process of movement through the soil by water is referred to as **percolation**. Water that enters the soil must go somewhere, some of it evaporates, some is absorbed by plant roots. Still more remains tightly adhered to the soil particles. Some remains held in the tiny spaces between soil particles (capillary spaces). Finally, the excess soil moisture percolates downward.

Eventually, the water that continues deeper into the soil will reach a zone of saturation. The level at which the soil becomes saturated more or less permanently is called the water table. Water contained below the water table is very stable and it is referred to as groundwater (see Chapter 4).

As we saw previously, water is a powerful solvent. As the rainfall strikes the soil surface, minerals (salts) are dissolved. Very tiny particles of insoluble materials are dislodged and held in suspension. If the water enters the surface water supply as runoff, the dissolved and suspended materials go with it. If the runoff is rapid, suspended particles can be quite large. This is what happens in the process of erosion, for instance. This is also how agricultural, industrial, and urban and domestic pollutants enter the surface water supply.

If the water enters the soil, it still carries the smaller suspended particles and most of the dissolved materials with it. As water continues to percolate through the soil, it dissolves still more soluble materials. Many organisms are very tiny and can move with the percolating water. The process by which soluble materials are dissolved and tiny particles are suspended in percolating soil water is called **leaching**.

Water Supply

The total amount of water on Earth has presented an interesting question for scientists for many years. Water is everywhere and it seldom remains still for long. Nevertheless, a number of interesting estimates have been made. Quantities of water are so vast that such estimates are expressed in cubic miles.

Reservoir

Water exists in a number of different kinds of locations. The term **reservoir**, as used here refers to a category of similar locations which hold water. The largest reservoir is the oceans. A very small reservoir (but important to us) is in the tissues of living organisms (biotic). The surface layer of the soil contains moisture that clings to soil particles and fills the smaller spaces between. This reservoir is referred to as soil moisture. Deeper into the soil, water saturates the spaces between soil particles and forms what is called groundwater. Large deposits of groundwater are termed aquifers (see Chapter 4).

Even though they are moving perceptibly, streams and rivers make up a major reservoir because as water moves downstream, more water enters to replace it. Freshwater lakes make up another important water reservoir. Saline lakes (which are called inland seas when they are particularly large) make up another very large reservoir.

One estimate of the total world supply of water is about 332 billion cubic miles. According to this estimate, the vast majority of the world's water is contained in the oceans—perhaps 97.4%. Of the remainder, most (two percent) is in the form of ice contained in the ice caps and glaciers of the world. Less than one percent of the world's water is distributed in all other reservoirs combined, shown in Table 8-1.

Over eighty percent of the world's ice is in the region of the Antarctic and another ten percent is in the Arctic. The remainder is included in permafrost, mountain ranges, and glaciers. The volume of water locked in ice equals the flow of all of the combined rivers and streams on Earth for about 900 years.

Some simple calculations show that the world supply of fresh water in rivers, lakes, soil moisture, and groundwater amounts to about two million cubic miles (about 0.6% of the total), with most of that contained in relatively deep deposits of groundwater which cannot easily be obtained for human use.

Table 8-1. Estimated Worldwide Water Supply*

Location	Amount (cubic miles)
Ocean	323,298,000
Atmosphere	3,000
Land	
Rivers	400
Lakes	24,000
Inland Seas (saline)	25,000
Soil Moisture	17,000
Groundwater	1,964,000
Ice Caps and Glaciers	6,586,000
Biota	300
TOTAL	331,917,700

*Source: Speidel, D.H., & Agnew, A.F. (1988). The world water budget. In Speidel, D.H., Ruedisli, L.C., & Agnew, A.F. (Eds). **Perspectives on water—Uses and abuses.** New York: Oxford University Press.

Flux

The world's total reservoir of water is of less importance to humans than its movement. **Flux** refers to the movement of water from one location to another or its transformation from one form to another. Flux also involves a period of time. For the purposes of this discussion, flux will be taken to mean the annual movement of water into or out of a reservoir. It occurs in the hydrologic cycle. As we have seen, the hydrologic cycle could be thought of as the engine of the ecosystem.

Each year, about 119 million cubic miles of water enter the atmosphere by evapotranspiration. About eighty-six percent of it comes from evaporation of ocean water. Eventually all of it returns to Earth in the form of precipitation. About seventy-eight percent falls back onto the ocean. The difference between the amount of evaporation from land sources (17,000 cubic miles) and the precipitation onto land areas (27,000 cubic miles) is the source of the fresh water that is available from the hydrologic cycle to support non-ocean life on the planet (see Table 8-2).

Figure 8-4. This water probably evaporated from the ocean surface and is falling back into the ocean. It could just as easily be redistributed to the land. (Photo courtesy Bill Camp.)

Table 8-2. Worldwide Annual Water Flux*

Flux	Amount (Cubic miles per yr)
Evaporation	
Ocean	102,000
Land	17,000
Precipitation	
Ocean	92,000
Land	27,000
Runoff to Ocean	
Streams	6,000
Underground Seepage from Land	3,000
Glacial Ice	600

*SOURCE: Speidel, D.H., & Agnew, A.F. (1988). The world water budget. In Speidel, D.H., Ruedisli, L.C., & Agnew, A.F. (Eds). **Perspectives on water—Uses and abuses.** New York: Oxford University Press.

Table 8-3. Water Runoff and Discharges from the Continents and Major Rivers *

Source	Discharge (Cubic miles per yr)
Continent	
Europe	614
Asia	2,986
Africa	816
Australia	573
North America	1,399
South America	2,644
Antarctic	476
River	
Amazon	902
Congo	301
Yangtze	165
Mississippi-Missouri	133
Yenisei	132
Mekong	129
Orinoco	129

*SOURCE: Speidel, D.H., & Agnew, A.F. (1988). The world water budget. In Speidel, D.H., Ruedisli, L.C., & Agnew, A.F. (Eds). **Perspectives on water—Uses and abuses.** New York: Oxford University Press.

You will notice that the annual flux balances. The total amount of evapotranspiration equals the amount of precipitation. The total amount of evaporation from the ocean (102,000 mi³) equals the amount of precipitation plus the runoff to the ocean, as shown in Table 8-2. These figures show some minor differences due to rounding and to differences in the processes by which the estimates were made.

The worldwide distribution of water is not even, however. Just in terms of runoff, rivers, streams, and groundwater discharge to the oceans ranges from 573 mi³ in Australia and 614 mi³ in Europe to 2,986 mi³ in Asia. The major mechanism for runoff is the great rivers. The Amazon River alone discharges more water (902 mi³ per year) than the entire runoff of Europe, see Table 8-3.

Residence Time

The length of time for a full turnover of the water in a reservoir is referred to as its **residence time**. This does not imply that every molecule of water in a reservoir will move once during the period of residence time. What it does mean is that the removal of water from the reservoir equals its total volume. Speidel and Agnew provide a simple formula for computing residence time:

$T = Q / \text{flux}$.
Where T = time in years
Q = quantity of water in a reservoir
Flux = annual rate of movement of water into or out of the reservoir

Using this formula and the data in Tables 8-1 and 8-2, we would estimate that the residence time for water evaporating into the atmosphere averages about 0.025 years: 3,000/119,000. To convert the residence time in years to days, simply use the daily flux (i.e., multiply by 365). Thus, atmospheric water residence time can be estimated at 3,000/119,000 × 365 = 9.2 days. Residence time for water captured in the world's ice pack would be estimated at about eleven thousand years. Residence time for groundwater would be estimated at 683 years.

Some scientists estimate that the residence time for water captured as ice in the Antarctic may be as long as 9,500 years and for groundwater it may actually be closer to 4,000 years if only the discharge to the sea is used as the flux.

Using the formula, we would estimate that the residence time for water in the world ocean system is about 3,169 (1,350,000,000/425,000) years, or over a million days. Interestingly, only the surface of the ocean is subject to evaporation and deep water practically never circulates to the surface. In the depths of the oceans, water molecules may remain for millions of years without evaporating.

Only the top two hundred inches of water are included in the actual calculations of quantity for ocean residence time. Using those figures, the residence time of water that undergoes evaporation from all of the planet's oceans is estimated to be about 255 years.

Water Demand

Estimates of worldwide use of water were made during the 1970s when water resources were on the front burner with many environmental scientists and world politicians. The period 1965–1974 was named by the United Nations as the International Hydrological Decade. As a result much of the data available on world water supply and use was collected at that time and has not been systematically updated. At that time, a number of estimates were made.

The United Nations Environment Programme (1988) placed total worldwide water use at 102 trillion

cubic feet in 1975, with about seventy-five percent going to agriculture, twenty percent to industry, and the remaining five percent to domestic and municipal use, as shown in Table 8-4. According to the United States Council on Environmental Quality and the Department of State, between 1975 and the year 2000, global use of water for industrial and domestic and urban use is projected to at least double. It should increase from the 1990 annual usage of 187 mi³ to between 430 and 550 mi³ annually by AD 2000.

Agriculture

As you can see, by far the largest use of water is for agriculture. The main agricultural use is in irrigation. The technology of irrigation is not new. The Hopi Indians of what is now Arizona used irrigation before the arrival of the Spaniards. Evidence of vast irrigation systems has been found in archaeological studies of the ancient Mayans, ancient Egyptians, and early Chinese, among many others.

Table 8-4. Worldwide Water Use in 1970 and 1975 mi³ per year***

Water Used	1970	1975	% Not Returned*
Municipal and Domestic	29	36	1 - 15
Industry	122	151	0 - 10
Agriculture	455	503	10 - 80**
TOTAL	606	690	

NOTES:

*Percentage of water withdrawn not returned directly to source.

**Majority returns to hydrologic cycle through evapotranspiration, part returns to soil water, remainder is converted into organic compounds.

***SOURCE: United Nations Programme (1988). Water demand. In Speidel, D.H., Ruedisli, L.C., & Agnew, A.F. (Eds). **Perspectives on water—Uses and abuses**. New York: Oxford University Press.

The most important form of irrigation is surface irrigation. Surface irrigation is done using either a flood technique or a furrow technique. In flood irrigation, shown in Figure 8-5A, the surface of the field must be level and the field is surrounded with berms (ridges or mounds of soil). The water is conducted to the field by ditches or berms and the entire surface of the field is flooded. In furrow irrigation, shown in Figure 8-5B, water is distributed throughout the field by furrows that lie between rows on ridges above the water level.

Sprinkler irrigation involves pumping water to the field in pipes and distributing it through a system of sprinklers, as shown in Figure 8-6. The water is applied to the tops of the crop first and it falls to the ground and infiltrates into the surface soil.

Much irrigation water comes from surface runoff and that often presents a problem. The heaviest runoff periods are normally when rainfall is heavy, but that is exactly the time when irrigation is not needed. During hot, dry weather, when irrigation is most needed, runoff supplies are likely to be at their lowest. In addition, irrigation water for very arid regions often must be moved over considerable distances. Surface irrigation canals in hot, dry climates and with slow-moving water produce a great deal of loss by evaporation. That further reduces already limited surface water supplies.

The most reliable and economical source of water for sprinkler irrigation has been groundwater. Unfortunately, groundwater tends to be replenished very slowly. As a result, groundwater levels have fallen dramatically in some very rich agricultural areas. In the central and southern Great Plains of the United States, the water table has been dropping by four to six feet per year. The water table is down in some places as much as four hundred feet in Texas, Arizona, and California since humans began to use the groundwater supply.

The result of such a lowering of the water table should be clear. It is more expensive to drill wells deeper. It is more expensive to pump the water out. We cannot continue to lower the water table forever —it will simply run out in those locations someday. Water shortages have occurred in many places al-

Figure 8-5. (A.) Flood irrigation in India. A canvas dam holds irrigation water back in the canal until one basin at a time is flooded. (Courtesy United States Agency for International Development in India.) (B.) Furrow irrigation of tomatoes in Sonoma County, California. Note the spiles (curved pipes) that siphon a uniform flow of water to each furrow. (Courtesy United States Bureau of Reclamation.)

A

B

A

B

C

Figure 8-6. Sprinkler irrigation systems at work. (A) A portable irrigation system. (Photo Courtesy Bill Camp.) (B) A semiportable irrigation system. (Photo Courtesy of Deborah M. Goetsch.) (C)Stationary irrigation system. (Photo courtesy Jack E. Ingles, Ornamental horticulture principles and practice, © 1985, Delmar Publishers, Inc.)

ready and will become more common and more severe in the future. Political decisions will become necessary to decide whether the scarce water will be used for irrigation on farms or for people in cities. The problems become more complex and more severe with time.

A further complication from conventional irrigation practices involves deposits of salts. Water taken from both surface and groundwater sources generally contains fairly high concentrations of dissolved salts. When the water is dispensed onto a field, it enters the top few inches as soil moisture.

As the plant roots and evaporation remove the excess soil moisture, all of the dissolved salts and minerals remain in the soil. The result, after many years of continued irrigation, is a buildup of salts that make the soil quite saline (salty). Most agricultural crops do not thrive in a saline environment (see Chapter 7). Thus, irrigation of a soil that is too dry for crop production can eventually produce a soil that is too saline for crop production.

A partial solution lies in so-called drip irrigation. This technique dispenses small quantities of water directly on the soil above the roots of the plants by means of a system of very small tubes attached to larger pipes.

This system deposits much smaller quantities of salt-laden water, thus reducing both the salinization problem and the water-use problem. This is a more expensive process, in terms of initial investment in the system and in terms of labor costs than the other two. On the other hand, it is less expensive in terms of water use and environmental impact.

Industry

More water is used by industry than any other natural resource. Only a small part of the water used by industry is actually consumed, however. Most of it is used as a medium for other processes. The United States Department of Agriculture estimates that U.S. industries alone withdraw about 40 billion gallons of water a day from the water supply. The water is used an average of two and one-half times. But only about 2.5 billion gallons per day are consumed (i.e., not returned directly into the water supply).

Industrial use of water is much more intensive in the developed world. In the United States, approximately fifty-seven percent of all water use in 1980 was by industry. In Canada, the figure was eighty-four percent. The percentage is much lower in the developing world. In 1980, industry's percentage of Mexico's total water use was about seven percent and in India it was only two percent.

Domestic and Urban Use

In the developed nations, we use water for much more than maintaining life. We bathe in it, water our lawns, supply water fountains for public decoration, and use it to carry away much of our waste products. In places where water is very cheap and plentiful, it is used quite recklessly. Where water is in short supply and is harder to get, it is a precious commodity that is hoarded and safeguarded (see Figure 8-7).

As with industrial uses, domestic use of water is a function of the degree of development of the country. In countries where water is drawn by hand, its consumption is very low. According to Postel, in India only six percent of total water consumption was for household and urban uses in 1980. For Mexico it was about five percent. For Poland, seventeen percent was used for domestic purposes and for the United Kingdom it was fourteen percent.

Figure 8-7. In many parts of the world, water is difficult to obtain. Potable water is a luxury that few outside the developed world take for granted. This farm woman in Ethiopia is carrying a five-gallon clay pot of water for her family to drink. The total weight is more than 22.5 Kg (fifty pounds). (Photo courtesy Roy Donahue.)

Major Sources of Water Pollution

For most of humanity's existence we have used water as a means of disposing of the things that we no longer want — our wastes. It is very simple, when the wastes are out of sight, they are gone. With the billions and billions of gallons of water on the planet, nothing people could do would ever ruin it. Such thinking has finally produced a major crisis. When people saw lakes becoming lifeless pools of filthy water and rivers become smelly, greasy streams of waste, they finally realized that something needed to be done.

Using the estimates for water removed and assuming fifty percent return of agricultural and ninety percent for other water users, over 423 cubic miles of used water are returned to the worldwide water supply annually. That converts to almost 147 trillion cubic feet. In that volume of water is contained massive quantities of materials added in the process.

There are many sources of water pollution. But as with water use, there are three broad categories that produce most of our water pollution: agricultural sources, industrial sources, and domestic and urban sources. For more details on the specific sources and nature of water pollution, see Chapter 9.

Agriculture

Farmers have found it necessary to expand agricultural production over the years. This has led to the development of intensive agricultural practices. One byproduct of those practices has been increased water pollution.

The most visible source of agricultural water pollution has been, and remains, erosion. Soil that has had its vegetative cover removed is subject to erosion. The result is muddy streams and lakes, along with silt deposits — usually where they are not wanted. Of course, silt deposits have also produced beneficial results in many places such as the Mississippi River Delta, the Mekong Delta in Southeast Asia, and the Nile Delta in North Africa. But the normal result we associate with erosion is depletion of productive capacity of the soil and pollution of surface waters.

Another pollutant that has become important in recent decades is agricultural chemicals. With scientific agriculture in the latter half of the twentieth century, has come pesticides and fertilizers. Farmers in the developed world found that fertilizers increased yields in rich fields and made poor fields capable of producing good crops. They found that insecticides could control damaging beetles, flies, grubs, worms, and other insect pests. They found they could increase production by using nematicides, herbicides, and fungicides. Without the modern pesticides and fertilizers that are used in the more productive agricultural countries, the present world population would be impossible to maintain. That is true simply because the present world food production could not have been reached without chemical agriculture.

Unfortunately, not all of the fertilizers, insecticides, fungicides, herbicides, etc. are used by the plants for which they are meant. In many cases, the chemicals are persistent. That means they last for long periods in the soil. In other cases, they are applied incorrectly or in heavier concentrations than is recommended. In either case, the result is that the excess chemicals end up in the water supply because in general they are water soluble.

Some pesticides are dangerous when they enter the water supply, others are relatively safe. Some are persistent, others biodegrade quickly. As a whole, the problem of pesticide runoff and leaching is a major one and is growing worldwide. The Food and Agriculture Organization projects a ten percent per year increase in pesticide use over the last quarter of this century in the world's less developed nations. That would mean a twelve-fold increase in pesticide use from 1975 to AD 2000 for those countries.

Persistent pesticides leaching into the surface water supply present a particular problem with fish and wildlife. Aquaculture is the intentional culturing of organisms in water for human use. Worldwide, aquaculture is becoming an important source of protein in the human diet. Aquacultural production of fish, mollusks, aquatic plants, shrimp, and crayfish are growing rapidly. In parts of Asia aquaculture provides a substantial portion of the food produced. Pesticide runoff in the surface water used for aquaculture is a serious threat (see Figure 8-8).

When excessive chemical fertilizers enter the water supply, the water becomes nutrient rich. The result of that is frequently eutrophication. An eutrophic lake soon becomes choked with algae. As the lake becomes more clogged with algae, some of

Figure 8-8. Aquacultural production of food and other products for human use is growing worldwide. Research done at universities like Virginia Tech in Blacksburg, Virginia is contributing to that growth. (Photo courtesy Bill Camp.)

the algae die and decompose, producing still more nutrients for the growing souplike mixture. In extreme cases, fish life in the lake can die and the lake can become almost like a bog. Chapter 9 discusses eutrophication in more detail.

Much of the nitrate increase in the groundwater supply is a direct result of the leaching of chemical fertilizers. It is a particular problem in the developed world where intensive agriculture is practiced. The United States Department of Agriculture estimates that fertilizer application rates will average 566 pounds per acre in Japan by the year 2000. The rates should reach 299 pounds per acre in Western Europe and 392 pounds per acre in Eastern Europe by then. Worldwide average fertilizer use is projected to rise from forty-nine pounds per acre to 129 pounds per acre in the twenty-five-year period ending in 2000 (Speidel, et al., 1988).

Another source of water pollution is from animal production. Large beef feedlots, swine production facilities, and poultry houses and processing plants produce vast amounts of waste products, mostly in the form of manure and other wastes. The most common method of disposal from such facilities is to spread the manure on the ground. If too much manure is spread in an area, the excess nutrients can be leached into the water supply as runoff or groundwater. Many of the same nutrient problems as from leaching fertilizer then occur.

Industry

Industrial pollution of the water supply can be divided into four major categories: thermal pollution, radioactive materials, organic chemicals, and inorganic chemicals.

Thermal pollution (warming) occurs when water is used for cooling in industrial processes. The warm water is then returned to the water supply. It may be just as clean as it was when it was removed, but, when warm water enters the water supply, it upsets the local ecosystem. The most common problems involve fish in the area immediately downstream. Some fish use temperature to trigger spawning. As they travel through the artificially warmed water, their biological mechanisms are disrupted.

Additionally, warmer water holds less dissolved oxygen. The water loses most of its oxygen supply as it removes the heat from the industrial process. Fish downstream from the reentry site can be killed. Finally, algae grows more profusely in warm water. Thus, just as in eutrophication, warmed streams can become algae-clogged.

Finally, thermal pollution causes increased evaporation. In parts of Europe where nuclear energy production involves massive thermal pollution, this can be a problem. Evaporated water returns to the hydrologic cycle. But it is no longer available for use at the point from which it was withdrawn. Thus, the evaporation of water is considered a consumption of part of the surface water supply. It is projected that as much as 7 cubic miles per year may evaporate from Europe's surface water supply by AD 2000 as a result of this. That would represent a one percent consumption of the total runoff from that continent.

Radioactive materials are used more each year in industry. Health workers use radioactive materials in many processes such as cancer treatment. Many electrical power generating plants use radioactive materials. The Chernobyl accident, in what was then the U.S.S.R, introduced large quantities of radioactivity into the world environment. Much of it entered the water supply. When radioactive wastes are improperly disposed of, sometimes they get into the water supply. When that happens they often enter the food chain and eventually find their way into plant and animal tissues, including those of humans.

Many organic and inorganic chemicals are used in manufacturing processes. Some are intentionally dumped into the water supply and others enter it accidentally. Of the massive quantities of water withdrawn annually for industrial purposes, about ninety percent is returned unchanged. Some is consumed and not directly returned to the water supply. Some is polluted only by increased heat. Yet the amount of chemical pollution is enough to endanger significant portions of the surface water and groundwater supplies.

Industrial pollution, as well as urban and domestic pollution in the most developed countries seems to be improving. Places such as the Chesapeake Bay and Delaware Bay of the United States once were in danger of becoming lifeless cesspools. Environmental legislation, international agreements, and popular activities since the 1960s have had a remarkable effect in many areas. Environmental movements have brought problems such as contamination of the groundwater supply and acid rain to the public's attention. The resulting national and international pressure has forced the developed nations to begin clean-up programs and to mandate limiting pollution by industries and urban centers.

Such positive steps have not generally been taken in the developing world. Underdeveloped countries can scarcely afford to build costly water distribution systems and water treatment plants. They need international investments badly. As a result, they believe that they cannot afford to require such technologies as smokestack scrubbers to remove acid emissions before they become acid rain. The people in the developing nations need more food, so the choice is starvation for their people or increased pesticide and fertilizer use, runoff, and leaching. In the real world, the results of such decisions can hardly be in doubt.

It is little wonder that many people in developing countries view environmental movements as attempts to keep them poor. The processes of exploitation and development led to wealth in the United States, Japan, Germany, Sweden, and other such developed countries. It is very simple to understand how leaders in poor countries look at the processes that led to such wealth and see a way out of poverty for their own countries by copying techniques used by wealthier nations.

Domestic and Urban

When the population density of an area allows for self-sufficiency, there is little problem with water pollution. Families take water from wells or surface water. They return waste products to the soil where it biodegrades and does little environmental damage. When urbanization followed the development of agriculture and with the rapid population expansion of the past few centuries, that situation changed. Today, large urban centers require central water distribution and elaborate sewerage systems.

Waste disposal technologies have improved in parts of the world, but in many countries, raw wastes are still being dumped into the water supply. The result is the spread of disease and the degradation of the ecosystem. Waste disposal will be discussed in more detail in Chapters 9 and 12.

Another important source of pollution of the water supply has only recently become a part of the public debate. Acidification of atmospheric water produces acid rain. In nature, the pH of rainwater is normally about 5.7. The pH of rainfall in areas affected by acid rain averages about 4.2 to 4.5 A rainfall with a pH of 2.4 was measured in Scotland in 1974. That is about the same pH as vinegar and will dissolve many organic and inorganic substances not normally soluble in water.

Results of Water Pollution

Water for human consumption comes from a number of different sources. In the developed countries most water for human use comes from groundwaters and surface waters. When necessary, it is treated. This treatment removes most contaminants by filtering and most organisms by various sterilization processes. The result is a safe, clean product for human consumption. Harmful pesticide and fertilizer contaminants are largely removed by this process. Pathogens are killed or filtered out. Most industrial contaminants and urban and domestic pollution are cleaned out before the water is released.

Water treatment requires sophisticated technology. It is relatively inexpensive in terms of cost per unit of water that is eventually produced. Nevertheless, it involves a large initial investment and requires highly skilled workers to operate and maintain. Sadly, in the less developed countries, funds for such investments are lacking. As a result, the water supply in rural areas throughout the less-developed world remains unpurified. In places where the water supply is contaminated by urban, industrial, and agricultural runoff and leachate, the water is unsafe to drink.

The result is widespread breakouts of human, plant, and animal diseases. The 1991 epidemic of cholera in South and Central America resulted from human fecal contamination of the water supplies of the affected countries. Water for human consumption there is drawn from shallow wells and surface runoff. Both sources of water are easily contaminated by runoff and leaching from the surface contaminants.

Severely acid rainfall leaches minerals from leaf surfaces. The result of this process is damaged and dying trees and other plants. The historic Black Forest of Germany as well as other important forests in Europe and North America are being severely affected by acid rain.

Buildup of acids in the surface water supply sometimes results in lakes and streams in which fish do not do well, or sometimes cannot even live. Buildup of acidity affects metal pipes in our plumbing systems. Lead in lead pipes and copper in copper pipes are dissolved more readily by acid water flowing through them. Percolation of acidified soil water results in increased leaching of partially soluble materials. In general, the acidification of the water supply could affect the entire ecosystem in ways that we cannot accurately predict.

Summary

Water is a critical part of the ecosystem. Its small molecule and structure make it a powerful solvent, a good conductor of electricity, and capable of transporting dissolved and suspended particles very efficiently. It exists in the solid, liquid, and gaseous states within a narrow range of temperatures.

To convert H_2O from a solid (ice) to a liquid (water), with no change in temperature, requires a relatively large amount of energy which is called the latent heat of melting. To convert liquid water at its boiling point to water vapor at the same temperature requires an even larger amount of energy, called the latent heat of evaporating. It is this phenomenon that produces a cooling effect to a can of soft drink placed in a container of ice. It is also this effect that produces a cooling effect under a shady tree on a still, hot day.

Water is ubiquitous, heterogeneous, and renewable. It is generally regarded as a common property. It is usually free for the taking. It is inexpensive to transport, treat, and dispense. As a result it is used in massive quantities.

If solar energy serves as the driving fuel of the planetary ecosystem, the hydrologic cycle (water cycle) could be thought of as its engine. Water evaporates, sublimates, and transpires from every part of the planet: oceans, land surfaces, ice surfaces, organisms, even from falling rain and snow. Water enters the atmosphere as water vapor and leaves the atmosphere as precipitation. More water evaporates from the oceans than is returned directly to them as precipitation. Conversely, more precipitation falls on the land than leaves it by evapotranspiration.

The difference between ocean evaporation and precipitation is about ten-thousand cubic miles of water each year. It is that ten-thousand cubic miles of water that is the entire source of the planet's fresh water supply for the year (except for trivial amounts taken from the oceans by desalinization, of course).

Water reaching the surface either evaporates, enters the surface water supply or ice pack, becomes soil moisture, or enters the groundwater supply. These are considered the main water reservoirs. The amount of water moving through a particular system is referred to as the system's water budget. The planetary ecosystem's annual water budget can be viewed as the total amount of water that evaporates in one year.

No matter where it goes, the water remains a part of the Earth's water budget because it is such a stable compound. The length of time that is required for a reservoir to return to the cycle an amount of water equal to its total volume is its residence time. The residence time for water in the atmosphere is less than ten days. The residence time of water in the Antarctic ice pack may be as much as 9,500 years.

Total worldwide water demand in 1975 was approximately 690 cubic miles. The human population is expected to grow between that year and the year 2000 to about 6.5 billion—a sixty-three percent increase. Assuming a level per capita water use (a very conservative guess) that would put total worldwide demand for water at 733 cubic miles per year by the end of the twentieth century.

For the planet as a whole, the major user of water is agriculture, primarily in the form of soil moisture for plants. Much of the agricultural use is for irrigation. But, agricultural use makes up only a small part of the water used in developed nations such as Japan,

Germany, and the United States. The other major uses of water are for industry and for domestic and urban functions.

Pollution of the surface and ground supplies of water is a growing problem. Increases in water pollution are a function of the expansion of technology, increasing population, and growing human wealth (consumption).

Agricultural pollution is increasing because of demands for more food. More intensive agriculture is simply more reliant on pesticides and fertilizers. Industrial pollution from the more developed countries is becoming less of a problem. Advances in technology and in environmental awareness are resulting in cleaner streams and lakes in countries like the United States. But, as underdeveloped countries institute development programs, they do not perceive that they can afford such "luxuries" as cleaner water and air when their people are constantly on the verge of starvation.

Even though water is an inexhaustible resource, clean water is not unlimited in a given location and during a given time span. Water pollution leads to a number of problems that must be solved if the world's water supply is to remain safe. Improved water distribution will be necessary if human populations are to continue to grow in places where supplies are limited.

DISCUSSION QUESTIONS

1. What are some characteristics of water that make it so unique and so important to life on Earth?

2. What is the hydrologic cycle and how does it work?

3. How much water exists on Earth? How is it distributed? How much fresh water is in the Earth's lakes and rivers?

4. What is the worldwide demand for water? What are the major user categories and how much does each use annually?

5. Why is it unlikely that agricultural use of water will decrease in the near future?

6. What factors have lead to the global water pollution problem today? Why is water pollution such a difficult problem to overcome?

ADDITIONAL ACTIVITIES

1. Collect beakers of water from several different sources in your community. Measure the pH of each. Filter each through a very fine filter (such as a coffee filter) and examine any residue. Examine some of the water under a microscope. Compare the kinds of particulate matter, living organisms, and pH you find.

2. Contact the director of your local water treatment facility. Find out the average annual output of the system. What filtration and sterilization procedures are used? How does the monthly output vary during a typical year?

3. Invite the manager of your water treatment facility to speak to your class.

4. Find a recent magazine article about water resources or water pollution. Read it and discuss the information in class. Is the article scientific, emotional, or political?

5. Find a small pond or lake in your community. Estimate its surface area and average depth. Use that data to estimate the water stored in it at the present time. Determine how much the water level changes in a typical year during wet and dry seasons. Estimate the variation in water content over the course of the year.

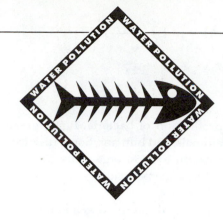

CHAPTER 9
Water Pollution

Courtesy United States Environmental Protection Agency.

Terms to Look for and Learn

Algae	Nitrate
Algal Bloom	Nitrite
Coliform	Pathogens
Copper Sulfate	Potable
Drain Field	Septic Tank
Eutrophication	Sodbuster Legislation
Green (Pond) Scum	Swampbuster Legislation
Lagoon	

Learning Objectives

After reading this chapter and participating in the activities, you should be able to:

- Discuss testing of water for potability.

- Discuss human and animal contamination of water.

- Discuss the contamination of water from sources other than human and animal wastes.

- Discuss nitrates as both a benefit and a pollution problem.

- Discuss oil as a water pollution source.

- Discuss the treatment of wastewater to prevent water pollution.

Overview

The United States contains 79,481 square miles of surface water, two percent of its total area. Water comes in three forms: liquid, solid, and gas. Liquid water is never pure in nature; solid water (ice) is often pure; and gaseous water (water vapor) is always pure. Water in the liquid form always contains ions and molecules of varying amounts of nearly all of the more than one hundred elements known. Liquid water is often the principal means of transport of bacteria, viruses, fungi, algae, and particulates such as clay, colloidal organic matter, and soluble salts.

Most elements are only slightly soluble, meaning that at any one time water contains only traces of them. Most elements are also harmless to the environment. A few are helpful by being essential for the growth and reproduction of bacteria, fungi, green plants, insects, animals, and humans. Some dissolved chemicals are harmful to most organisms; these include lead, mercury, some pesticides, and nitrates in high concentrations.

Water is the universal solvent as well as the principal transport for polluting wastes. Who knows if our water supply is fit to drink?

Several federal agencies are involved in global monitoring and data collection of water quality. Each brings unique expertise to the U.S. Global Change Program, and each plays an important part in advancing the nation's knowledge of global conditions and trends. These federal agencies and their environmental responsibilities are displayed in Table 9-1.

Table 9-1.   Major Federal Agencies Involved in Global Monitoring and Research Programs, Including Water Quality

Agency	Area of Expertise	Agency	Area of Expertise
National Oceanic and Atmospheric Administration (NOAA)	Climate Studies Weather Prediction Ocean Studies Data Management Atmospheric Studies Stratospheric Ozone	U.S. Geological Survey (USGS) (continued)	Water Resources Information and Data Management Land Surface Geology Geography
Department of Energy (DOE)	Climate and CO_2 Precipitation Chemistry Atmospheric Chemistry Ecosystems Research Geosciences	Minerals Management Service (MMS)	Arctic Species Ocean Circulation and Heat Flux
Department of the Interior (DOI) U.S. Geological Survey (USGS)	Ecosystem Modeling Land and Water Use Remote Sensing Studies Hydroclimatology Hydrologic Systems, Processes, and Modeling Spatial Data Collection	Fish and Wildlife Service (FWS)	Wetlands Biological Diversity Wildlife Fish Threatened and Endangered Species Habitat

Table 9-1(Continued).   Major Federal Agencies Involved in Global Monitoring and Research Programs, Including Water Quality

Agency	Area of Expertise	Agency	Area of Expertise
National Park Service (NPS)	Ecological Research Man and the Biosphere Program Vegetation Classification Populations and Communities Natural Resources Earth History	National Aeronautics and Space Administration (NASA)	Earth Observing System Atmospheric-Surface Interactions Climate Systems Greenhouse Effects Remote Sensing Weather Cloud-Radiation Processes Hydrological Cycle Components General Calculation Models Biosphere-Atmosphere Interactions Sea Ice and Ice Sheets
Bureau of Land Management (BLM)	Fire Weather/Climate Wilderness Management Rangeland Wildlife and Fish Threatened and Endangered Species		
Environmental Protection Agency (EPA)	Air Quality Climate Scenarios Acidic Deposition Atmospheric Research Terrestrial Ecosystems Quality Assurance Water Quality Effects on Ecosystems Pesticides Transport and Ecosystem-Level Effects Toxic Substances Research Environmental Monitoring and Assessment Program Human Health Research/Policy Analysis Domestic Regulatory Actions	National Science Foundation (NSF)	Atmospheric Chemistry Biological Diversity Climate Studies Earth History Ecosystems Research Geosciences in Polar Regions Ocean Studies Regional Geography Social and Economic Research Solar-Terrestrial Studies

Water for Farm and Home

It is obvious that muddy and foul-smelling water is not fit for drinking (**potable**). Not so obvious is water with **pathogens** (disease-causing organisms). Your local County Extension Director will have information on how to take a water sample and where to send it for bacteriological and chemical testing to determine its potability.

Testing for bacteria is for the purpose of detecting animal or human contamination. The test is specifically for **coliform** bacteria concentration. These bacteria occur in the intestinal tract of humans and animals. Their presence may indicate serious disease organisms, such as many forms of diarrhea, themselves pathogenic. Please note that technically, coliform bacteria include these genera: *Escherichia, Klebsiella, Enterobacter, and Citrobacter.*

Legal maximum drinking water standards are presented in Table 9-2, and recommended maximum standards are displayed in Table 9-3.

Chemical tests for water for farm and home should include all of the legal maximum standards and as many of the recommended maximum elements as possible. Federal laws mandate maximum concentrations of the following ten chemicals in drinking water: arsenic, barium, cadmium, chromium, fluoride, lead, mercury, nitrate nitrogen, selenium, and silver. The same laws recommend maximum concentrations for drinking water for these ten chemicals: chloride, copper, fluoride*, iron, manganese, sodium, sulfate, total alkalinity as calcium carbonate equivalent, total dissolved solids, and zinc.

About 22,500 infant children under a year of age per year in the U.S. are exposed to drinking water with nitrate nitrogen (NO_3-N) levels above ten milligrams per liter. This is above the maximum contaminant level set by the U. S. Environmental Protection Agency. Use of drinking water from other sources is recommended for these infants.

* The RECOMMENDED maximum concentration for fluoride in drinking water (2.0 ppm) is lower than the MANDATED maximum concentration (4.0 ppm).

Table 9-2. Required* Maximum Drinking Water Standards**

Chemical	Maximum Permitted (milligrams per liter) ***
Arsenic (As)	0.05
Barium (Ba)	1.00
Cadmium (Cd)	0.01
Chromium (Cr+6)	0.05
Fluoride (F)	4.00
Lead (Pb)	0.05
Mercury (Hg)	0.002
Nitrate Nitrogen (NO_3-N)	10.00
Selenium (Se)	0.01
Silver (Ag)	0.05

*Notice the difference between Required maximums (Table 9-2) and Recommended maximums (Table 9-3).
Source: U.S. Environmental Protection Agency, in cooperation with the U.S. Department of Agriculture, U.S. Army Corps of Engineers, and the U.S. Department of the Interior (undated). **Process design manual: Land treatment of municipal wastewater.
***Parts per million (ppm) is the same concentration as milligrams per liter (mg/l).

Table 9-3. Recommended* Maximum Drinking Water Standards**

Chemical	Maximum Permitted (milligrams per liter) ***
Chloride	250.00
Copper	1.00
Fluoride	2.00
Iron	0.30
Manganese	0.05
Sodium	20.00
Sulfate	250.00
Total Alkalinity as Calcium Carbonate equivalent	400.00
Total dissolved solids	500.00
Zinc	5.00

*Notice the difference between Required maximums (Table 9-2) and Recommended maximums (Table 9-3).
**Recommended by the U.S. Environmental Protection Agency.
***Parts per million (ppm) is the same concentration as milligrams per liter (mg/l).

Pollution Violations of Rivers

The number of pollution violations of rivers are itemized by kind of pollution in Table 9-4. All five causes of pollution have decreased over the fourteen-year period reported. The number of violations for fecal coliform dropped thirty-nine percent. Dissolved oxygen violations were lower by sixty percent. Phosphorus violations decreased twenty percent. Dissolved lead varied from no change in 1975 to a decrease of more than eighty percent from 1980 to 1988. Finally, the violations for excess cadmium in U.S. river waters was from one in 1975 to less than one in 1988.

The authors compliment the U.S. Environmental Protection Agency for these good records on the reduction of pollution violations of rivers in the United States. Please note that the standards are established by the U.S. Environmental Protection Agency, but the U.S. Geological Survey samples and tests the river waters for pollution.

Reducing Pollution of Rivers

Certainly industries, cities, and households are major contributors to our water pollution problem. But,

Table 9-4. National Water Quality of Rivers—Violation Rate, 1975-1988*

Year	Fecal Coliform Bacteria >200 per 100 milliliters	Dissolved Oxygen >5 milligrams per liter	Total Phosphorus >1 milligram per liter	Dissolved Lead >50 micrograms per liter	Dissolved Cadmium >10 micrograms per liter
			Total Number of Violations Per Year		
1975	36	5	5	<1	1
1980	31	5	4	5	1
1981	30	4	4	3	1
1982	33	5	3	2	1
1983	34	4	3	5	<1
1984	30	3	4	<1	<1
1985	28	3	3	<1	<1
1986	24	3	3	<1	<1
1987	23	2	3	<1	<1
1988	22	2	4	<1	<1
Increase/ Decrease, 1975-1988	-39%	-60%	-20%	no change to > -80%	<1%

*SOURCE: U.S. Geological Survey.
**Violation level set by U.S. Environmental Protection Agency.

farmers and ranchers must also accept part of the blame for the pollution of streams, rivers, and lakes. They have been using increasing amounts of energy, pesticides, and fertilizers with potential to pollute. According to the U.S. Environmental Protection Agency:

- Nearly two-thirds of U.S. surface water pollution is caused by sediment from croplands.

- Ground waters are becoming polluted by forty-six pesticides in twenty-six states. As a further complication, many ground waters are being depleted faster than they are renewed.

Suggestions for reducing pollution hazards to streams, rivers, and lakes include :

- Control erosion and sedimentation on farms, ranches, construction sites, and surface mining sites by following the guidelines in Chapter 5 of this book.

- Use pesticides according to suggestions found in Chapter 11.

- Use fertilizers and organic residues as proposed in Chapter 12.

- Dispose of toxic wastes according to rules laid down in Chapter 14.

- Maintain woodlots and other forested areas by excluding or limiting livestock and by harvesting trees selectively rather than clearcutting as described in Chapter 6.

Federal Clean Water Laws and Their Enforcement

The first Federal Water Pollution Control Act was passed by Congress in 1972. It authorized the U.S. Environmental Protection Agency to set national water quality standards and delegated enforcement to the states. The Clean Water Act was authorized by Congress in 1977, to be administered by the U.S. Environmental Protection Agency and the Department of Defense. Amendments to both of these acts were passed in 1981, 1987, and 1988.

The U.S. Army Corps of Engineers was authorized to enforce wetland regulations and the U.S. Coast Guard to enforce oil spill regulations. There were provisions in these amendments for citizens to bring suit against other citizens or companies.

Within the U.S. Environmental Protection Agency (known by many people as the EPA), the Office of Enforcement and Compliance Monitoring enforces pollution laws governing water, air, hazardous wastes, toxins, and pesticides. Through ten U.S. EPA regional offices, enforcement personnel cooperate with U.S. attorneys, state attorneys general, district attorneys, as well as with other federal and state agencies to enforce nine major statutes and a number of minor ones. A criminal unit, established in 1982, consists of eight attorneys in the Washington, D.C. office, at least one attorney in each of the ten regional offices, and forty-nine investigators, most with criminal justice experience.

As of 1989, in addition to the Environmental Protection Agency, the following federal agencies were actively enforcing environmental laws, including clean water laws. These included:

- The Council on Environmental Quality

- The Occupational Safety and Health Administration

- The Department of Defense

- The Army Corps of Engineers

- The United States Coast Guard

- National Oceanic and Atmospheric Administration

- National Marine Fisheries Service

- Department of State

- Fish and Wildlife Service
- Department of Commerce
- Department of Interior, Surface Mining Reclamation
- Department of Agriculture
- Department of Treasury, Customs Service
- Nuclear Regulatory Commission
- Department of Transportation

Water Pollution: State Laws

By 1988 all fifty states had established water pollution control laws that prescribed and set criminal fines for unlawful acts resulting in water pollution. No longer do violators of water pollution regulations receive a "slap-on-the-wrist" — there are now crimes with fines and jail sentences.

Although the various fifty states' laws for water pollution vary, each set of state laws must be as strict as the federal laws set by the U.S. Environmental Protection Agency. Some state laws on water quality are stricter than federal laws.

Pennsylvania has a model state law regulating water pollution which handles about two thousand violations a year, and the U.S. Environmental Protection Agency handles about fifty state violations in Pennsylvania. Other states with well-developed water pollution laws are Minnesota, Massachusetts, and New Jersey.

Despite the relatively small number of criminal prosecutions, environmental crimes stand out as the major change in environmental enforcement over the past twenty years. A growing public reaction to those who knowingly damage the environment and endanger human life has been reflected in tougher enforcement provisions in environmental laws. Many things that had been just misdemeanors have been raised to felonies. Million-dollar fines are no longer uncommon, and corporate executives serve jail terms for knowingly violating the environmental laws.

Civil enforcement actions often are taken against large companies capable of absorbing even million-dollar penalties as just another cost of doing business. Civil judgments may carry little public stigma, but, in contrast, criminal sanctions can serve as effective deterrents. A jail sentence for the Chief Executive Office or for the President of the guilty company is one cost of doing business that cannot be passed on to the consumer.

The goal of environmental enforcement is partly to punish polluters and poachers. But the goal is by means of timely and appropriate actions against significant violators, to discourage others. The goal is not to close down the industries that built the American economy. Rather, the goal is to motivate them, out of enlightened self-interest, to clean up their own houses — and keep them clean.

Federal and state legislators have empowered environmental enforcers in both the public and private sectors. Environmental enforcers range from the city policeman to the federal judge and from the individual citizen to large public-interest groups.

Today environmental enforcers are armed with an array of techniques: high-tech surveillance, environmental audits, contractor debarment, and citizen suits. These tools are far more socially powerful than those available in 1970 when the National Environmental Policy Act was passed which established the U.S. Environmental Protection Agency.

Cleaner Water from Less Soil Erosion Sediment

Title XII of the 1985 Food Security Act authorized the establishment of the Conservation Reserve Program. Its objectives are to retire thirty-four million acres of highly erodible cropland to reduce erosion and sedimentation and thereby improve water quality. An estimated 422 million tons of soil will be saved, valued at $3.5 to $4 billion. Additional benefits were projected:

- Less flood damage
- Lower costs of sediment removal
- Lower water treatment costs
- Increased recreational fishing

Erodible land retired from cultivation is to be planted to grass or trees for a contract period of ten years. The land owner who signs the contract will receive fifty percent of the cost of establishing the vegetation and yearly rental payments to offset loss of income from crops.

Other provisions of Title XII of the 1985 Food Security Act included a shift from all-voluntary water pollution control. It implemented an unprecedented system of incentives and penalties for agricultural nonpoint pollution.

Sodbuster provisions of the 1985 Food Security Act prohibit cultivated crops from being grown on sloping lands that are stabilized by grasses without an approved farm/ranch plan. **Swampbuster** provisions of the same Act prohibit the draining of wetlands to produce cultivated crops.

Nitrates as Pollutants

Nitrates are essential in the environment because they are necessary for growth and reproduction of green plants. Nitrates may also be harmful because they can cause sickness or death of small babies and domestic livestock.

Nitrates get into the water environment primarily from these five sources:

- Chemical fertilizers containing nitrogen
- Animal wastes (feces and urine)
- City sewage and septic tank effluents
- Municipal and factory wastes
- Nitrification of organic matter in the soil (Figure 9-1) (see Chapter 4)

The reader should note that many meats and fish are treated with nitrates and nitrites to enhance color and lengthen shelf-life. They are not destroyed by cooking.

For humans, the U.S. Public Health Service has established 45 parts per million of nitrate (NO_3) (or 10 ppm* of nitrate nitrogen, NO_3-N) as the threshold of nitrate toxicity in drinking waters. In the absence of further research information, veterinarians have accepted the same level for toxicity to domestic animals.

Nitrate poisoning may affect humans and animals in these ways:

- Vitamin utilization interference—in suspected instances of nitrate poisoning, large doses of vitamin pills are usually given.

- Antibody production impairment—this hazard may result in an animal or human being more susceptible to pathogens.

- Abortion—abortion in cattle has been suspected as being caused by the animal drinking water high in nitrates.

The disease of infants commonly called "blue baby disease," but technically known as methemoglobinemia, has been associated with nitrates in drinking water since 1945. Since that time, two thousand cases have been reported in North America and Europe and between seven and eight percent have died. Babies less than three months of age appear most susceptible. Reporting of the disease is not mandatory, however, and some estimates place the incidence ten times more than the cases reported.

Nitrate is itself not harmful in the usual amounts ingested. In the stomach of the infant, bacteria reduce the innocuous nitrate (NO_3-) to harmful nitrite (NO_2-). The **nitrite** then converts hemoglobin in the blood to methemoglobin and thus reduces the ability of the hemoglobin to carry oxygen throughout the body. The result is suffocation as observed by a blue

* Parts per million (ppm) is the same concentration as milligrams per liter (mg/1)

Figure 9-1. Concentration of total **nitrogen** (N) in U.S. streams. Annual mean values are in milligrams per liter. (**Source**: CEQ analysis of data from the United States Geological Survey's National Stream Quality Accounting Network as reported in "The Seventh Annual Report of the Council on Environmental Quality," Sept., 1976, page 268.)

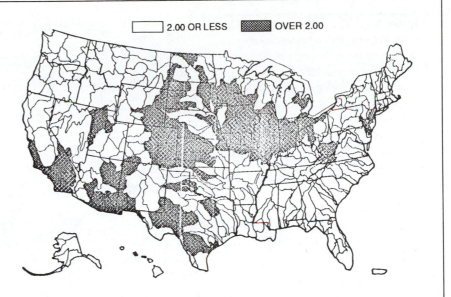

2.00 OR LESS   OVER 2.00

appearance of the skin, similar to carbon monoxide poisoning (see Chapter 13).

Most states and the federal government have passed laws regarding pollution abatement, including the release of solid wastes into the environment. In addition, the federal government, through the Environmental Protection Agency, is making loans and grants to municipalities for the upgrading of their sewage disposal systems.

So far, no laws exist that regulate the location of cattle feedlots. However, various amounts of money have been made available by the federal government through the Rural Environmental Assistance Program of the Agricultural Stabilization and Conservation Service to share the cost with livestock owners for the construction of lagoons below feedlots for pollution abatement.

Phosphorus and Eutrophication

Algae are the principal aquatic plants in the nation's lakes. Algae are a group of single-celled or multi-celled, primitive, green plants that grow in water or in very damp places. Levels of algae are increasing as a result of increases in nutrient elements essential for plant growth, as illustrated in Figure 9-2.

Although algae require sixteen elements for plant growth, the elements most frequently limiting growth are carbon (C), nitrogen (N), and phosphorus (P). Some authorities claim that algae require twenty-one elements to survive. Those include the same sixteen required by higher plants plus aluminum, cobalt, silicon, sodium, and vanadium.

Algae utilize C, N, and P in the approximate ratio of 106:16:1. Carbon is available to algae from the biodegradation of organic matter and from atmospheric carbon dioxide (see Chapter 13).

Nitrogen may be a principal element limiting growth of algae. But certain algal species, especially some blue-green algae, have the capacity to utilize (fix) atmospheric nitrogen. They can then use the nitrogen for manufacturing their proteins. Since that discovery, increased nitrogen from anthropogenic (human caused) sources of nitrogen in effluents have been cited less often as a major cause of increased algal growth.

By the process of elimination, phosphorus must be guilty of being the principal element causing

Figure 9-2. Concentration of algae in U.S. streams. Annual mean values (total algae) are in numbers of cells per milliliter of stream water. (**Source**: CEQ anaysis of data from the United States Geological Survey's National Stream Quality Accounting Network as reported in "The Seventh Annual Report of the Council on Environmental Quality," Sept., 1976, page 267.)

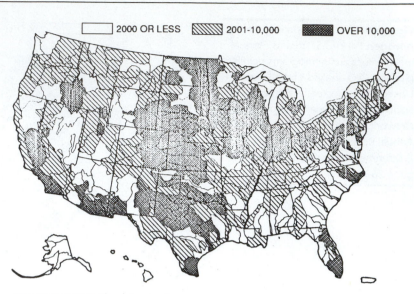

2000 OR LESS   2001-10,000   OVER 10,000

Alaska, Hawaii, and Puerto Rico not drawn to scale

eutrophication of lakes. Eutrophication refers to the build-up of plant nutrients in a body of water and normally results in the growth of excessive plants, particularly algae. The threshold level of phosphorus (P) in waters for the growth of algae has been established between 0.02 and 0.10 part per million. Some environmentalists have predicted that this much phosphorus gets into waters from sources in nature alone.

The amount of phosphorus in the environment has increased slowly from phosphorus fertilizers. Phosphorus concentrations are low in animal manures, and the amount that gets into lakes from feedlots is minor.

Detergents in the 1970s were responsible for about seventy percent of the phosphorus in the waterways of our environment. Some states banned the sale of detergents containing phosphates, hoping to solve the problem of eutrophication of lakes.

As of mid-1972, the United States Government, Department of Health, Education, and Welfare, decided that phosphates in detergents were the *least harmful* in the environment and should continue to

be used until a substitute could be found that was less harmful and/or less polluting.

If the manufacturer and the consumer need any defense for using phosphorus-containing detergents, examine the following:

- They soften hard water by immobilizing iron, calcium, and magnesium and dispersing scum deposits.

- They help to control pathogenic bacteria.

- They emulsify oils and greases for easier removal by the wash water.

- They are nontoxic, noncorrosive, nonflammable, safe for use on all kinds of natural and synthetic fabrics, and can be effectively removed in sophisticated (tertiary) sewage treatment plants.

In defense of phosphates in the environment, they are not harmful to plants or humans. In fact, phosphorus is one of the essential and frequently limiting elements for maximum development of plants, animals, and humans (see Chapter 5). If phosphorus from all detergents were *eliminated*, the nation's waters would continue to receive 680 million pounds of phosphorus (P) a year from other sources (Ferguson, 1968).

One of the authors of this book made a household check on the detergents being used in his home in April, 1991. Here is what he found:

- "Cold Water All," "Low in Phosphorus," actually contained 4.8 percent phosphorus which is the equivalent of 5 grams per wash assuming usage of one-third cup per wash. "The organic surfactants in this formula are biodegradable in lakes and streams." Lever Brothers Company, New York, N.Y. 10022.

- "Dove," "Ingredients: The surfactants in Dove Liquid are biodegradable. This formula contains no phosphorous*" (sic). "If you have any questions or comments on Dove, please call us toll-free at 1-800-544-2002." Lever Brothers Co., New York, N.Y. 10022.

Algae Indicate Eutrophication

One of the first indications of eutrophication of waters is **algal bloom.** A "bloom" means a very large concentration of algae, perhaps five hundred individuals or more per milliliter of water. There are at least twenty thousand distinct species of algae that are classified into five groups as follow.

- Blue-green algae (blue-green bacteria) (*Cyanophyta*) (classified by some as Cyanobacteria) are single-celled plants whose nuclear material is scattered throughout the central part of the cell. The green chlorophyll is diffused throughout the outer edge of the cell. A blue pigment is also present and often a red pigment. Cell reproduction is by fission (simple division). Blue-green algae cover rocks with a slimy gelatinous mass that smells like a pig pen, tastes like fish, and looks like pollution. Some

species of blue-green algae are capable of fixing atmospheric nitrogen.

- Green algae (*Chlorophyta*) have pigments that are principally chlorophyll confined to definite bodies called chloroplasts. There is a well-organized nucleus. The vegetative and reproductive cells have appendages for locomotion, known as flagella.

Figure 9-3. Algal bloom and sludge at the edge of Lake Michigan at Chicago was made more noticeable by a man who accidentally stepped in the polluted algae with his left foot. (Courtesy United States Environmental Protection Agency.)

*Phosphorus is misspelled. One of the authors informed Lever Brothers of the mistake. He received a "thank you" note and an explanation that they were in the process of printing new labels.

- Diatoms are classified as *Chrysophyta*. The pigments are confined to definite bodies, with more yellow or brown pigment, respectively, than chlorophyll. The cell wall of diatoms is composed mostly of silica. (Diatomaceous earth is used as a scouring agent.)

- Euglenoids (*Euglenophyta*) are one-celled algae with one to four flagella for locomotion. There is a definite nucleus and the grass-green chlorophyll is confined to plastids.

- Dinoflagellates (*Pyrrophyta*) include many diverse, pigmented, and one-celled algae. Brown pigments predominate, although chlorophyll is present.

Algal Control

Enriched waters in livestock watering troughs, farm ponds, streams, and lakes may contain so many algae (**"green scum," "pond scum"**) as to be a nuisance, as shown in Figure 9-4. Such waters may be enriched by factory wastes, animal manures, municipal sewage, septic tank effluent, chemical fertilizers, and surface water runoff from fertile soils. Moreover, water from some wells, lakes, and streams may be so fertile that algae will grow luxuriantly with no additional nutrients.

If the cause of the enrichment is known and can be controlled, this should be done before applying any treatment. The usual treatment to control algae consists of applying **copper sulfate** ($CuSO_4$), available at many drug stores under the trade name of "bluestone" or "blue vitriol" at a cost usually of less than two dollars per pound. Caution must be used in its application because less than 1/2 ppm of copper sulfate will not be effective, and above 2 ppm may kill certain species of fish and animals, including humans (Klussman, 1971).

Precautions on the household use of copper vessels may be repeated here. (1) One of the authors drank lemonade (pH 2.0) mixed in a *copper* pitcher, became sick at his stomach and vomited, whereas his wife and three children were not affected. (2) Most

Figure 9-4. Algae on the bank and floating in the water is an example of serious pollution of the water environment. (Courtesy United States Environmental Protection Agency.)

house plants will not grow in soil in a *copper* pot unless the pot is first lined with plastic.

Concentration of the copper sulfate in treated water is the critical factor in determining whether only the algae are killed. The most difficult part of the problem of correct concentration is in determining the volume of water to be treated.

The final concentration of copper sulfate in the treated water will be in units of *parts per million*, on a weight basis. One part per million of copper sulfate would be the concentration when:

One pound of copper sulfate is dissolved in one million pounds of water (16,020.51 cubic feet)

To determine the pounds of water in a pond, wade or use a boat and a pole to determine the average depth of water in feet. Then measure the average width and the average length of the pond in feet. Cubic feet of water are obtained by multiplying the average depth (feet) × average width (feet) × average length (feet). Multiply the total cubic feet × 62.42 (pounds of water per cubic foot) = pounds of water in pond.

Example:

Average dimensions of pond (**feet**)
   Depth—4 ft.
   Width—30 ft.
   Length—400 ft.
4 × 30 × 400 = 48,000 cubic feet of water.
48,000 cu. ft. × 62.42 lb. per cu. ft. = 2,996,160 lb.
(**total cubic feet**) (**pounds per cubic foot**) (**total pounds**)

For purposes of this kind of calculation, the 2,996,160 pounds can be called 3,000,000 pounds of water. If the desired concentration of copper sulfate to kill algae is one part per million, three pounds of copper sulfate would be the required amount.

Actual recommendations of copper sulfate in Texas for the control of algae in ponds vary as follows:

- For acid to neutral waters (pH 7.0 and below)—0.5 to 1.0 part per million.

- For alkaline waters (pH above 7.0)—2.0 parts per million.

In the example given, the actual amount of copper sulfate that should be used would depend on the pH of the water:

- For acid to neutral waters—1.5 to 3.0 pounds of copper sulfate.

- For alkaline waters—6.0 pounds of copper sulfate.

In making such decisions, there are two points that should be taken into account:

- To calculate the *gallons* of water, these factors may be useful:

  1 cubic foot = 7.48 gallons.
  1 gallon = 8.345 pounds.

- The control of aquatic plants other than algae is more complex and depends upon the species of plants to be killed. Write to your state agricultural university for this information or contact the county extension director.

Wastewater Treatment for Water Pollution Control

Treatment of wastewaters may be by a central municipal treatment plant, a septic tank with drain field, or a lagoon. Only the last two techniques fit the subject of this book.

A septic tank is a large, usually underground tank that receives raw sewage from a home or business. The **drain field** is the soil area used to receive the biodegraded liquid from the septic tank. The selection of soils for establishing a drain field for a septic tank is in contrast to soils for a lagoon. A drain field must have a medium-textured soil or a fine-textured soil that permits water to percolate through it at a moderate rate to filter out and adsorb the pollutants.

In contrast, a soil for a lagoon must not permit *any* percolation (Figure 9-5).

The principal pollutants of major concern in wastewater treatment are nitrogen and phosphorus, disease-producing organisms (pathogens), heavy metals, and trace organics. The pathogens of concern are bacteria, viruses, protozoa, and helminths (intestinal worms). The heavy metals of environmental concern in wastewater are cadmium, copper, chromium, lead, mercury, selenium, and zinc.

Septic Tanks and Drain Fields

Three septic systems for rural homes are diagrammed and explained in Figure 9-6. Systems 1 and 3 are similar except that No. 3 is recommended for a smaller lot. The entire drain field is excavated and made into a drain bed instead of drain lines, as in technique No. 1. No. 2 has a motor and blade mixer

Figure 9-5. The clay loam soil in Figure 9-5A would make an ideal site for a septic tank drain field (infiltration rate > 1 in./hr.); whereas, the clay soil in Figure 9-5B would be desirable for building a lagoon for manure disposal (infiltration rate < 0.1 in. per hr.)(Courtesy Texas A & M University.)

A

B

1 Septic Tank & Soil Absorption Field (Trench)

Sewage bacteria break up some solids in tank. Heavy solids sink to bottom as sludge. Grease & light particles float to top as scum. Liquid flows from tank through closed pipe and distribution box to perforated pipes in trenches; flows through surrounding crushed rocks or gravel and soil to groundwater (underground water). Bacteria & oxygen in soil help purify liquid. Tank sludge & scum are pumped out periodically. Most common onsite system. Level ground or moderate slope.

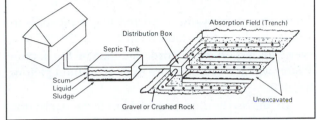

Figure 9-6. Septic systems for rural homes. (**SOURCE**: United States Environmental Protection Agency. "Small Wastewater Systems." FRD-10, 1980.)

2 Aerobic System & Soil Absorption Field

Air and wastewater are mixed in tank. Oxygen-using (aerobic) bacteria grow, digest sewage, liquify most solids. Liquid discharges to absorption field where treatment continues. Can use same treatment & disposal methods as septic tank. Maintenance essential. Uses energy.

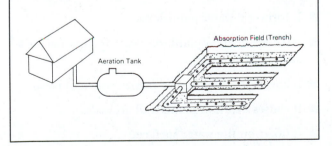

3 Septic Tank & Soil Absorption Field (Bed)

Similar to Sketch 1 but smaller field. Total field excavated. Used where space limited. Nearly level ground.

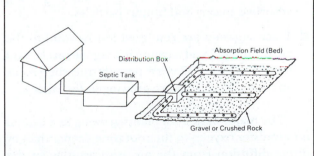

in the **septic tank** which mixes air with the effluent. The aerobic decomposition reduces odors. Also more pathogens are destroyed more quickly. The mixer need not be run constantly but can be set for any recommended cycle.

Lagoon Technology

A **lagoon** is a shallow lake constructed to reduce the concentrations of hazardous substances in wastewaters. When adapted aquatic vegetation is established in the lagoon, its efficiency is increased. For example:

- Levels of coliform bacteria decline rapidly due to sedimentation, filtration, absorption (physiological uptake), adsorption (physical adhesion), and the lethal effect of sunlight.

- Pathogenic bacteria are reduced by predation, sedimentation, absorption, adsorption, and die-off because of an unfavorable aquatic environment, ultraviolet light, and a water temperature unfavorable for bacterial cell reproduction.

- Heavy metals, especially copper, zinc, cadmium, mercury, and selenium are effectively precipitated and adsorbed by algal cells, reeds, and water hyacinths. Heavy metals are also stabilized by being adsorbed to clay and humus particles.

- Trace organics are rendered less harmful to the environment by adsorption in a lagoon on clay and humus particles. This is especially true when reeds and water hyacinths are growing in the lagoon.

The major benefit of plants growing in a lagoon is to transfer oxygen to the root zone deeper than by direct diffusion from the air. Also assisting in the efficient cleansing of wastewaters by lagoons are green leaves of submerged plants. These provide substrate for attached predacious microbes.

Aquatic vegetation effective in wastewater lagoons are water hyacinths, cattails, reeds, rushes, bulrushes, and sedges. All of these are distributed throughout the world except water hyacinths, which are restricted to tropical and semitropical environments.

Oil Pollution of Water Environment

There were twenty-one reported major oil spills in and around the fifty states from 1968 to 1988. However, the largest on record was that of the Exxon Valdez which released more than 10 million gallons in 1989.

In the early morning of March 24, 1989, the Exxon Valdez hit a submerged rock near Valdez, Alaska, and spilled more than 10 million gallons of crude oil. This was the largest accidental oil spill in history. The total fines and clean-up costs were estimated at one billion dollars at the time. When the final decision was made, the Exxon Corporation was not required to pay any fines based on the facts that the spill had been properly reported and clean-up efforts were undertaken promptly.

Oil in the water environment is difficult to clean up because it reacts in the following ways.

- Volatile components in oil evaporate and form insoluble tar balls that sink.

- It forms oil-water emulsions.

- It is consumed by plankton (water plants, bacteria, and small animals).

- It covers wildlife, higher plants, soil, and rocks.

- It causes death of plants and animals.

- It floats on the water surface.

Fifteen months after the oil spill, a group of environmental authors revisited Green Island which had been severely affected at first by the Exxon Valdez oil spill. Their conclusions are summarized as follow.

■ Nearly all the beach oil and tar of 1989 was naturally removed or buried.

■ The recolonization of intertidal organisms, especially barnacles and rockweed (*Fucus* sp.) plants, was underway.

■ Changes in the intertidal community, especially delayed mortality probably due to the oil spill, were still occurring in 1990, so that *both* recovery and oil spill effects were interacting.

■ Remobilization of buried oil may cause additional damage, although the risk in the study area would not be as great as leakage of unweathered oil sealed in fine sediments. (Juday & Foster, 1991).

The general tone of the fifteen-month, after-the-fact study could be characterized as, "We thought it would be worse." By the time of the writing of this book in 1992, little environmental effect still remained from the massive spill.

Summary

While some contaminated water is easy to detect by its appearance or smell, other contaminants are less obvious. In fact, it is necessary to perform very precise chemical analyses to determine whether water is suitable for human and animal consumption. Water can be contaminated by organisms such as coliform bacteria, or, it can be contaminated by inorganic chemicals like lead and mercury. Just because the water is cool, looks clear, and smells fresh does not mean that it is potable.

The pollution of our surface waters has lessened since the environmental movement of the early 1970s. That does not mean that water pollution is no longer a major environmental problem in this country — it certainly is. What it does mean is that local regulations and state and federal laws as well as public pressure from increased environmental awareness have reduced the severity of the problem in many ways. If the situation is to improve further, continued efforts are necessary on the part of concerned citizens as well as government agencies at all levels.

Water pollution is not just an urban or industrial problem. Farmers and ranchers have contributed to water pollution as well. By far the largest polluter of surface water is sediment from soil erosion. Pesticide and fertilizer residues get into the surface and ground water supplies. There are many things that farmers and ranchers should be doing to reduce agricultural pollution of the water supply.

One major source of water pollution is nitrates. Nitrates themselves are not particularly harmful. In fact, nitrates are essential for plant growth. But, once ingested, they can be reduced in the intestines to nitrites. Nitrites do have harmful effects on animal life, including humans, by reducing oxygen transport in the blood.

Another source of both problems and benefits in the water supply is phosphorus. Excess phosphorus in surface water has been blamed for heavy growth of algae, which in extreme cases results in eutrophication. Phosphorus is probably not the real problem, but it took much of the blame for water problems in the 1970s.

The major governmental agency concerned with the water supply is the United States Environmental Protection Agency, but many other federal and state agencies are involved. In addition to numerous federal laws affecting water quality, all fifty states have laws on the subject.

Wastewater, particularly sewage, must be treated before it is returned to the environment. In rural settings, that usually means individual septic systems for homes. In suburban and urban settings, it usually means wastewater treatment plants.

Oil spills are sometimes major catastrophes for ecosystems. Oil covers the water and floats onto the shore. Birds and fish die in massive numbers. Such spills certainly produce major news items and result in powerful emotions. It appears though, that intermediate term (2–3 years) damage to the aquatic ecosystems from oil spills is largely corrected by natural forces. Long-term damage from oil spills appears to be minimal.

DISCUSSION QUESTIONS

1. How do we determine if water is potable from the standpoint of organic contamination?

2. What is the single biggest polluter of the surface water supply in the U.S.?

3. What are some steps that we should take to further reduce the pollution of our surface and groundwater supplies?

4. What are some of the most common chemical contaminants of water? Where do they come from?

5. Is the problem of contamination of the water supplies in the United States getting better or worse? What are the indications of this? Why has the situation changed in the past several decades?

6. What are the harmful and beneficial effects of nitrogen compounds in our water supply?

7. Why is there so much controversy over the effects of phosphorus in the water supply?

8. What can be done to control the excessive buildup of algae in surface water (both prevention and cure)?

9. How should wastewater be treated to reduce its negative effects on the environment?

10. What are the short-term and long-term results in the environment of oil spills?

ADDITIONAL ACTIVITIES

1. Visit a wastewater treatment plant or invite a speaker from the plant to visit your class. Find out what mechanical and chemical procedures are used to remove the contaminants from the water. What happens to the water after it has been processed?

2. Try to find a eutrophic lake or pond in your community. Ask your chemistry teacher to help determine the concentrations of nitrogen and phosphorus in the water. Then try to find a clear lake or pond and repeat the chemical analyses. What are the differences? What are the implications?

3. Develop an inventory of the household products in your home that contribute to the pollution of the water supply.

4. Find articles about the Exxon Valdez oil spill from magazines published in 1989 and 1990. How is the spill described? Discuss the nature of the descriptions in class. To what extent is the discussion centered on actual damage to the ecosystem rather than on the "humane" or emotional aspects of the spill?

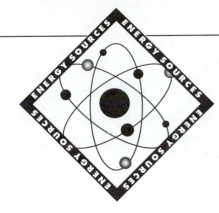

CHAPTER 10
Energy Sources and Use

Courtesy Institute of Energy Conversion, University of Delaware.

Terms to Look for and Learn

Anthracite Coal	Kinetic Energy
Atomic Energy	Lignite Coal
Bituminous Coal	Low-level Nuclear Waste
Chemical Energy	Methanol
Electrical Energy	Natural Gas
Energy	Nuclear Waste
Ethanol	OPEC
Geothermal Energy	Potential Energy
Heat Energy	Sub-bituminous Coal
High-level Nuclear Waste	

Learning Objectives
After reading this chapter and participating in the activities, you should be able to:

■ Discuss the forms of energy usable to humans.

■ Outline energy production trends in the United States.

■ Discuss alternative energy sources.

■ Discuss nuclear waste disposal and the problems associated with it.

■ Discuss oil and coal as major energy sources for human consumption.

157

Overview

Energy (Latin *energi*; Greek *energeia*) is the capacity for overcoming resistance, producing motion, or of doing work. Energy sources for humans include petroleum, coal, gas, running water, wind, sunlight, nuclear fuel, and foods and feeds from which electricity, heat, or mobile power can be produced. Some scientists classify energy into these six categories: potential, kinetic, heat, chemical, electrical, and atomic.

■ **Potential energy** is stored energy ready to be put to work, such as water behind a dam.

■ **Kinetic energy** is the energy of motion. An example is flowing water.

■ **Heat energy** is a measure of the relative activity of atoms in a substance.

■ **Chemical energy** is the energy essential for ions to form compounds.

■ **Electrical energy** consists of electrons that readily move through a metal or, as with lightning, move through particulates in the atmosphere.

■ **Atomic energy** is the energy that holds the nucleus of an atom together.

Energy can be neither created nor destroyed — only transformed. With suitable techniques, all six of these categories of energy may be interchanged, and therefore transformed.

The principal sources of energy produced in the United States, 1950–1988, are listed in Table 10-1. Please note that in 1950, the rank of sources of energy were coal, crude oil, natural gas, and water power, respectively. Energy sources from nuclear power were zero in 1950. Total U.S. energy production increased ninety-four percent from 1950 to 1988.

Table 10-1. Energy Production in the United States by Source, 1950–1988* **(in quadrillion British Thermal Units)****

Year	Coal	Crude Oil	Natural Gas	Water Power	Nuclear	Other	Total
1950	14.06	12.27	6.23	1.42	0.00	0.010	33.98
1960	10.82	16.39	12.66	1.61	0.01	<0.005	41.49
1970	14.61	22.91	21.67	2.63	0.24	0.010	62.07
1980	18.60	20.50	19.91	2.90	2.74	0.110	64.76
1988	20.94	19.52	17.19	2.32	5.68	0.240	65.89
Percent of 1988 Total	31%	30%	26%	4%	9%	<1.00%	100%

*SOURCE: United States Department of Energy.

**In the United States and France, a quadrillion is 1 followed by 15 zeroes; in Great Britain and Germany, it is 1 followed by 24 zeroes.

Trends in sources of energy production in 1988 compared with 1950 were as follow.

- Energy from coal decreased ten percent.

- Energy from natural gas increased eight percent.

- Energy from crude oil increased six percent.

- Energy from water power remained the same.

- Energy from nuclear power sources increased from zero in 1950 to nine percent of all energy sources in 1988.

The United States could become self-sufficient in energy, but to do that, Congress would have to permit offshore drilling for oil on the Florida and California sea coasts and in the Alaskan interior. Or, we could burn more coal or use more nuclear energy. We would have to be willing to accept the effects of increased drilling, coal burning, and nuclear waste on the ecosystems in those regions—damage to wildlife habitats, increased likelihood of oil spills, radioactive waste, and so on. Alternative energy sources may make the goal of energy self-sufficiency more attainable in the future.

Energy Conservation

Is anyone really interested in the efficient use of energy or is our national concern with energy conservation just talk? It appears to be just talk and here are some examples to illustrate that conclusion.

- We ride an airplane when the use of an automobile is more energy efficient.

- We drive an automobile when it would be more energy efficient to use a bus. Using a train would be even more energy efficient.

- Our government promotes highway construction instead of railroads even while we proclaim that energy is a major national concern.

- Riding a bus is less efficient of energy than walking, but bicycling is the most efficient use of energy (see Table 10-2).

One-fourth or more of all new homes in the United States are mobile homes. Thin walls, poor insulation, and very little design for air circulation make heating and cooling of the homes extremely inefficient. Almost all homes could save twenty-five percent of their heating and cooling fuel bills by better insulation in ceilings, walls, and floors; tight-fitting storm windows; weather

Table 10-2. A Comparison of the Energy Efficiency in Various Modes of Human and Freight Transport*

Mode of Human Transport	Energy Efficiency (British thermal units per passenger mile)
Bicycle	200
Walking	300
Bus	2,650
Railroad	2,900
Automobile	5,750
Airplane	8,400

Mode of Freight Transport	Energy Efficiency (British thermal units per ton per mile)
Pipeline	450
Railroad	670
Waterway	680
Truck	3,800
Airplane	42,000

*SOURCE: Oak Ridge National Laboratory, Oak Ridge, Tennessee.

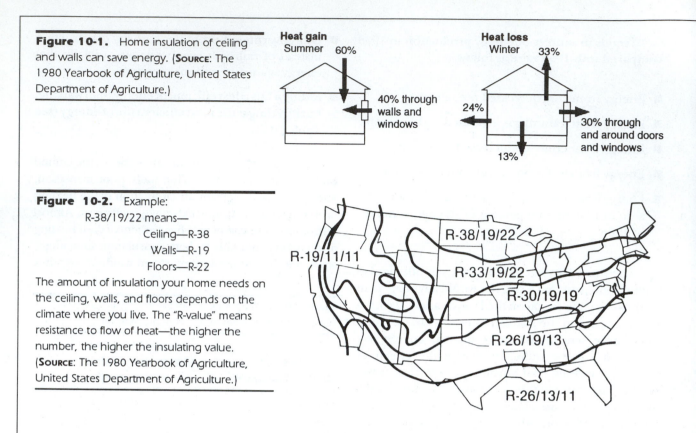

Figure 10-1. Home insulation of ceiling and walls can save energy. (**Source**: The 1980 Yearbook of Agriculture, United States Department of Agriculture.)

Heat gain
Summer 60%
40% through walls and windows

Heat loss
Winter 33%
24%
30% through and around doors and windows
13%

Figure 10-2. Example:
R-38/19/22 means—
Ceiling—R-38
Walls—R-19
Floors—R-22

The amount of insulation your home needs on the ceiling, walls, and floors depends on the climate where you live. The "R-value" means resistance to flow of heat—the higher the number, the higher the insulating value. (**Source**: The 1980 Yearbook of Agriculture, United States Department of Agriculture.)

R-19/11/11
R-38/19/22
R-33/19/22
R-30/19/19
R-26/19/13
R-26/13/11

stripping; and periodic tightening of loose boards (see Figures 10-1 and 10-2).

Newer Energy Technology

Energy production techniques now in the research and testing stage but predicted to be available for mass use sometime in the near future include solar collectors; fusion; and more readily available, safer nuclear energy.

Solar Collectors

Solar power is by far the greatest potential source of energy we have. Each day the sun floods Earth with at least 100,000 times as much energy as humans use. Solar energy can be captured for human use by either active or passive systems. Most solar energy is collected as simple heat. Such solar systems will be discussed under ancient energy sources.

For many years heat from the sun has been used to generate steam. The steam is used to drive turbines that generate electricity. Of even greater potential are solar cells that convert sunlight energy directly into electricity. Small amounts of electricity are generated by solar cells in our everyday lives—solar powered calculators are an example.

Larger solar panels have been perfected and used successfully on everything from solar cars to space ships, but their costs are not yet competitive with other sources of energy on the Earth. Solar scientists at the University of Delaware at Newark designed and built a house in 1973 whose entire roof consisted of cadmium sulfide solar cells encapsulated in plastic. This represented an early attempt to produce electricity directly from the sun (see Figure 10-3).

Fusion

Fusion power is predicted some day to produce from a gallon of seawater enough fuel—deuterium—to have the energy of 300 gallons of gasoline. This concept makes beautiful headlines today to relax an entire world on the need to use energy efficiently, but will cause frustration tomorrow when the hard facts are known. *Scientists can do this now, but the energy output is less than one percent of the energy input.* Until the efficiency is more than one hundred percent, the process is simply trading energy. Fusion power produces some radioactive waste and tritium but less waste heat than existing fission atomic power.

Nuclear Energy

Nuclear power has been controversial in the United States, but in many parts of the world, nuclear power supplies most of the electricity generated.

The first nuclear chain reaction of a uranium isotope, U-235, was accomplished on December 2, 1942. This was followed by the first field testing of an atomic bomb at Alamorgordo, New Mexico, on July 16, 1945. To win World War II with Japan, the United States dropped two atomic bombs, one on Hiroshima on August 6, 1945 and the second on Nagasaki on August 9, 1945 (see Figure 10-4).

Coal as Energy

By far the greatest source of readily available energy, given our present technology is coal. There is enough coal in the United States in known reserves to last five

Figure 10-3. **Solar One**, the first house to convert sunlight into electricity and heat, was designed and built as an experiment to determine the technical feasibility and cost of using solar energy. (Courtesy Institute of Energy Conversion, University of Delaware.)

Figure 10-4. Initial test of the atomic bomb in New Mexico on July 16, 1945, from a distance of 6 miles. **Above**: The start of the explosion. This small cloud later rose to a height of 40,000 feet. **Center**: Multicolored cloud from the explosion. Black areas were brighter than the sun itself, according to observers. **Below**: A later stage of the development of the cloud. (Courtesy United States Atomic Energy Commission.)

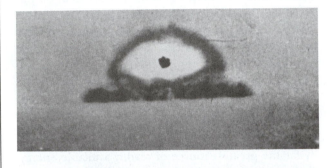

hundred years. In 1950 coal production comprised forty-one percent of all energy sources, however, this percentage had dropped to thirty-one percent by 1988.

Generating electrical energy from coal requires burning it. Coal has always been known, and rightly so, as a dirty energy source. Burning coal produces large volumes of visible smoke. At one time, people thought the solution was to build taller smokestacks and continue to burn coal. The smoke then spread out over a larger area and the visible pollution was less concentrated. Today we still use very tall smokestacks to make the pollution appear less severe. Smokestacks as tall as 750 to one thousand feet are presently in use in the United States.

But there are two much more important problems with burning large volumes of coal. Burning high-sulfur coal puts large amounts of sulfur compounds in the atmosphere. Many scientists believe that results in the formation of large amounts of sulfuric acid in the atmosphere. The result of that is highly acidic water in the atmosphere which falls as "acid rain."

But, an even more important long-term problem with coal-burning is not the smell, or smoke particles settling on the surface. The big culprit is the carbon that is added to the atmosphere from coal burning, as you will learn in the chapter on global temperatures (Chapter 20).

The word *coal* is used to apply to various stages of fossilized herbs, trees, and shrubs. In northern Europe, peat is also burned as a fuel. Peat is still being formed in continuously wet places where vegetation dies and is preserved because of a low level of oxygen for decomposition.

Present-day coal in the United States started with green plants perhaps 250,000 years ago. Throughout the centuries, with varying heat and pressure, the organic residues formed what is known today in the United States as:

- **Lignite** — soft and very dirty coal with thirty to forty percent water and which turns to dust and may ignite spontaneously when exposed to the atmosphere.

- **Sub-bituminous coal** — a commercial product with characteristics between that of lignite and bituminous coal.

■ **Bituminous coal** (soft coal) — the most abundant kind of coal of commercial value.

■ **Anthracite coal** (hard coal) — coal with the greatest heating value per pound. Some anthracite coal is so hard and firm that some specimens can be carved and polished into jewelry.

Coal is not far from U.S. centers of population; it occurs in commercial amounts for mining in three geographic regions of the United States:

■ Appalachian (Eastern)—coal contains large amounts of sulfur and, therefore, adds to acid rain when burned.

■ Interior Region (Iowa, Missouri, Arkansas).

■ Northern Great Plains and Rocky Mountain Region.

U.S. coal production from 1970 to 1989 is displayed in Table 10-3. Production of sub-bituminous coal increased faster than any other kind. This was followed by an increase in lignite and bituminous coal. Anthracite coal decreased in production. Total coal production, 1965–1989, increased fifty-nine percent.

Crude Oil

Wars have been fought over crude oil including the U.S. war with Iraq in 1990-91. The upper graph in Figure 10-5 shows that the four countries with the *largest* crude oil reserves were involved in the Iraqi war with the United States, which ranks ninth in world crude oil reserves (lower graph in Figure 10-5).

Table 10-4 indicates that the United States imported twenty-five percent of its total crude oil consumption in 1960. This has increased every year except one since then. In 1980 imports were forty-seven percent of consumption but fell to thirty-nine percent in 1985. Most startling of all was the fact that in 1989 the United States imported fifty-three percent of its crude oil for consumption.

Table 10-3. Coal Production in the United States, by years, 1970–1989[*] **(in million tons)**

Year	Lignite Coal	Subbituminous Coal	Bituminous Coal	Anthracite Coal	Total
1965	**	**	512.1	14.9	527.0
1970	8.0	16.4	578.5	9.7	612.7
1975	19.8	51.1	577.5	6.2	654.6
1980	47.2	147.7	628.8	6.1	829.7
1985	72.4	192.7	613.9	4.7	883.6
1989	89.6	227.2	656.3	3.5	974.1
Increase/ Decrease, 1970–1989	+1,020%	+1.285%	+14%	-64%	+59%

[*] **Source:** United States Department of Energy.
[**] Included with bituminous coal in 1965.

Figure 10-5. **Above**—The United States ranks ninth of all countries of the world in crude oil reserves. **Below**—However, the United States is first by far in oil consumption. Is anyone concerned?

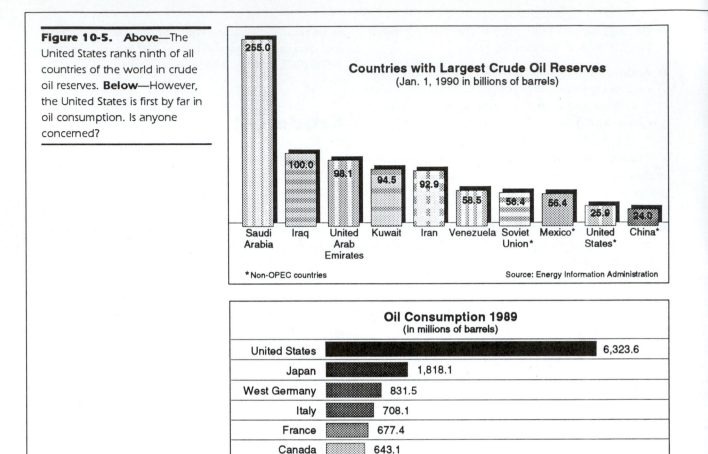

Countries with Largest Crude Oil Reserves
(Jan. 1, 1990 in billions of barrels)

*Non-OPEC countries

Source: Energy Information Administration

Oil Consumption 1989
(In millions of barrels)

United States	6,323.6
Japan	1,818.1
West Germany	831.5
Italy	708.1
France	677.4
Canada	643.1
Britain	634.0

Source: Energy Information Administration

World Production

World production of crude oil is dominated by the oil cartel known as the Organization of Petroleum Exporting Countries (**OPEC**). The objective of this cartel is to control world oil prices by managing worldwide production of crude oil. Representatives of **OPEC** meet periodically to agree on production goals for each member nation.

World crude oil production has been dominated by the OPEC countries since 1960. In that year they produced forty-two percent of world crude oil. However, in 1989 they accounted for thirty-eight percent.

Note: The OPEC countries include these thirteen: Algeria, Ecuador, Gabon, Indonesia, Iran, Iraq, Kuwait, Libya, Nigeria, Qatar, Saudi Arabia, United Arab Emirates, and Venezuela.

The rank in production in 1989 was OPEC, U.S.S.R., United States, China, Mexico, United Kingdom, and Canada, respectively.

Crude oil production in the United States remained relatively level from 1960 to 1989 (see Table 10-5).

Table 10-4. Crude Oil Production and Consumption in the United States, 1960–1989[*]

Year	Crude Oil Production (in quadrillion British Thermal Units)	Crude Oil Consumption	Crude Oil Imported: (Percentage of Total consumption)
1960	14.93	19.92	25
1965	16.52	23.25	29
1970	20.40	29.52	31
1975	17.73	32.73	46
1980	18.25	34.20	47
1985	18.99	30.92	39
1989	16.12	34.21	53

NOTE: "Production" listed as "Crude Oil," and "Consumption" as "Petroleum."

SOURCE: Energy Information Administration, U.S. Department of Energy.

Alternative Fuels

The average miles per gallon of gasoline for passenger cars in the United States has been increasing steadily since 1970, from 13.5 to twenty miles per gallon in 1988. Even though car mileage is increasing satisfactorily, the resulting air pollution must be reduced.

Motor vehicles are responsible for more than sixty percent of all air pollution in the United States. They emit ninety percent of all toxic carbon monoxide (CO), sixty percent of all hazardous hydrocarbons (HC), fifty percent of acid-producing nitrogen oxides (NO_x), ten percent of all acid-producing sulfur dioxide (SO_2), and seventeen percent of all suspended particulates (mostly dust) according to the U.S. Department of Commerce. Are there cleaner fuels?

The advantages and disadvantages of automobile fuels alternative to gasoline are detailed in Table 10-6. The principal alternative fuels are methanol (methyl alcohol, CH_3OH), ethanol (ethyl alcohol, C_2H_5OH), liquified petroleum gas (LPG), compressed natural gas, oxygenated fuels, and reformulated gasoline.

Table 10-5. World Production of Crude Oil, 1960-1989 (in millions of barrels per day).[*]

Year	OPEC Countries	U.S.S.R.	United States	China	Mexico	United Kingdom	Canada	Other	Total World
1960	8.70	2.91	7.04	0.10	0.27	**	0.52	1.42	20.96
1965	14.34	4.79	7.80	0.23	0.32	**	0.81	2.01	30.30
1970	23.41	6.97	9.64	0.60	0.49	**	1.26	3.50	45.87
1975	27.15	9.47	8.38	1.49	0.71	0.01	1.43	4.14	52.78
1980	26.99	11.46	8.60	2.11	1.94	1.62	1.44	5.20	59.35
1985	16.63	11.25	8.97	2.51	2.75	2.53	1.47	7.54	53.65
1989	22.63	11.36	7.61	2.76	2.51	1.79	1.56	9.24	59.46

[*]Energy Information Administration, U.S. Department of Energy.
[**]Less than 5,000 barrels a day.

NOTE: Although the U.S.S.R. no longer exists, the countries that made it up are still major energy producers and consumers.

Table 10-6. Advantages and Disadvantages of Alternative Fuels Compared to Gasoline[*]

Alternative Fuel	Advantages Compared to Gasoline	Disadvantages Compared to Gasoline
Methanol (CH_3OH, also known as methyl alcohol, carbinol, wood alcohol, wood spirit)	May reduce ozone-forming hydrocarbon emissions by up to forty percent May reduce ozone-forming hydrocarbon emissions by up to ninety percent when one hundred percent methanol is used Eliminates benzene and other toxic emissions	May increase formaldehyde emissions Requires significant costs for new production and distribution systems Reduces vehicle driving range and is corrosive to engine parts Difficult to start vehicle in cold temperatures
Ethanol (C_2H_5OH also known as ethyl alcohol)	Reduces ozone-forming hydrocarbon and toxic emission similar to methanol Reduces carbon dioxide emissions	Emits more acetaldehyde Would cost consumers substantially more without federal tax exemption Requires vehicle modifications estimated at three hundred dollars per vehicle
Liquified Petroleum Gas	Produces an estimated fifty percent fewer hydrocarbons, which have less ozone-forming potential May reduce carbon monoxide emissions by an estimated twenty-five to eighty percent	Reduces vehicle driving range and causes refueling inconveniences Required pressurized fuel tanks restrict vehicle cargo space Increases new car cost by up to one thousand dollars
Compressed Natural Gas	Reduces hydrocarbon emissions by an estimated forty to ninety percent Reduces carbon monoxide emissions by an estimated fifty to ninety percent Reduces emissions of benzene and other toxic pollutants	Emits more nitrogen oxides Requires new distribution system Reduces vehicle driving range and causes refueling inconveniences Requires large, heavy, pressurized fuel tanks May increase vehicle costs by up to two thousand dollars

[*]United States Government Accounting Office (1990). **Air pollution: Air quality implications of alternative fuels**. GAO/RCED-90-143. Washington, D.C.

Table 10-6 (Continued).   Advantages and Disadvantages of Alternative Fuels Compared to Gasoline[*]

Alternative Fuel	Advantages Compared to Gasoline	Disadvantages Compared to Gasoline
Oxygenated Fuels	Reduces oxygenated emissions by an estimated twelve to twenty-two percent	Reduces carbon monoxide emissions of volatile organic compounds
	Increases gasoline octane levels, thus decreasing the need for harmful additives such as benzene	Oxides to form toxic chemicals, such as formaldehyde
		May increase nitrogen oxide emissions
		May contribute to auto fuel system problems, reduced fuel economy
		Increases the cost of gasoline
Reformulated Gasoline	Deliverable to consumers through existing distribution system	May require purchasing significant amounts of new refinery equipment and constructing new refinery units
	Estimated to reduce ozone and air toxics produced by automobile emissions	Will not provide emission benefits comparable to fuels such as pure methanol and compressed natural gas
	Requires few, if any, vehicle modifications	Extensive reformulation could result in substantial price increases to consumers

The Federal Clean Air Act Amendments of 1990 require companies and government entities with fleets of ten or more vehicles to begin by 1998 to replace part of their vehicles with alternative-fueled vehicles. Such vehicles can be fueled with "natural gas, electricity, ethanol, methanol, hydrogen, propane, diesel, and reformulated gasoline." (*The Wall Street Journal,* July 24, 1992)

Methanol (CH₃OH) (Methyl Alcohol, Wood Spirit, Wood Alcohol)

Methanol was formerly made from the distillation of wood but now made by the oxidation of hydrocarbons, including most organic materials. It is a poison-ous, highly flammable, and explosive liquid. Blindness is one of the acute toxic effects. Death can result from drinking as little as one cup full.

As an alternative to gasoline, cars can be adapted to run on pure methanol. The advantages of pure methanol fuel are: ten percent to forty percent more miles per gallon, up to twenty-five percent more horsepower, and ninety percent fewer smog-producing hydrocarbons and cancer-causing chemicals.

Ethanol (Ethyl Alcohol) from Farm Products

Ethyl alcohol (C_2H_5OH), also known as **ethanol**, is diluted and used as a motor fuel blend. As a motor fuel additive, ten percent anhydrous (without water) ethanol is mixed with ninety percent unleaded gasoline and sold as "gasohol." Ethanol can be made by yeast fermentation of cereal grains, sugarcane, sugarbeets, and potatoes. About 2-1/2 gallons of anhydrous ethyl alcohol can be made from one bushel of a cereal grain. The residues are valuable for use as foods and feeds.

Sales of gasohol in the United States increased from about one billion gallons in 1980 to eight billion in 1986. Consumption of gasohol is predicted to double during the next ten years (Windholz, The Merck Index, 1983; *The 1980 Yearbook of Agriculture*).

Natural Gas

Natural gas (methane, CH_4), the cleanest of the power sources (except for water power), releases only carbon dioxide and water when burned. The simplest chemical equation is:

$$CH_4 + 2O_2 \longrightarrow CO_2 + 2H_2O$$

methane gas + oxygen → carbon dioxide + water

Production of natural gas in the United States increased from 1960 to 1970 but has been decreasing almost every year since then. Consumption was greatest in 1970 but has been erratic since that year (see Table 10-7).

Water Power

The production and use of water power energy in the United States is presented in Table 10-8. Production reached its peak in 1975, then declined. Use also reached its maximum in 1975, declined in 1980, peaked in 1985, and again decreased in 1989. When

Table 10-7.   Natural Gas: Energy Production and Consumption in the United States, 1960-1989[*]

Year	Energy Production from Natural Gas	Energy Consumption from Natural Gas
	(in quadrillion British Thermal Units)	
1960	14.12	12.39
1965	17.66	15.77
1970	24.18	21.79
1975	22.01	19.95
1980	22.20	20.39
1985	19.16	17.85
1989	19.77	19.36

[*]**Source**: U.S. Department of Energy.

Table 10-8.   Water Power: Production and Use in the United States, 1960-1989[*]

Year	Water Power Production	Water Power Use
	(in quadrillion British Thermal Units)[**]	
1960	1.61	1.66
1965	2.06	2.06
1970	2.62	2.65
1975	3.15	3.22
1980	2.90	3.12
1985	2.94	3.36
1989	2.75	2.86

[*]**Source**: U.S. Department of Energy.
[**]In the United States and France, a quadrillion is 1 followed by 15 zeroes; in Great Britain and Germany, it is 1 followed by 24 zeroes.

use exceeds production, that means that energy from water power was imported from Canada. The maximum imported was in 1985 (see Figure 10-6).

The North American Electricity Reliability Council reported in September, 1988, that the United States needed an *annual* growth rate of *new* electrical capacity of two percent. In fact, annual sales of electricity have increased by four percent from 1979 to 1990. In 1990, hydroelectric power supplied from four to six percent of U.S. electricity and was not predicted to increase (Ray & Guzzo, 1990).

Another scientist, W.E. Matson, agrees with the above prediction but states that there is some potential for small-scale hydroelectric developments on small rivers. Preventing such developments are the growing popularity of laws establishing Federal Wild and Scenic Rivers.

Figure 10-6. "The clean look" is the best way to describe a dam, a lake, beautiful clouds, and electricity generated from the power of water falling through a turbine (The Morrow Point Dam in Colorado). (Courtesy Bureau of Reclamation, United States Department of the Interior.)

Nuclear Energy

People are afraid of the word "nuclear." Those who remember 1945 recall that the United States became a nuclear power. Those not old enough to remember 1945 may recall these more recent nuclear incidents.

■ Three Mile Island nuclear accident on March 28, 1979, at which time a combination of human errors and equipment failures caused a partial core meltdown in Middletown, Pennsylvania. The plant shut itself down and no one was injured.

■ In April, 1986, the Chernobyl nuclear plant about sixty miles from Kiev, Soviet Union, released clouds of radiation over many countries in Europe. The Soviet government was slow in announcing this accident to the world.

Large numbers of Soviet citizens died from exposure to radiation right away and even larger numbers died from long-term effects over the following years. We can expect increased deaths from radiation-linked health problems (mostly cancer) for decades to come. The radioactive cloud dropped fallout across much of Europe where the effects were more subtle but still quite real.

It became clear later, that the technology used in the Soviet power plant had been outmoded and safety training and procedures had been inadequate by western standards. Nevertheless, this nuclear accident was a great catastrophe and a tragedy that has led to still more fears of nuclear energy as a major power source.

The number of nuclear power reactors in the United States increased from sixty-one in 1976 to 110 in 1989. During the same period, the percentage of U.S. electric generating power from nuclear energy increased from 9.4 % to 19.0 % (see Table 10-9 and Figures 10-7 and 10-8).

Table 10-9. U.S. Nuclear Power Plant Operations in the United States, 1976–1989[*]

Year	Nuclear Power (no. of reactors)	Nuclear Power (% of electricity generated)
1976	61	9.4
1977	65	11.8
1978	70	12.5
1979	68	11.4 [**]
1980	70	11.0
1981	70	11.9
1982	77	12.6
1983	80	12.7
1984	86	13.6
1985	95	15.5
1986	100	16.6 [***]
1987	107	17.7
1988	108	19.5
1989	110	19.0

[*]SOURCE: U.S. Department of Energy.
[**]Three Mile Island nuclear accident, March 28, 1979.
[***]Chernobyl nuclear accident, April, 1986.

The Three Mile Island nuclear accident in 1979 appears to have slowed the development of nuclear power the following year. Was this hysteria or prudence? There was no such slow-down after the Chernobyl nuclear incident in 1986 in the U.S.S.R. The U.S. Environmental Protection Agency plays a major role in any federal response to nuclear accidents. It coordinates and participates in environmental monitoring during and after emergencies. It also maintains emergency mobile monitoring teams that can be deployed rapidly to a nuclear accident site.

In 1989 nuclear power supplied nineteen percent of the U.S. electricity. Without these nuclear plants, the United States would have used an additional 275 million tons of dirty coal *or* another 800 million barrels of crude oil. With coal or oil the emission of sulfur dioxide and nitrogen oxides would have increased by many millions of tons and acid rain would have worsened (*The Wall Street Journal*, August 23, 1990).

Nuclear power is claimed to be cheaper than other sources of energy, with a saving of $115 billion on the U.S. foreign exchange oil bill over the period of 1960–1990. However, the problem of safe disposal of **nuclear waste** has not been adequately addressed to solve it, either politically or technically.

Safe Nuclear Waste Disposal

The U.S. Atomic Energy Commission had at one time considered a plan to send nuclear wastes in a permanent orbit around the sun. More mundane proposals include:

- Above-ground storage in bunkers on U.S. government land, as shown in Figure 10-9.

- Underground storage in deep, unfractured, igneous, nonporous rock such as basalt and granite.

- Underground storage in salt beds.

At Lyons, Kansas, twenty-five miles northwest of Hutchison, the geologic materials consist of deep and thick salt beds that appear to be the most feasible site for radioactive waste disposal. Two other sites in the United States also have thick salt deposits, one in western New York State and the other under Detroit, Michigan.

Of the three locations with suitable depths of salt beds, the site at Lyons, Kansas, appears the most suitable, at least in terms of weather. Annual rainfall in Lyons is twenty-eight inches; Detroit, thirty-one; and in western New York, about thirty-five inches. Water tables are high in the Detroit area, intermediate in western New York, and very deep in Lyons, Kansas. Furthermore, evaporation rates are the high-

Figure 10-7. Nuclear power reactors in the United States.

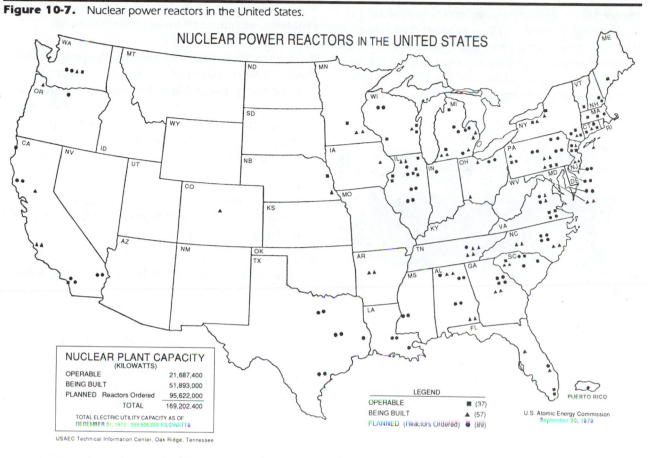

NUCLEAR POWER REACTORS IN THE UNITED STATES

NUCLEAR PLANT CAPACITY
(KILOWATTS)

OPERABLE	21,687,400
BEING BUILT	51,893,000
PLANNED Reactors Ordered	95,622,000
TOTAL	169,202,400

TOTAL ELECTRIC UTILITY CAPACITY AS OF
DECEMBER 31, 1972 399,606,000 KILOWATTS

USAEC Technical Information Center, Oak Ridge, Tennessee

LEGEND

OPERABLE	■	(37)
BEING BUILT	▲	(57)
PLANNED (Reactors Ordered)	●	(89)

U.S. Atomic Energy Commission
September 30, 1973

Figure 10-8. The world's first nuclear commercial power station was built in 1957 at Shippingport, Pennsylvania, on the Monongahela River by the Westinghouse Corporation. (Courtesy United States Environmental Protection Agency.)

Figure 10-9. Radioactive waste storage tanks at the U.S. Atomic Energy Commission's Savannah River Plant in South Carolina. The tanks are in reality a tank within a tank which will be encased in concrete three to four feet thick. All known precautions are taken to prevent pollution by spills. (Courtesy United States Atomic Energy Commission.)

est in Lyons. A salt bed is predicted to be the best "sink" for such wastes because:

■ In the past, water has not been penetrating salt; otherwise, the readily soluble salt would not exist. (The salt beds were deposited under an ocean, but that was several million years ago.)

■ In the future, water is not expected to move through the salt to carry the nuclear wastes into the water table and into wells, rivers, or lakes where they would endanger people and animal life.

■ The salt (NaCl) itself constitutes a safety factor. Salt resists corrosion and flows slowly under the influence of heat. (The nuclear wastes are constantly giving off heat.) The hot and flowing salt moves only a short distance and then precipitates to seal the edges of the flow space.

■ Salt beds should be at least two hundred feet thick, and the depth to the top of the bed from the surface of the soil should be between five hundred and two thousand feet. Any deeper than two thousand feet would be technically suitable but too expensive to reach.

In terms of the salt deposits at Lyons, potential problems encountered include the following:

■ The American Salt Company which operates a salt mine fifteen hundred to eighteen hundred feet from the proposed site for disposal of nuclear wastes, does not leave pillars to support the cavity when salt is removed. The cavity *could* collapse and form a deep surface lake that could possibly disturb, at least psychologically, the local residents.

■ Existing deep holes nearby, drilled in search of gas and oil, cannot effectively be plugged. In the future, water may leak from these into the proposed radioactive waste burial site.

■ The American Oil Company pumped 175,000 gallons of water into the salt bed with the hope that the water would return to the surface as saturated brine as a means of recovering the salt more cheaply. Instead of surfacing, however, the water disappeared. The fear is that the water may have moved into the proposed "sink" and could possibly carry harmful nuclear wastes into usable water supplies if nuclear wastes were buried there.

The proposed disposal site near Lyons, Kansas, was abandoned because of the objection of the local residents, and a newer proposal to use the potassium and sodium chloride salt mines near Carlsbad, New Mexico, was being considered.

In truth, the best scientific evidence indicates that the hazards to public safety were almost (but not quite) zero. Nevertheless, people were afraid of the very slight hazards. When a scientist or engineer says that he or she is 99.99% certain that a thing (like an accident) will not occur, people react not to the near-certainty but to the 0.01% risk. They do not understand that zero risk is unattainable in any human activity. As a result, they react emotionally to a risk more perceived than real.

There are two contrasting kinds of nuclear wastes that must find safe disposal sites: low-level and high-level. The dividing line is any radioactive waste emitting more than 100 nanocuries per gram.

> *Note:* A Curie is that amount of any radioactive substance that has 3.7×10^{10} disintegrations per second and a nanocurie is 10^{-9} Curie (0.000000001 Curie.) This is one billionth of a Curie.

■ **Low-level nuclear waste**—slightly contaminated items such as rags, paper, discarded protective clothing, wood, and spent resins used in water purification systems.

■ **High-level nuclear waste**—spent fuel rods removed from the nuclear reactors after the fissioning process is complete. These rods are highly radioactive and must be isolated from the environment for several hundred years.

The experience of the Tennessee Valley Authority with spent fuel rods will be quoted here without change:

> Originally, the nuclear industry believed that spent nuclear fuel would be shipped to a reprocessing facility, where unburned uranium and fission-produced plutonium would be recovered for use in new fuel assemblies. Any high-level waste remaining after processing would be prepared for underground disposal. However, in 1977 President Carter ordered a halt to development of the reprocessing industry because of the potential for proliferation of nuclear weapons. This decision left the nuclear industry without an established policy for disposing of spent nuclear fuel and with an urgent need to provide greater storage capacity (Tennessee Valley Authority, 1983).

In the United States, low-level nuclear wastes have been disposed of by shallow land burial, under the U.S. Low-Level Waste Policy Act of 1980 and the U.S. Nuclear Regulatory Commission in 1982. Nuclear waste disposal must follow these guidelines:

> Concentration of radioactive material which may be released to the general environment in ground water, surface water, air, soil, plants, or animals must not result in an annual dose exceeding an equivalent of 25 millirems to the whole body, 75 millirems to the thyroid, and 25 millirems to any other organ of any member of the public. Reasonable effort should be made to maintain release of radioactivity in effluents to the general environment as low as is reasonably achievable. (Source: U.S. Geological Survey).

> *Note:* REM stands for "Roentgen Equivalent Man" and is a unit measure of biological damage to humans. A millirem is one-thousandth of a REM.

In a study of several existing low-level nuclear waste disposal sites, the U.S. Geological Survey found (see Figure 10-10):

■ Burial sites that are satisfactory in arid climates were not suitable in humid climates.

■ Waste containers in the burial trench may decompose and cause a collapse of the clay cap, thus permitting animals and surface water to reach the nuclear waste.

■ A shallow water table must be avoided as a nuclear waste disposal site. A fifty-foot-deep water table is most desirable.

■ A deep and productive loam to clay loam soil makes a safer burial site than a sand or fractured rock site.

Figure 10-10. Locations of commercial and major Department of Energy low-level radioactive waste burial sites in the United States. (**Source**: United States Geological Survey, Washington, D.C.)

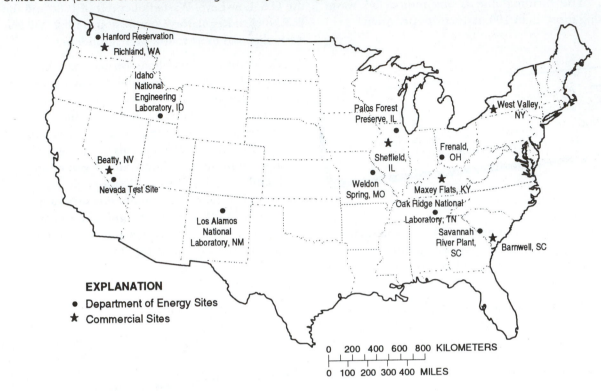

Ancient Power Sources

As of 1988 there had accumulated in the United States almost fourteen million cubic feet of high-level nuclear waste and no government policy or approved plan to safely dispose of it. However, the U.S. Geological Survey has made recommendations on its safe disposal at these five sites:

■ Cypress Creek Dome—Mississippi

■ Richton Dome—Mississippi

■ Deaf Smith Site—Texas

■ Yucca Mountain Site—Nevada

■ Hanford Site—Washington State

Ancient power sources that occasionally find champions to promote them are solar, wood, wind, and geothermal.

Geothermal Energy

In some parts of the world, heat within the Earth is an important source of energy, called **geothermal energy**. Particularly where there is a lot of volcanic activity, there are reservoirs of hot groundwater. In

Iceland, for instance, hot water is taken from geo-thermal reservoirs and piped directly into homes, factories, and businesses to provide heat. A geother-mal plant operated by Pacific Gas and Electric Company in California uses steam from geysers. It generates part of the electrical energy used by the city of San Francisco.

The economical use of geothermal energy appears to be limited, though. Most parts of the Earth's surface do not have hot springs. Where there are hot springs, the quantity of water is often limited or the temperature is really only warm.

The normal thermal gradient (geothermal gradient), the rate at which temperature increases with depth in the Earth, averages about 150° F per mile. Since water boils at 212° F, water at a depth of about 1.41 miles (212/150) would be at the boiling point.

Hot springs (thermal springs) are evidence that there is a geothermal gradient and that groundwater can be so heated. There are more than one thousand thermal springs in the western United States, forty-six in the Appalachian Mountains, six in the Ouachita Mountains of Arkansas, and three in the Black Hills of South Dakota. Geothermal resources occur in thirty-seven states.

> *Note:* Natural springs are classified as thermal springs when temperature of the water is at least 15°F higher than the mean temperature of the ambient air.

Solar Energy

Solar power is "free" energy from the sun. Every square yard of the Earth on a sunny day receives about one thousand watts of energy. To use it to generate electricity by means of solar collectors is relatively expensive. But, for many thousands of years, humans have collected solar energy directly as heat. Such heat can be used immediately or stored in water, oil, or some other medium. The energy is then used to warm our homes or businesses elsewhere or later as it is needed.

Newer technologies to collect, store, and move solar energy as heat can provide limited sources of energy. Passive solar collectors for homes and businesses can help in heating and in providing hot water. However, it is doubtful that solar energy collected as heat will provide a major energy source in the developed world.

Wind Power

Wind power is also free, but it has always had severe limitations. For thousands of years, people have harnessed wind in sails to produce movement. Later, windmills were constructed. Wind could then be used to draw water from wells and pump it into troughs or pipe systems. It could also be used to drive simple machinery. But such uses have been limited to when the wind is blowing steadily from a given direction and a windmill has been erected.

Newer technologies have produced more efficient windmills. Large windmill farms have the potential to supplement more traditional electrical generation systems. But, it is unlikely that wind power can produce significant amounts of energy for an energy-dependent world.

Wood

For thousands of years, wood has been the number one energy source for cooking food and heating the home. In most developing countries it is still the number one source of fuel for cooking but the supply is getting low (see Figure 10-11).

Figure 10-11. A Peace Corps Volunteer in Mali, semiarid western Africa, is discussing the need for firewood with the village leader. Planting trees is the logical answer. (Courtesy United States Peace Corps.)

Summary

Energy can be defined as the capacity to overcome resistance, produce motion, or do work. There are six general categories of energy used by humans: potential, kinetic, heat, chemical, electrical, and atomic energy. Most of the energy used by humans originated as solar energy and was stored as chemical energy in coal, crude oil, or natural gas. Solar energy stored in water as potential energy is used to turn turbines and generate electricity. Total energy production in the United States almost doubled between 1950 and 1990.

There is little reason to believe that energy conservation is taken seriously in this country. In fact, there may be little reason to conserve energy in the long run. The real effort might better be directed at converting energy generation from fossil fuels (coal, oil, and gas) to new energy sources. The real problem is not energy sources or potential. The real problem is the impact on the environment of how that energy is generated.

There is enough coal in the United States in known reserves to last five hundred years. But coal is known as "the dirty energy." Coals' reputation for being dirty came from the smoke and smell that results from burning. But, the real environmental damage from burning coal comes from the sulfur and carbon emissions into the atmosphere. There are four major categories of coal: lignite, sub-bituminous, bituminous, and anthracite.

The single energy source that is most critical in the world of the 1990s is oil. Control of the world supply of oil (and consequently its price) is one of the most pressing political problems of today. Regardless of the speeches and the emotions, the Gulf War between the United States and Iraq in 1990–91 was over oil — who owns it and who controls it.

The most important environmental problems center around energy. Coal and oil are burned to generate electricity and heat. Worldwide, cars produce more air pollution than any other source. The pollution comes from burning fuels which produce carbon compounds, nitrogen oxides, and sulfur dioxide. A number of alternatives are already available to power our cars and our electricity generating plants.

When we refer to alternative energy sources, we are generally talking about anything other than oil and coal. There are a number of alternative energy sources that are, or could be important. These include water, wind, solar, alternative fuels (ethanol, methanol, alcohol, and others), geothermal, fusion, and nuclear energy.

Solar, geothermal, fusion, wind, and tidal energy generation are basically nondamaging to the environment. Nuclear power generation has the potential to produce great amounts of energy, but nuclear power is very controversial because of the radioactive waste it produces. Indeed, for each of these there are severe limitations to the amount of energy that can be captured and stored with current technology.

The alternative energy source with the greatest short-term potential is clearly nuclear energy. If the political problems associated with it could be overcome, there would still be legitimate safety and waste disposal concerns. Nuclear waste disposal has received great engineering and scientific attention over the past few decades. Still, the problems with waste disposal remain the main stumbling block to expansion of nuclear power as a worldwide energy source.

DISCUSSION QUESTIONS

1. What are the six broad categories of energy?
2. What have been the energy production trends in the United States over the past half century?
3. What have been the trends in energy production in the United States in recent decades?
4. Why do you suppose the United States imports so much of its energy from other countries?

5. What is OPEC and what is its primary purpose? How does it work to achieve that purpose?
6. What are some of the more important alternative energy sources and what are their short-term potentials for solving this country's energy problems?
7. How can nuclear waste be safely disposed?
8. What are some of the problems associated with nuclear waste disposal?
9. What are the ancient power sources?

ADDITIONAL ACTIVITIES

1. Make a collection of energy sources for display. Include as many different kinds of coal, wood, charcoal, fuel oil, etc. as you can gather.
2. Arrange for the manager of your local electric utility office to visit your classs to talk about energy sources in your area.
3. Construct a working model of a passive solar energy collector. Plans should be easily available in your school library.
4. Organize a class debate on whether to allow a nuclear power plant to be planned to serve your community.
5. Organize a class debate on whether drilling for oil should be allowed in the Arctic National Wildlife Refuge. As an alternative, assume that an oil deposit had been discovered under your school. Debate whether to allow drilling on the school grounds.

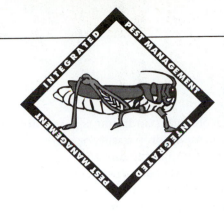

CHAPTER 11

Integrated Pest Management

Courtesy Cornell University.

Terms to Look for and Learn

Avicide	IPM
Biological Control	Nonpoint Source Pollution
Calibration	Pesticide
Cultural Pest Control	Point Source Pollution
Economic Threshold	Rodenticide
Genetic Control	Sex Attractants
Insecticide	Transgenic
Interplanting	Ultrasonic Pest Repellers

Learning Objectives

After reading this chapter and participating in the activities, you should be able to:

- Explain integrated pest management.

- Discuss the relationships between agriculture pest control practices and the environment.

- Discuss the pros and cons of banning pesticides with potentially harmful effects on animal life.

- Discuss alternative pest control techniques.

- Perform simple calculations for pesticide applicator calibrations.

Overview

In a very real sense, pest management is one of the most critical aspects of the human fight for survival. Without effective management of damage from insects, rodents, fungi, bacteria, and other pests, human populations could not be maintained even remotely near present levels.

People often talk about "exterminating" pests, such as mice or cockroaches. Extermination implies complete elimination of the species. Not only is exterminating such pests impossible, but from an ecological standpoint it would be undesirable. The most practical and overall beneficial solution would be to manage the populations of these organisms so the damage they do will be acceptable.

Any number of pest management techniques have been used by humans over the past several thousand years. Only in the last few decades has large-scale chemical pest management become possible. Unfortunately, when synthetic chemical pesticides became available beginning in the 1940s, many people believed they could abandon all other pest management strategies and rely strictly on chemicals. That has proven not to be true. On the other hand, many people would now have us abandon chemical pesticides because some problems with side effects from pesticide use have developed. Neither position is very wise. In this chapter, a variety of pest management techniques will be examined, including the judicious use of synthetic pesticides.

Pests and Human History

A pest can be defined as "anything, such as an insect, animal, or plant that causes injury, loss, or irritation to a crop, stored goods, an animal, or people." (Herren & Donahue, 1991, 344). Clearly, this definition implies that the idea of pest is a human concept. We think of other organisms as having competitors, parasites, or predators. We think of humans as having pests. From an ecological standpoint, a pest is nothing more than a competitor, parasite, or predator of humans or something that humans use.

Throughout history, people have had to contend with pests. Insects and rodents take our food; fungi and bacteria cause our food to rot in the field or in storage; fleas, mosquitos, mites, ticks, and other insects irritate us with their bites and spread disease. The Bible speaks of plagues of locusts that descended on the fields and produced great famines. The Black Plague of the Dark Ages was spread by infected fleas.

Native Americans taught the early European settlers in this country how to plant corn and numerous other local crops. Legend has it that they taught the settlers to plant seven corn seeds in every hill. An old proverb grew from that legend, "Two for mice, two for crow, two to rot, and one to grow." Although it leaves out insects, the meaning of the proverb is clear and based on thousands of years of human history — humans must be prepared to lose most of the food they try to produce to a range of pests.

Indeed, this reality placed severe limitations on the human population until recent years. It was the mechanization of agriculture, advances in plant genetics, the invention of synthetic fertilizers, and the development of chemical pesticides that made possible the huge increases in human population that have occurred over the past century.

Our ancestors found that environmental sanitation and personal hygiene were important pest control measures. They found that cats were an effective rodent control measure and that dogs were effective against many small predators. They discovered that sulfur had some effect as an insecticide for crop pests. They found that some plants were less likely than others to suffer insect damage and that the time of planting was critical in minimizing insect damage. In short, they used a variety of biological, cultural, and chemical pest control measures. Thus, early human attempts to control pests were not so very different from what we are now calling Integrated Pest Management.

Elimination or Management?

Ultimately, there is no real solution to the problems of pests. In the planetary ecosystem, there are literally millions of species of plants and animals. All of those species must compete for niches. Each species forms parts of many interconnected food webs. Humans are no different. Even if one insect could be completely eliminated, some other pest would simply move into the vacated food niche that had been occupied by the eliminated species. We can, however, hope to manage the populations of various pests at such levels that they do not cause unacceptable harm.

Three very important problems can develop when pesticides are used as the primary means of pest management: (1) pests can eventually develop resistance to specific pesticides, (2) populations of insects can rebound (resurgence) to even higher levels after pesticide applications, (3) and other (secondary) pests can arise to take the place of the original pest. These problems give rise to what some environmentalists have called the "pesticide treadmill." What these people are referring to is the tendency for a given pesticide to become gradually less effective in controlling a given pest. (The reason for that tendency will be explained shortly.) The result of that tendency is that the pesticide must be used more often or in heavier concentrations to get the same results.

> In reality, the same problems are true for biological controls, genetic manipulations, and all other pest management techniques. A more accurate and less biased term would be the "pest control treadmill."

Development of Resistance

As seen in Chapters 1, 2, and 3, and as noted in Chapter 19, all organisms adapt to changes in their environment by genetic modification across generations. Smaller organisms tend to have shorter life spans. Thus, the smaller an organism, the more quickly it can generally adapt. What that means is that pests such as insects, fungi, and bacteria are very adaptable. An example will help explain the effects of that rapid adaptability.

Assume that some widespread species of insect, such as a cockroach, has become a pest. Some ingenious chemist will surely develop an **insecticide** (insect killer) that is very effective against cockroaches and sell it to pest control companies because people will pay to get rid of such obnoxious insects. Unfortunately, it is very rare that an insecticide is either completely effective or completely safe. But, let's assume that this is an unusually lethal compound and it is ninety-nine percent effective at the recommended application rate. What that means is that for every one thousand cockroaches to which the insecticide is applied, about ten will probably survive. It should come as no surprise that those ten surviving insects will be the one percent that have a resistance to the insecticide. Without the 990 other cockroaches to compete for food and living space, these resistant cockroaches multiply rapidly (remember the Logistic Curve?). That particular insecticide will be less effective on this next generation of cockroaches. In effect, a new "super cockroach" will eventually develop which is not at all susceptible to that insecticide.

This phenomenon occurs in all forms of pest control. The result is invariably the same. No species will become a serious pest unless it occurs in large numbers. Such a species is almost impossible to completely eradicate. Use of any single control measure will at best produce only partial results. The surviving organisms will simply reproduce successive generations that are more resistant to the pesticide or other control measure.

Resurgence

A resurgence occurs when the specific pest comes back in even larger numbers after a pesticide has been applied or some other pest control technique has been used. That does not seem to make sense, but the process is simple to illustrate. Lady beetles eat aphids and help keep their population under control. When a broad-spectrum insecticide is applied to a garden to control the remaining aphids, the lady beetles are also eliminated. The surviving aphids reproduce much more quickly than the surviving beetles. The result may very soon be even more aphids than were present before. In this example it is said that the aphids experienced a resurgence.

Secondary Pest Development

A third problem that sometimes occurs when an attempt is made to eliminate one pest is that a different organism simply replaces the original pest. An example of this effect occurs on cotton. Spraying the cotton with an insecticide to control boll weevils results in a reduction in the number of insect predators as well as the target species (boll weevils). The result is that the number of serious pests of cotton has increased from six to sixteen.

Economic Threshold

Just as in many other aspects of environmental science, economics must be considered in pest management. We have seen that eradication of a pest organism is probably neither possible nor desirable. Thus, the question becomes, "How much population reduction of a given pest is enough?"

Pest management practices have associated costs. For instance, applying an herbicide to reduce the weeds in a field of soybeans requires money to purchase the pesticide. It also requires labor on the part of the farmer. It requires the use of equipment, fuel, and other expenses. Before applying the herbicide, the farmer should consider how heavy the infestation of weeds must be to cause more economic damage to the soybean crop than it costs to apply the herbicide. If leaving the weeds in the field will cause less damage than applying the herbicide will cost, it makes no economic sense to use the chemical.

The point at which the damage from a pest exceeds the cost of controlling the pest is said to be the **economic threshold**. To determine when the economic threshold is reached in a crop or livestock enterprise requires constant monitoring of all potentially serious pests.

So-called "insect scouts" gather samples of specific insects in a given area to estimate population levels. Other sampling techniques are used to estimate densities of mites, nematodes, and other pests. If insect scouts have (1) information on the density of a given pest required to reach the economic threshold and (2) the actual density of the pest in an area, the farmer can avoid unnecessary pesticide applications.

Insurance and Cosmetic Applications

All too often pesticides are applied for "cosmetic" or "insurance" reasons. Some examples should clarify what these two terms mean in relation to pesticide use. A homeowner may find even one dandelion in the lawn unacceptable. The homeowner would then apply herbicides much more often and at much higher rates than are really needed. Proper mowing and good cultural techniques would control the weeds without herbicides, but the herbicide is used as a cosmetic to improve the lawn's appearance.

The same homeowner may find several Japanese beetles on the roses. When that happens, the homeowner may elect to spray the roses with an insecticide. Reasoning that if one tablespoon of the insecticide per gallon of water is good, then two tablespoons must be better—the homeowner doubles the dosage rate. Then reasoning that if one spraying per week is good, then two per week must be better, the homeowner doubles the frequency of application. The extra dosage and application are done "for insurance."

Cosmetic and insurance application of pesticides produce much of the damage to the environment that results from pesticides. These are examples of what we mean when we refer to misuse or overuse of pesticides.

Pesticides

A **pesticide** is any substance used to kill organisms unwanted by humans. These include insecticides (insects), herbicides (higher plants), algicides (algae), fungicides (fungi), bactericides (bacteria), nematicides (nematodes), **avicides** (birds), molluscacides (slugs and snails), viruscides (virus), and **rodenticides** (rodents). A more general term for chemicals that are used to kill organisms is biocide.

Nebel described as "first generation" pesticides those early pesticides which were based on inorganic sources, such as sulfur, and compounds of heavy metals such as mercury, lead, and arsenic. Other early insecticides that could also be considered first generation include a number of naturally occurring toxic substances such as salt and petroleum. These were the only pesticides available in large quantity until the 1930s.

Second generation pesticides began to be developed in the 1800s in the form of synthetic organic compounds. The first important second generation pesticide was dichlorodiphenyltrichloroethane (DDT). It was first synthesized in the 1880s, but it was not until 1938 that DDT became recognized as a potentially important insecticide.

DDT was extremely toxic to insects, yet seemed to have no effect on larger animals such as birds, mammals, and humans. It was cheap to manufacture, selling for as little as twenty cents per pound at retail. It could be produced in huge quantities, and was simple to transport and apply. It was so effective that crop production in treated fields increased dramatically. Mosquitoes, the carrier of tropical diseases such as malaria, could be effectively controlled for the first time in history. It appeared that a solution to one of humanity's oldest scourges, insects, might be at hand. The remarkable properties of DDT and the other second generation pesticides that have followed have had massive effects on the human condition and on the environment.

Classifying Pesticides

The U.S. Environmental Protection Agency uses the following list to code pesticides for registration.

- Fiber crops, such as cotton and hemp.

- Specialized field crops, such as tobacco.

- Crops grown for oil, such as castor bean and safflower.

- Ornamental shrubs and vines, like mistletoe.

- General soil treatments, such as manure and mulch.

- Household and domestic dwellings.

- Processed nonfood products, like textiles and paper.

- Fur and wool-bearing animals, such as mink and fox; laboratory and zoo animals; pet sprays, dips, collars, litter, and bedding treatments.

- Dairy farm milk-handling equipment.

- Wood production treatments, for example railroad ties, lumber, boats, and bridges.

- Aquatic sites, including swimming pools, diving boards, fountains, and hot tubs.

- Uncultivated nonagricultural areas, such as airport landing fields, tennis courts, highway rights-of-way, oil tank farms, ammunition storage depots, petroleum tank farms, sawmills, and drive-in theaters.

- General indoor/outdoor treatments, in bird roosting areas, or mosquito control.

- Hospitals, including syringes, surgical instruments, pacemakers, rubber gloves, bandages, and bedpans.

- Barber shops and beauty shops.

- Mortuaries and funeral homes.

- Preservatives in paints, vinyl shower curtains, and disposable diapers.

■ Articles used on the human body, such as human hair wigs, contact lenses, dentures, and insect repellents.

■ Refuse and solid waste sites, home trash compactors, and garbage disposals.

■ Specialty uses, such as mothproofing and preserving specimens in museums.

Each state and the United States government, mainly through the Environmental Protection Agency (EPA), have set standards for pesticide registration, handling, and use. Some practices that were *suggested* for safe use in the past are now *required* by law. These include recordkeeping, storage and disposal procedures, reentry intervals, filling, and mixing methods. All the new standards are designed to help make pesticide use safer for people and the environment.

The state pesticide laws cannot overrule or conflict with federal laws. Both federal and state laws and regulations apply to any person using pesticides within a state.

Early recognition of possible harm to the environment by pesticides was emphasized in 1939 by A.E. Michelbacher, an agricultural scientist at the University of California at Berkeley. He published a report which he titled, "Integrated Control." No doubt this was the scientific background for the term "Integrated Pest Management."

"Integrated Pest Management" (**IPM**) is: "An economical approach to pest management in which all available necessary techniques are systematically consolidated in a unified program, so that pest populations can be managed in such a manner that economic damage is reduced and adverse side effects are minimized" (Herren & Donahue, 1991, 245).

A summary of the relationships between pesticide control systems used in agriculture and pollution of the environment was objectively stated in 1971 by a task group at the University of Illinois. The general conclusions of that report are still accurate.

■ Available evidence does not indicate that present levels of pesticide residues in human food and environment produce an adverse effect on human health.

■ Present methods of regulating the marketing and use of persistent pesticides appear to accomplish the objectives of providing the user with a properly labeled product and holding the amounts of residue in food at a low level. They do not, however, appear to ensure the prevention of environmental contamination.

> Persistent pesticides remain stable, and therefore toxic, for extended periods of time. Nonpersistent pesticides biodegrade (decompose) rapidly into nontoxic compounds once they are applied. Some persistent pesticides remain stable for many years. Some nonpersistent pesticides decompose into harmless compounds in only a few days.

■ Residues of certain persistent pesticides in the environment have an adverse effect on some species of wild animals and threaten the existence of others.

■ Persistent pesticides are of special concern when their residues possess—in addition to persistence—toxicity, mobility in the environment, and a tendency for intensification in the food chain.

■ Usage of certain persistent pesticides in the United States should be restricted to specific essential uses that create no known hazard to human health or to the quality of the environment, and that are unanimously approved by the Departments of Health, Education, and Welfare; Agriculture; and the Interior; and by the Environmental Protection Agency. (This jurisdiction now resides solely in the U.S. Environmental Protection Agency.)

■ Action should be increased at international, national, and local levels to minimize environmental contamination where the use of persistent pesticides remains advisable.

■ Suitable standards for pesticide content in food, water, and air must be developed to protect the public from undue hazards, yet recognize the need for optimal human nutrition and food supply.

- Public demand for attractiveness in fruits and vegetables and statutory limits on the presence of insect parts in processed foods have invited excessive use of pesticides. (We should note that in recent years, there has been a growing public acceptance of "organically-grown" fruits and vegetables, even though the appearance is often not as nice as that of conventional produce.)

- There should be more federal support of research on all methods of pest control, the effects of pesticides on human health and the ecosystems, and improved techniques for prediction of effects on humans.

- Incentives should be provided to encourage industry to develop more specific pest control chemicals.

- In the United States, no decrease in the use of pesticides is expected in the foreseeable future. On a world basis, increased use is probable.

- More participation is needed in international cooperative efforts to promote safe and effective use of pesticides.

By the time that report was published, a number of insect species had begun to develop resistance to DDT. We saw earlier how that process occurs through natural genetic selection. In addition, there was evidence that a build-up of organochlorines, primarily DDT, in animal tissues had begun. Persistent pesticides retain their chemical structures and toxicity for extended time periods. In addition, there is a tendency for those compounds to accumulate in fatty animal tissues because they are soluble in water but not in oils. The higher up the food chain, the higher the concentrations of those compounds became. Although no evidence of harmful effects of DDT were ever demonstrated for mammals, high levels of DDT in birds were associated with problems in egg production. No direct causal relationship was ever established, but the circumstantial evidence was actually quite substantial. Many people concluded that DDT

was causing reproductive failure in some birds, most notably in eagles.

Because of the growing environmental controversies, that 1971 report from the University of Illinois came at a time when pest management and environmental concerns were in great turmoil. As of January 1, 1971, the U.S. Environmental Protection Agency banned DDT, one of the most effective and economical pesticides available at the time. Hurt most by the ban were the cotton growers who had been using eighty-six percent of the DDT used in the United States for the control of the cotton bollworm and the cotton boll weevil. Approximately thirteen percent of the DDT had been used on soybeans and peanuts, and the remaining one percent on vegetable and fruit crops. The ban on DDT also was to blame for tussock moth damage to Douglas fir trees in the Pacific Northwest. This tree species furnishes more building lumber than any other species.

Soon thereafter other organochlorines besides DDT were outlawed. They were aldrin, dieldrin, chlordane, endrin, and heptachlor. These pesticides were long-lasting, cheap, effective on the target species, and so harmless to humans that no deaths had ever been documented. Then why were they banned? Because they accumulated in the food chain of wildlife and caused thin egg shells of wild birds.

These long-lasting, effective, and relatively nontoxic organochlorines were officially replaced by a group of organophosphorus insecticides and carbamates. Many of the substitute organophosphorus and carbamates are extremely toxic to humans, other warm-blooded animals, and to wildlife. A number of human deaths have been directly attributed to use of these pesticides over the years, and countless deaths of small mammals and birds have occurred.

More sophisticated methods of application are required to apply these newer pesticides that are toxic to animals other than the target species. Prescription-type sales to licensed pesticide custom applicators will gradually replace over-the-counter sales to the public. Compulsory insurance against damage to people, animals, fish, wildlife, and the neighbors' crops is predicted.

In common language, banning pesticides like DDT was an environmental trade-off where humans lost and the nonhuman parts of the environment won. In at least this case, the purely anthropogenic (human-centered) perspective was set aside and a broader environmental perspective was taken.

Pesticides that have been restricted or taken off the market, their present use, and their environmental concerns are detailed in Table 11-1.

Although many pesticides have been restricted in use and others have been taken off the market, there is serious concern because many are still polluting our environment. Seventy-four different pesticides have been reported in groundwaters in thirty-eight states in the United States. Of these, forty-six pesticides in twenty-six states have been traced to normal agricultural (**nonpoint**) **sources**. The others are from known **point sources** such as from a factory sewer (see Chapter 9). Point source pollution can be traced to one or more specific sources. Pollution that cannot be traced to single or multiple points is called nonpoint source pollution.

To interpret these data, one report states: "It is important to note, however, that one or two detections of a pesticide at trace levels, does not indicate a problem in itself. However, a particular pesticide detected at numerous locations at concentrations equal to or greater than Health Advisory Levels is reason for concern" (Williams, Holden, Parsons, & Lorber, 1988).

Eighteen pesticides have been found in ground water at levels equal to or greater than those established Health Advisory Levels. Of these eighteen, seven of the pesticides have been banned or their use severely restricted as of December, 1988 (see Table 11-1). An estimated 150,000 people in the United States drink water from rural domestic wells with at least one pesticide above health-based limits (United States Environmental Protection Agency, 1992).

Pesticide control techniques other than the use of traditional chemical techniques will be discussed next. The discussion will be under the headings of preventive, genetic, cultural, biological (sex attractants and insect eradication by irradiation of males).

Note: The last section in this chapter will be on calculations for formulating chemical pesticides.

Preventive Control of Pests

Many insect, disease, and weed species have been "imported" from abroad. At all entry points in the United States there are specialists working for the Animal and Plant Health Inspection Service in the U.S. Department of Agriculture whose job it is to confiscate and destroy such pests. Even so, many get through the cracks.

Preventive control of pests in the garden can be practiced by following these suggestions.

- Buy seeds or plants with resistance to as many diseases as possible. For example, seeds labeled "V-F-N" indicates *resistance to* verticillium wilt, *fusarium* wilt, and *nematodes*. But "resistance to" does not guarantee "freedom from" (see Chapter 5).

- Plant "early maturing" varieties to *avoid* maximum insect build-up.

- Spray or dust with a pesticide as a last resort.

- Avoid a concentration of any one species or crop cultivar. Insects locate host plants by their odor.

- Prepare a weed-free seed bed.

- Rotate the location of crops.

- Do not compost diseased plants.

- Fertilize and lime based on a soil test.

- Keep plants growing vigorously by carefully watering, weeding, manuring, and fertilizing the soil.

- If possible, plant in rows running north and south to reduce shading by the taller plants.

Table 11-1.   Pesticides Restricted or Taken Off the Market[*]

Pesticide	Use	Environmental Concerns
Aldrin	Insecticide	Oncogenicity[**]
Chlordane (Agricultural uses: termiticide uses suspended or cancelled)	Insecticide/Termiticide	Oncogenicity, reduction in nontarget and endangered species
Compound 1080 (Livestock collar retained, rodenticide use under review)	Coyote control; Rodenticide	Reductions in nontarget and endangered species; no known antidote
Dibromochloropropane (DBCP)	Soil Fumigant–fruits/vegetables	Oncogenicity, mutagenicity; reproductive effects
DDT and related compounds	Insecticide	Ecological (eggshell thinning); carcinogenicity
Dieldrin	Insecticide	Oncogenicity
Dinoseb (in hearings)	Herbicide/Crop desiccant	Fetotoxicity; reproductive effects; acute toxicity
Endrin (Avicide use retained)	Insecticide/Avicide	Oncogenicity; teratogenicity; reductions in nontarget and endangered species
Ethylene Dibromide (EDB) (Very minor uses and use on citrus for export retained)	Insecticide/Fumigant	Oncogenicity; mutagenicity; reproductive effects
Heptachlor (Agricultural uses; termiticide uses suspended or cancelled)	Insecticide	Oncogenicity; reductions in nontarget and endangered species
Kepone	Insecticide	Oncogenicity
Lindane (Indoor smoke bomb cancelled; some uses restricted)	Insecticide	Oncogenicity; teratogenicity; reproductive effects; acute toxicity; other chronic effects
Mercury	Microbial Uses	Cumulative toxicant causing brain damage
Mirex	Insecticide/Fire Ant Control	Nontarget species; potential oncogenicity
Silvex	Herbicide/Forestry, rights-of-way, weed control	Oncogenicity; teratogenicity; fetotoxicity
Strychnine (Rodenticide use and livestock collar retained)	Mammalian predator control; rodenticide	Reductions in nontarget and endangered species
2,4,5-T	Herbicide/Forestry, rights-of-way, weed control	Oncogenicity; teratogenicity; fetotoxicity
Toxaphene (Livestock dip retained)	Insecticide–Cotton	Oncogenicity; reductions in nontarget species; acute toxicity to aquatic organisms; chronic effects on wildlife

[*]Source: U.S. Environmental Protection Agency.
[**]Definitions: Oncogenicity — Causes tumors
Mutagenicity — Causes mutation
Carcinogenicity — Causes cancer
Teratogenicity — Causes major birth defects
Fetotoxicity — Causes toxicity to the unborn fetus
Avicide — Killer of birds

■ Before applying *any* chemical insecticide, determine the economic threshold of the target insect species. An economic threshold is the number of insects that justify treatment. This number can be based on insects per plant, per square yard, or per sweep net sample. The threshold number is always based on economics. In other words, at what point does the insect begin to cause as much damage to the crop as the insecticicde will cost to apply? In a home garden, the gardener makes this decision; on large acreages, the farmer usually hires an insect scout to determine the economic threshold. When many farmers have many acres of the same crop, they often coordinate scouting and the use of pesticides because insects often move from field to field.

Under preventive control, some people have even used ultrasonic repellers to try to drive pests away. These are mechanical devices that make high-pitched sounds that the manufacturers claim repel insects, moles, and others pest. In the late 1970s, the U.S. Environmental Protection Agency spent $100,000 on research testing **ultrasonic pest repellers.** Their conclusion: *Worthless.* In addition, the Denver Wildlife Research Center of the U.S. Fish and Wildlife Service found pesky animals apparently enjoying the ultrasonic sounds. On the other hand, a few species of birds can be scared away by their electronically reproduced distress calls.

Preventive control has recently been practiced in California by rigid control of imported citrus fruit to prevent the reintroduction of the Mediterranean fruit fly shown in Figure 11-1. The idea of this approach is to quarantine potential sources of infestation. Quarantines can be used either to confine a pest to an affected area to prevent further spreading or to prevent the introduction of a pest into an unaffected area.

Many of the most damaging pests result from either accidental or intentional introduction of some animal or plant into an area. Customs laws are designed to prevent the accidental introduction of pests that may be contained in fruits, vegetables, or other materials being transported from one part of the world to another.

A scientist at the University of Rhode Island tried some preventive control remedies usually recom-

Figure 11-1. A Mediterranean fruit fly is a very serious pest on citrus. (Courtesy United States Agricultural Research Service.)

mended by organic gardeners. Here are some of his results in the form of questions and answers.

Q. If I place a cardboard collar around my young vegetable plants, will this protect them from cutworms?

A. Yes, to a limited degree. If the cutworm is one of the species that burrows into the soil, adequate control will probably be achieved. But if it is of a species that lays its eggs on the plant, the collar will be useless.

Q. Do marigolds, **interplanted** (marigolds alternated with vegetables plants) among vegetables, repel insects?

A. No. To our knowledge, they do not satisfactorily reduce insects or nematodes. In fact, we can even *grow* nematodes on marigolds!

Q. If I place unleached wood ashes around my cabbage and broccoli, will it prevent cabbage maggots?

A. Yes, but if you use too much the plants will be damaged by potassium toxicity. Write to the University of Rhode Island Soil Testing Laboratory, East Farm, Route 108, Kingston, RI 02881 and request a fact sheet which discusses the pros and cons of wood ashes as a soil amendment.

Q. If I grind up Colorado potato beetles and spray them in a water solution on my potato plants, will they prevent other beetles from attacking? (see Figure 11-2).

A. Emphatically, NO. We tested this home remedy with disastrous results. Within one day of spraying, the poor plants looked as if they had been singed by a flame thrower. Within three days, *the plants were dead.*

Q. If I spray my vegetables with garlic, will I deter insects?

A. We recently tried hot pepper, garlic, and onions on beans in an effort to repel the Mexican bean beetle. The experiment had questionable results, so we will try it again next year. (Our findings may not have been overwhelming, but the smell that still haunts the spray tanks is.)

Q. If I interplant petunias with potatoes, will this deter Colorado potato beetles?

A. No (But this was our prettiest experiment.) (Wallace, 1982).

Prevention is the only practical means of control of ticks that may carry Lyme disease or Rocky Mountain spotted tick fever to humans. These tick species are usually more numerous in woods where deer live. *Ticks must be avoided.* Before going into the woods during tick seasons (spring and summer), wear protective clothing and a hat. Spray all clothing, hat, and shoes with some insect repellent containing DEET*. Light-colored clothing will aid in seeing and removing all ticks not attached to the skin. If a tick is attached to the skin, remove it gently with tweezers and apply an antiseptic (University of Rhode Island, 1991) (see Figure 11-3).

*WARNING! DEET should not be used by pregnant women!!

Figure 11-2.   Colorado Potato Beetle (**Leptinotarsa decemlineata**, family Chrysomelidae) is a very serious phytophagous (plant eating) insect throughout the United States, Canada, and Europe. Its principal food preference is leaves of the Irish potato plant, but it also eats other solanaceous plants such as tomatoes and eggplant.

1. Adult.
2. Larva.
3. Egg mass.

(Courtesy Ohio State University.)

Figure 11-3.   In the South, both the Lone Star tick (left) and the black-legged tick (right) have been found to harbor the bacterium that causes Lyme disease. In the past 5 years, the spread of Lyme disease has made it the most common tick-borne illness in the United States.

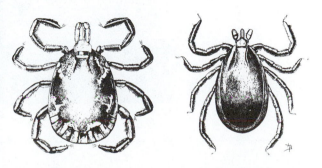

Genetic Control of Pests

Some plants can be selected and/or bred to be resistant to certain insects, diseases, and/or insecticides (see Figure 11-4). This is known as **genetic control.** More recently, it has been confirmed that interplanting even different cultivars of the same plant species reduces insect damage. This may be called "genetic diversity." For example, in California it has been reported that:

■ Mixing wheat varieties can reduce certain rusts.

■ Four broccoli cultivars that were interplanted reduced the number of aphids as compared to solid plantings of the same cultivars (monocultured). The difference was greater in fall-planted than in spring-planted broccoli (Altieri & Schmidt, 1987).

The lacewing larvae are relatively tolerant of insecticides such as chlorinated hydrocarbons, pyrethroids, and many microbials; however, most individuals are easily killed by organophosphates and carbamates (such as Sevin™). If some individual lacewings can be found in nature to be resistant to Sevin (a carbamate), could these be reproduced and distributed in areas where Sevin™ was used? The answer is yes. This research was reported for the first time in California in 1986 (Grafton-Cardwell & Hoy, 1986).

Figure 11-4. Downy mildew-resistant lima beans at left growing beside two varieties that are susceptible. Developing vegetable varieties with resistance to serious diseases is an important phase of the research conducted in vegetable improvement at Beltsville and elsewhere. (Courtesy Agricultural Research Service, United States Department of Agriculture, Beltsville, Maryland.)

Genetic Control of a Disease-carrying Tick

Bovine babesiosis is a tick-borne protozoan disease that causes cattle fever, anemia, loss of appetite, and a drop in milk production. It is transmitted by the bite of a tick, genus *Babesia*. The tick and its associated disease also affect humans. As of March, 1990, there was no protective vaccine available for humans.

A new integrated pest management technique being developed to control bovine babesiosis consists of:

■ Crossing two species of *Babesia, B. annulatus* and *B. microplus* to produce *sterile* male hybrid ticks; however, female hybrids are fertile.

■ Mass-producing large numbers of males and females of both species and crossing them.

■ When the *sterile* male hybrids mate with normal females, the eggs are infertile (will not hatch).

■ Although hybrid females are fertile, their male offspring are *sterile.*

As of early 1990, this nonchemical technique of *Babesia* tick control was being tested on a field scale in the Virgin Islands.

A Cotton Resistant to Insects?

Of all extensively cultivated crops, cotton seems to require the most pesticides. The concerned public, however, may be wondering when the biotechnologists will be able to develop a cotton transgenic cultivar resistant to insect damage. Monsanto Chemical Co. has developed such a cotton and it successfully passed its first field tests in 1990 in Mississippi, Texas, Arizona, and California.

Scientists spliced into a cotton plant a gene from a bacterium that is the natural enemy of the cotton bollworm and the cotton pink bollworm. The "new" cotton plant contained the same toxin as the bacterium and its sap killed the larvae of the two species of cotton bollworms (*The Wall Street Journal*, 1990). An organism (the cotton) that has genes from another organism (the bacterium) transferred into its cells, is known as **transgenic.**

Can pesticides now be withheld from cotton plants? Not to date; unfortunately, it will take several more years of field testing to determine:

■ If the "new" cotton will produce economic yields of cotton lint.

■ If total yields are economical, will the *quality* of the cotton fiber be satisfactory.

Assuming that the "new" cotton resists the bollworms and has satisfactory lint yields and lint quality, can all pesticides be withheld from cotton? The most serious insect pest of cotton is still the cotton boll weevil (see Figure 11-5). This breakthrough will not affect the boll weevil. Moreover, there is every reason to assume that the pests will simply develop resistance to the transgenic toxin, just as they would to synthetic pesticides.

Cultural Control of Pests

In general, plants and animals are less susceptible to pest damage when they are healthy. Proper cultural practices to produce healthy, vigorous plants and animals will go a long way to minimizing pest damage. Cultural pest control refers to the use of planting, cultivation, pruning, fertilization, and other plant managment processes to manage pest populations and damage. For instance, plants that are stressed by damage to their root systems from tillage are more susceptible to any number of diseases and other pests. Another example would be trees that receive mechanical injury, which are more subject to further damage by insects. After all, one of the functions of tree bark is to prevent insect and disease invasions.

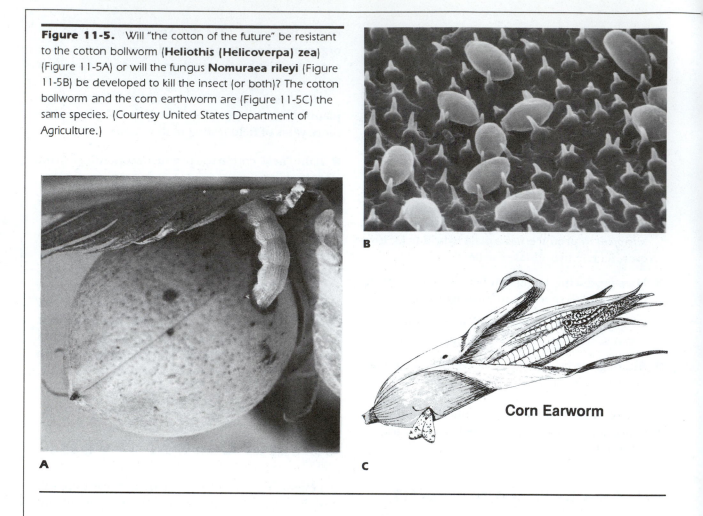

Figure 11-5. Will "the cotton of the future" be resistant to the cotton bollworm (**Heliothis (Helicoverpa) zea**) (Figure 11-5A) or will the fungus **Nomuraea rileyi** (Figure 11-5B) be developed to kill the insect (or both)? The cotton bollworm and the corn earthworm are (Figure 11-5C) the same species. (Courtesy United States Department of Agriculture.)

Corn Earworm

A   B   C

Cultural pest control techniques include:

- Modifying the planting and/or harvest dates to *avoid* pest abundance.

- Rotating crops to reduce pest species build-up in the soil and in crop residues.

- Mixed cropping to confuse insect pests by mixing odors.

- The use of scarecrows or noise-makers to control birds.

- Drainage of wet spots and empty containers to reduce the mosquito population.

- The burial or desiccation (drying) of garbage.

- The spreading of animal manures on fields to control fly populations.

- Tillage to eliminate weeds.

- Use of smother techniques such as black plastic covers, crop residues, or use of aggressive plant covers to control weeds.

- Destruction of weeds before they bear seed.

Biological Control of Pests

Biological control of pests of plants, animals, and people includes the use of one living plant or animal to control another unwanted plant or animal. One of the first successful biological controls of a pest in the United States was in California in 1887–1889. A few hundred vedalia lady beetles were imported from Australia and New Zealand to control cottony cushion scale of citrus. Within two years the pest was no longer an economic threat. Since 1991 more than three hundred pests are managed by biological agents. Other kinds of biological control agents are:

- Predators—These include a fish to control a noxious water weed, a beetle to control the cotton bollworm, also known as the corn earworm, and lady beetles to control aphids (see Figure 11-6).

- Parasites—Among the "good" parasites are species of Tachinid flies and many species of wasps (see Figures 11-7 and 11-8). Also, see Figures 11-9 and 11-10.

- Sex Attractants—These include pheromones (sex odors).

- Bacteria—These include *Bacillus thuringiensis,* soil bacteria for pest control.

- Irradiation of male insects—An example of this is gamma radiation of the screwworm fly.

- Fungal control of a fungus.

The European lady beetle (*Coccinella septempunctata*) is also known as the seven-spotted lady beetle. All lady beetles eat aphids (see Figure 11-8). In 1953, scientists brought many of the beetles into the United States to serve as insect predators. Twenty years later the species was established in New Jersey. By 1989, it had become established in all regions of Arkansas. This large lady

Figure 11-6. A predacious beetle (**Calosoma sayi**) that feeds on cotton bollworm (also known as corn earworm). (Courtesy University of Arkansas.)

Figure 11-7. Two species of Tachinid flies that in nature parasitize pests on valuable agricultural crops. **Left—** Imported from Europe to help to control the European corn borer. **Right—**Parasitizes the Mexican bean beetle larvae. (Courtesy National 4-H Service Committee, Washington, D.C.)

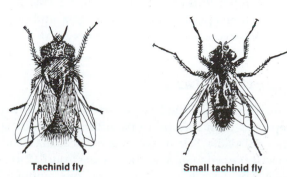

Tachinid fly   Small tachinid fly

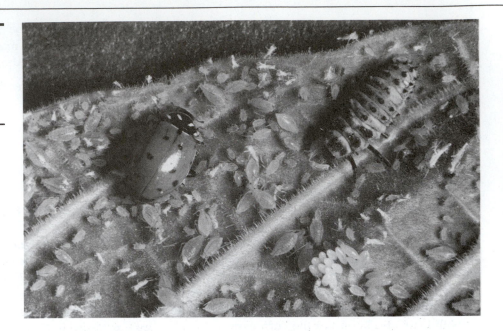

Figure 11-8. Biological control of insects. Here an adult and a larva ladybird beetle are feeding on aphids in New York State. (Courtesy Cornell University.)

beetle has been known to eat more than two hundred aphids a day. Studies have also shown that this European lady beetle co–exists with native species of lady beetles and does not replace them as some scientists had predicted (Kring & Bush, 1991).

Genetic control of pests may also include the selection and propagation of predacious insects that have greater resistance to pesticides. An example is the green lacewing (*Chrysoperla carnea*), a common nectar-feeding, predacious insect. It lays its eggs on many orchard and field crops. When larvae hatch they feed on aphids, mites, larvae, psylla, and eggs of many insects of economic importance.

Parasitic Wasps

Three species of stingless parasitic wasps have been imported from Europe to parasitize the blotch leaf miner of alfalfa. Scientists estimate that alfalfa farmers no longer need to spray and are saving from $12 to $16 million per year in insecticides. Also, one cannot fully calculate the environmental benefits involved, but before the parasites were introduced, farmers used heavy applications of pest control chemicals on nearly two million acres of alfalfa. In some instances this resulted

in insecticide residues on alfalfa leaves and in cow's milk, and kill-off of honeybees and other beneficial insects. The insecticides probably contributed to some contamination of soil and water.

The three species of wasps parasitic on the alfalfa blotch leaf miner insect are **Dacnusa dryas**, **Chrysocharis punctifaciens**, and **Miscogaster hortensis**.

The cost-effectiveness to the farmer of such biological means of pest control is matched by the economy of its development by public-funded research. For example, the cost of the U.S.D.A. program to import enemies of the alfalfa weevil, including twenty years of exploration and research, was less than $1 million. This was only a fraction of one year's savings in insecticide costs by alfalfa farmers. More important, control via natural enemies of a pest is self-perpetuating; the control agent continues to reproduce and attack the pest, season after growing season.

Figure 11-9. A parasitic wasp (**Apanteles rubeoula**) attacking the imported cabbage worm larva and laying its eggs inside the worm. (Courtesy United States Department of Agriculture.)

Two other species of parasitic wasps are shown in Figures 11-9 and 11-10.

Fish Eat Noxious Water Weed

Carp, a rough fish, eat the noxious water weed hydrilla (*Hydrilla verticillata*). This weed is a nuisance in ponds and lakes in southern and southwestern United States. It is very difficult to kill with any chemicals without poisoning the fish. However, carp have been credited with cleaning about four hundred miles of irrigation canals that had been choked with hydrilla.

In this particular case, there is no danger of the carp reproducing and driving game fish away. The carp used were specially-produced fish with an abnormal three sets of chromosomes. That made them sterile, so they cannot reproduce. This is a prime example of scientific biological control of a weed pest.

A Fungus to Control a Fungus

A "bad" fungal disease of lettuce and peanuts, *Sclerotinia minor*, can now be controlled by a "good" fungus, *Sporidesmium sclerotivorum*. This discovery has been described as an accident but such accidents happen only to highly educated and alert scientists who know

Figure 11-10. This parasitic wasp (**Microplitis croceipes**) is laying eggs (parasitizing) in the tobacco budworm. (Courtesy United States Department of Agriculture, Agricultural Research Service.)

and care about people and their environmental enemies. The "good" fungus was identified as a species of *Sporidesmium* previously unknown to science.

A good cultural practice in plant production, gardening, and landscaping is the use of composts, organic fertilizers, and manures. Such organic materials are rich in microorganisms to begin with and encourage even more microorganism growth when applied to growing plants. Active and organically diverse conditions such as these tend to naturally suppress the growth of many pathogens. Of course, you would not want to put diseased plant materials into a compost pile because that would simply inoculate the pathogens to the next crop.

Field Bindweed Control by a Mite

Bindweed (*Convolvulus arvensis*, family Convolvulaceae) is a perennial, extensive-rooted vine and a worldwide pest. It is a weed that reduces yields of crops such as corn and wheat, especially in the West and Midwest. It is difficult to control with cultivation and herbicides. The flowers are pink to whitish, one inch across, and the plant is so vigorous and attractive it is used for beauty in hanging baskets. In crop fields it is classified as the fifteenth worst weed in the United States.

Agricultural research scientists from the U.S. Department of Agriculture are now successfully using a mite shipped from the Biological Control of Weeds Laboratory in Rome, Italy to control field bindweed. The mite has the scientific name of *Aceria malherbae*. Will the mite kill all field bindweed in the Midwest and West? Probably not. Chemical herbicides will likely be necessary for many years, but, the mite will reduce the amount of chemical herbicide that farmers have to use.

Sex Attractants (Pheromones)

Pheromones are hormones that produce sexual attraction between insects. Such hormones are used to control insect populations in two ways: trapping and confusion.

Insects find "friendly" foods and sex partners by odors. In years past it has been the odor of the female insect that attracted the male. But in 1987 scientists synthetically duplicated a **sex attractant** (pheromone), odor of the male papaya fruit fly to lure female fruit flies to a sticky surface and to their death. As of 1987,

only one other synthetic male pheromone was being used to attract female boll weevils into a trap. At that time it was discovered that the synthetic male pheromone attracted both female and male boll weevils. This pheromone is now used in cotton fields by insect scouts to trap the cotton boll weevil. They use the count of weevils caught to determine the economic threshold of insects and when, if ever, to use chemical insecticides.

The second technique is to release pheromones over a large area. That produces a mass confusion among the males. They search in vain for females unable to determine where the real females are.

Insect Eradication by Irradiation of Males

Control of the screwworm was conceived many years ago by Agricultural Research Service entomologist Edward F. Knipling, who thought insects could be used for their own self-destruction (see Figure 11-11). He proposed to rear insects in huge quantities. He would sterilize the males with gamma radiation from cobalt-60, and then release them from aircraft into infested areas to breed with normal females. Such matings would produce no offspring, so that if enough sterile male insects were released, the species would be annihilated.

Knipling's first chance to test his theory against screwworms came in 1954 on the island of Curacao off the coast of Venezuela in South America. He chose the screwworm as the target species because of its monogamous breeding habits. In six months Knipling's new method wiped out the small island's screwworm population.

Success of the sterility principle, as the new method was called, brought a clamor for similar help from Florida cattle producers. The U.S. Department of Agriculture designed a "fly factory" in Sebring, Florida, that could rear fifty million flies a week at peak capacity. When the campaign got into full swing, five-day-old pupae were sexually sterilized by exposure to radioactive cobalt-60, packaged four hundred to a box, and dropped over infected areas from small

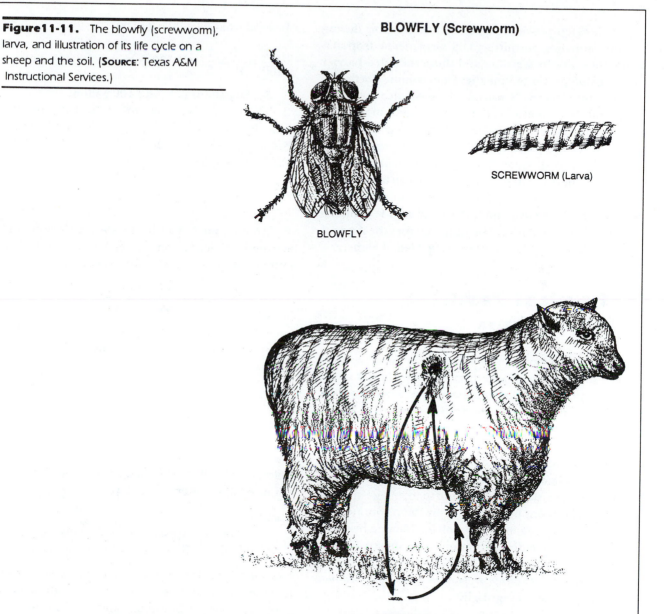

Figure11-11. The blowfly (screwworm), larva, and illustration of its life cycle on a sheep and the soil. (**Source**: Texas A&M Instructional Services.)

BLOWFLY (Screwworm)

SCREWWORM (Larva)

BLOWFLY

LIFE CYCLE

airplanes. In only a year and a half, the screwworm pest was eradicated from Florida.

The Florida effort was a prelude to the large-scale campaign that was to unfold in Texas and the Southwest. A "fly factory" was housed at an abandoned Air Force base near Mission, Texas. It reared over 100 million flies a week—an effort that required nearly sixty tons of ground meat, six thousand gallons of blood, and seventeen thousand gallons of water. Total costs for the three-year campaign were $12.5 million.

By the end of 1964, the screwworm was eradicated from Texas. It was later eradicated from small areas of Arizona and California that were not included in the original program, thus making the U.S. screwworm-free.

Mexico provided excellent cooperation during the campaign, permitting U.S. airplanes to drop sterile flies over its territory and thus establish a barrier zone that prevented the flies from coming north into the United States. Now a joint U.S.–Mexico campaign is underway to eradicate the screwworm in Mexico, from Texas to the isthmus of Tehuantepec.

Over the years, Knipling's sterility principle has also been used against such other serious insects as the Mediterranean fruit fly, oriental fruit fly, and pink bollworm. For his dedication to agricultural science and a cleaner environment, Dr. Edward F. Knipling was the first scientist to be inducted into the Agricultural Research Service's Scientific Hall of Fame.

Chemical Pesticide Calculations

If everyone who mixed and used a chemical pesticide did so properly, a large part of the pollution of the environment by pesticides would be eliminated. No longer can society tolerate the myth that "If the correct amount of a pesticide will kill a few insects, a little more will kill all of them." The "little more" usually is the amount that pollutes the environment.

Calculations will be presented for how to mix wettable powders, liquids, and percentage mixing, and calibrations on how to apply the recommended amounts of pesticides. This will be followed by useful facts for pesticide calibration. The source for this section is the United States Department of Agriculture's *Pesticide Application Training Manual*, which is updated regularly.

Wettable Powder Mixing

You may be given directions to add 2 pounds of pesticide in 100 gallons of water but you wish to fill a 300 gallon tank. Since you know that 300 gallons is 3 times 100 gallons, you simply add 3 times 2 pounds, or 6 pounds in 300 gallons. If you wish to mix only 20 gallons of finished spray you must use some simple arithmetic. Follow these steps:

1. Find what part 20 is of 100

$$\frac{20}{100} = \frac{2}{10} = \frac{1}{5} \quad \text{OR 20 goes into 100 five times.}$$

So, 20 gallons is 1/5 of 100 gallons.

2. Therefore, you must add 1/5 of 2 pounds of pesticide to your finished spray. One pound contains 16 ounces; 2 pounds contain 32 ounces. 1/5 of 32 oz. = 32 divided by 5

$$\frac{5}{32} = 6.4 \text{ oz.}$$

Another way to handle the above situation would be to figure that at 2 pounds per 100 gallons, 0.2 (two tenths) pound is required for every 10 gallons.

$$\frac{2 \text{ lb}}{100 \text{ gal.}} = \frac{0.2 \text{ lb.}}{10 \text{ gal.}}$$

0.2 of 16 ounces = 3.2 ounces, so every 10 gallons requires 3.2 ounces. Twenty gallons would require 6.4 ounces.

Liquid Mixing

Liquids are mixed in the same manner. If your directions call for 2 pints emulsifiable concentrate per 100 gallons, 300 gallons will take 6 pints, and 20 gallons will take 0.2 (two-tenths) of 2 pints.

Example
>   2 pints (1 quart) = 32 fluid ounces.
>   0.2 of 32 = 32 x 0.2 = 6.4 fluid ounces.

A used plastic two-cycle engine oil container with fluid measurements stamped on the side makes a good measure for liquid amounts up to one quart. The top should be cut off above the 1 quart mark for easy filling and pouring.

Percentage Mixing

Sometimes you will find directions telling you to make a finished spray of a specific percentage, for instance a 1% spray for ants. The pesticide may be formulated as a 57% emulsifiable concentrate. To make a 1% finished spray you would add 1 part of pesticide to 56 parts of water. For example, 1 fluid ounce in 56 fluid ounces (1-3/4 quarts) of water.

When mixing percentages you should remember that 1 gallon of water weighs about 8.3 pounds and 100 gallons weigh about 830 pounds. Thus, to make a 1% mix of pesticide in 100 gallons of water you must add 8.3 pounds of active ingredient of pesticide to 100 gallons of water. The following formulas may be used for reference.

> *Note:* One gallon of kerosene weighs 6.6 pounds. One hundred gallons of kerosene weigh 660 pounds.

Formula for Wettable Powder Percentage Mixing

To figure the amount of wettable powder to add to get a given percentage of active ingredient (actual pesticide) in the tank:

$$\frac{(gallons\ of\ spray\ wanted) \times (\%\ active\ ingredient\ wanted) \times 8.3\ (pounds/gallon)}{(\%\ active\ ingredient\ in\ pesticide\ used)}$$

Example

How many pounds of an 80% wettable powder are needed to make 50 gallons of 3.5% spray for application by mist blower?

$$\frac{50\ (gallons\ wanted) \times 3.5\ (\%\ wanted) \times 8.3\ (lbs/gal.)}{80\ (\%\ active\ ingredient)}$$
$$= \frac{1452.5}{80} = 18.1\ pounds\ 80\%\ wettable\ powder$$

Formula for Emulsifiable Concentrate Percentage Mixing

Use the following formula to figure the amount of emulsifiable concentrate to add to get a given per-

centage of active ingredient (actual pesticide) in the tank.

$$\frac{(gallons\ of\ sprayed\ wanted) \times (percent\ active\ ingredient\ wanted) \times 8.3\ (lbs/gal.)}{(lbs.\ of\ active\ ingredient\ per\ gallon\ of\ concentrate) \times 100}$$

Example

How many gallons of a 25% emulsifiable concentrate (2 pounds pesticide per gallon) are needed to make 100 gallons of 1% spray?

$$\frac{100\ (gal.\ wanted) \times 1\ (\%\ wanted) \times 8.3\ (lbs/gal.)}{2\ (lbs.\ active) \times 100}$$
$$= \frac{830.0}{200}$$
$$= \frac{830}{200} = 4.15\ gal.\ of\ 25\%\ emulsifiable\ concentrate$$

Calibration

Often the label will give mixing instructions in terms of quantities of pesticide to be used per one thousand square feet of area to be treated. Examples are in pesticide treatments for lawns or other turf, or per acre as in commercial vegetables. In this case you will have to determine and adjust the amount of liquid your sprayer applies over a given area. This is called **calibrating** the equipment. When the equipment is calibrated you can add the proper amount of pesticide to give the recommended dosage per area.

Even if you have the right mixture in your spray tank, you can still apply the wrong amount of pesticide. You need to know at what rate your equipment is applying the pesticide to the target. When you have found the rate per minute of your machine, you have calibrated your machine. Then you can decide how long (minutes, seconds, miles per hour) you must spray to get the right dosage on each target.

You need to know more than the delivery rate when a small amount of pesticide must be applied accurately and evenly to a relatively large area; for example, treating a field with an herbicide at the rate of one-half (0.5) pound per acre of actual pesticide (active ingredient).

This job is usually done with a tractor- or truck-mounted sprayer equipped with a folding boom having several nozzles. In order to apply the pesticide evenly and accurately your sprayer must move at a constant speed over the field and be pumping at a constant pressure. Each nozzle must be clean, at the proper height from the ground, and applying the same amount of spray. Choose the speed, pumping pressure, and nozzles that you wish to use. Then your sprayer must be calibrated to find the total amount of spray it will apply to an acre.

One accurate method of calibrating the sprayer does not require any arithmetic. Simply mark out an acre, then fill the spray tank with *water* and spray the acre as if applying the pesticide. Measure the amount of water needed to refill the tank. That is the rate per acre. If it takes 9.9 gallons to refill the tank, then you are spraying at the rate of 9.9 gallons per acre. Finally, adjust the speed or pressure until the desired application rate is achieved.

An acre is a square area 209 feet on each side. An area 100 feet by 435 feet is about an acre. Any similar area marked off totalling 43,560 square feet is an acre.

Gallons Per Acre — In Less Time

You may not wish to spray over a whole acre to calibrate the equipment. It is not necessary if a few basic facts are known and simple multiplication is used. One method of calibration is shown below, but there are several others. The facts you must know for this method are:

- One acre equals 43,560 square feet.

- The distance the sprayer moves in one minute at the speed and throttle setting being used.

- The width of the spray boom.

- The pumping rate of the sprayer.

Example

Suppose the sprayer's boom is twenty feet wide, that it travels 440 feet in one minute, and that it pumps two gallons each minute.

Figure the area that the sprayer covered in one minute (the distance traveled × boom width).

Then figure how many minutes it would take to spray one acre. One acre = 43,560 square feet. Divide 43,560 square feet per acre by 8,800 square feet per minute.

Finally, figure the amount of spray pumped per acre:

$$\text{2 gallons per minute} \times \text{5 minutes per acre}$$
$$= \text{10 gallons per acre.}$$

You now know how many gallons of spray per acre the equipment will apply. Next, determine the amount of pesticide to be put in the tank to apply the correct dosage. Two more facts need to be known.

- How many gallons the sprayer tank holds.

- The recommended amount of formulation desired to use per acre. Recommended doses will be listed on the label and in extension publications.

OR

The amount of active ingredient per gallon of formulation and the amount of active ingredient desired to apply per acre. Look on the label.

Example #1

Suppose the tank holds 80 gallons of ready-to-use spray (finished spray) and you wish to apply 1 pint of formulation on each acre. You have already found that the sprayer applies 10 gallons per acre. First, find the number of acres one tank load will spray. Divide 80 gallons (what the tank will hold) by 10 gallons per acre.

$$\frac{\text{80 gallons per tankful}}{\text{10 gallons per acre}} = \text{8 acres per tankful}$$

Then find the amount of formulation that must be added to the tank in order to spray 8 acres with 1 pint per acre. Multiply 1 pint per acre by 8 acres (the number of acres one tank load will cover).

1 pint per acre × 8 acres per tankful = 8 pints per tankful.

Eight pints must be used in each tankful.

Example #2

Suppose the formulation of a pesticide contains 4 pounds active ingredient per gallon and 1/2 pound of active ingredient per acre is desired. The tank covers 8 acres as was determined earlier.

First, find how much formulation in pints is needed to apply 1/2 pound active ingredient per acre. It is known there are 4 pounds of active ingredient in 1 gallon. One gallon = 4 quarts; therefore, there are 4 pounds in 4 quarts and 1 pound in 1 quart. One quart = 2 pints, so, there is 1 pound in 2 pints and 1/2 pound in 1 pint.

$$\frac{4 \text{ lbs}}{1 \text{ gal}} = \frac{4 \text{ lbs}}{4 \text{ qts}} = \frac{1 \text{ lb}}{1 \text{ qt}} = \frac{1 \text{ lb}}{2 \text{ pts}} = \frac{1/2 \text{ lb}}{1 \text{ pt}}$$

So, 1 pint contains 1/2 pound active ingredient. Then, figure how much formulation is needed per tankful. Multiply 1 pint per acre times 8 acres per tankful.

1 pint per acre × 8 acres per tankful = 8 pints per tankful.

So, 8 pints of concentrated pesticide will be added to the spray tank in order to apply 1/2 pound active ingredient per acre.

Useful Facts for Pesticide Calculations

- 1 gallon of water weighs about 8.3 pounds.

- 100 gallons of water weigh about 830 pounds.

- 1 pound = 16 ounces = 453.6 grams.

- 1 pint = 16 fluid ounces = 946 milliliters = 0.946 liters.

- 1 quart = 32 fluid ounces = 946 milliliters = 0.946 liters.

- 1 pound wettable powder per 100 gallons = 1 level tablespoon full per gallon (approximately).

- 1 pint emulsifiable concentrate per 100 gallons = 1 level teaspoon per gallon.

Summary

Pesticides are substances that kill unwanted organisms. The most common pesticides around the home are bactericides and insecticides. Many homes have rodenticides. In the lawn and garden, insecticides and fungicides are commonly used. In agricultural and industrial use, all kinds of pesticides are found.

In the United States, pesticides are regulated at both federal and state levels. Each state has its own laws and regulations regarding pesticide storage, handling, application, and use. At the federal level, the office with primary responsibility for pesticide regulation is the Environmental Protection Agency.

In the past few decades a large number of pesticides have been restricted or entirely banned in the U.S. In many cases, pesticides are restricted or banned because of potential or feared human toxicity. In other cases, pesticides with little effect on humans have been banned because of their reported effects on nonhuman animals and plants. One notable example is DDT which was banned in the United States in the 1970s largely because of charges that it caused a softening of the egg shells of such birds as bald eagles. DDT and related insecticides are remarkably effective against many insects, including mosquitoes which carry malaria, one of the most serious diseases of humans in the tropics. It is also cheap and has no known effects on humans. Because of those properties, DDT is still being used in developing countries to control human diseases. In those regions, the choice is human survival or environmental damage to birds. Unfortunately, as we have seen, overreliance on a single pesticide often results in resistant pests. Where DDT alone is used to control mosquitoes, the insects eventually develop more resistance. It is essential that a more balanced approach be used, in which the pesticide is just one part of the control program.

Integrated Pest Management (IPM) refers to pest control systems that use multiple techniques to manage pest populations. One desirable result of IPM is that the environment is often less affected than with the use of chemical pesticides alone. Integrated Pest Management is not new. In fact, throughout history, people have sought to use all kinds of techniques to prevent pests from damaging crops and livestock. With the advent of chemical pesticides, mostly in the latter part of the twentieth century, mankind came to rely on chemical controls almost totally. That was apparently a serious mistake, as we have come to realize the harmful effects of many synthetic pesticides on the environment. Also understood is the ability of pests to develop genetic resistance to almost any pesticide. As a result, governments and scientists have begun to emphasize an integrated approach to pest control in recent years. Extreme requirements are sometimes placed on pesticide handling for the specific purpose of making it more difficult to use pesticides. One specific purpose of some regulations is to discourage pesticide use by making it complicated.

There are many ways to control pest populations besides using chemical pesticides. Some of the more important alternative measures include prevention, genetic techniques, cultural practices, biological control, sex attractants, and irradiation. One of the most important considerations is maintaining healthy, vigorous plants and animals. The importance of plant and animal health in pest management cannot be overstated. In fact, many scientists have begun to use the term Plant Health Care (PHC) instead of the more traditional term Integrated Pest Management to emphasize the central role of basic health management.

One important consideration in pesticide use is following directions. The pesticide label always provides application rates and other important information. If you use pesticides, it is CRITICAL that you follow the label directions. In fact, it is ILLEGAL if you do not follow the label directions, either intentionally or through ignorance. In agricultural chemicals, one very important factor is that the chemical be applied at the recommended rate. Because agricultural and industrial pesticides normally come in concentrated form, they must be mixed to provide the proper concentration. Procedures to follow for mixing wettable powders and liquids are discussed in this chapter. In both cases, label instructions may specify that certain volumes of pesticide be combined with certain volumes of other liquids. Sometimes the directions may specify that the finished mixture or solution should be a certain percentage concentration.

Another important factor is the accuracy of the applicator. That means the machinery that will be used to apply the pesticide must be accurately calibrated.

DISCUSSION QUESTIONS

1. What is integrated pest management?
2. What effect has agricultural pest management had on the environment?
3. How severe is the problem of pesticide contamination of the groundwater reservoir?
4. What are some preventive pest control techniques that gardeners can use?
5. What are some examples of genetic pest control measures?
6. What cultural pest control measures appear to be effective?
7. What are some of the more common biological pest control measures?
8. How were screwworms controlled by irradiation?
9. Why is calibration so important in pesticide application?
10. What is economic threshold and why is it important in pest management?

ADDITIONAL ACTIVITIES

1. Visit a local garden center or a store that carries garden supplies. Make a list of all of the pesticides carried and group them as insecticides, fungicides, etc. Select one insecticide, one herbicide, and one fungicide. For the three pesticides selected, make a complete list of the safety precautions and potential human health hazards listed on the labels.

2. If you have a pest control problem in your own home or garden, consult your agricultural extension agent to help you plan an integrated pest management system for the pest. Be sure to not rely strictly on chemical control except as a last resort.

3. Make a complete inventory of all pesticides in your home. Be sure not to forget materials that may be listed as "disinfectants." Determine if each pesticide is being used in accordance with its directions.

4. Visit a local golf course, public park, or other area that has a groundskeeper. Try to determine what kinds of pest control measures are in use. Does there seem to be an overreliance on chemical pest control?

CHAPTER 12

Fertilizers, Manures, and Human Effluent

Courtesy United States Department of Agriculture, Soil Conservation Service.

Terms to Look for and Learn

Desiccation
Effluent
Essential Elements
Human Effluent
Nitrate Nitrogen

Nitrogen
Phosphorus Pentoxide
Potassium Oxide
Septage
Sewage Sludge

Learning Objectives

After reading this chapter and participating in the activities, you should be able to:

■ Discuss the potential positive and negative effects of chemical fertilizers in the environment.

■ Discuss the effects of animal manures in the environment.

■ Outline environmentally safe ways of disposing of animal manures.

■ Discuss the effects of human effluent in the environment.

■ Discuss the safe treatment and disposal of human effluent in the environment.

Overview

When the soil does not have all of the essential elements in available forms and in balanced proportions for optimum plant growth, fertilizers and/or

Table 12-1. Elements Essential for the Growth of Crop Plants, Their Chemical Symbol, Source, and Average Percentage Composition*

Essential Element	Chemical Symbol	Principal Source	Average Percentage Composition of Plants (dry-weight basis)
Carbon	C	Air	44%
Hydrogen	H	Water	6%
Oxygen	O	Air/Soil	45%
Nitrogen	N	Air/Soil	2%
Phosphorus	P	Soil/Fertilizer	0.5%
Potassium	K	Soil/Fertilizer	1.0%
Calcium	Ca	Soil/Lime	0.6%
Magnesium	Mg	Soil/Lime	0.3%
Sulfur	S	Organic matter	0.4%
Iron	Fe	Soil	0.02%
Boron	B	Soil	0.005%
Manganese	Mn	Soil	0.05%
Copper	Cu	Soil	0.001%
Zinc	Zn	Soil	0.01%
Nickel**	Ni	Soil/Fertilizer	Undetermined
Molybdenum	Mo	Soil	0.0001%
Chlorine	Cl	Soil	0.015%

*SOURCE: Adapted from Donahue, R.L., Follett, R.H., and Tullock, R.W. (1990). **Our soils and their management**, 6th ed. Danville, IL: Interstate Publishers.

**In their summary report of 1991, the USDA-ARS Plant, Soil, and Nutrition Laboratory (p. 5) stated that nickel was found to be an essential element for plants as well as for animals.

organic residues are needed. The sixteen essential nutrient elements for plant growth are listed in Table 12-1. All sixteen **essential elements** must be supplied to the growing plants from air, water, and soil. The average dry-weight composition in crops of each element is also given. Please note that oxygen composition is greatest, followed by carbon. These two make up a total of eighty-nine percent of the plant's weight. Next are hydrogen at six percent and nitrogen at two percent. Even though the other twelve plant nutrients occur in plants in small amounts, each of the sixteen elements is as essential as the other and no one element can replace or substitute for another.

Fertilizers

Fertilizers are used by farmers to increase plant growth and profit. No one can argue with this objective. The disagreement is with some scientists and pseudoscientists who proclaim: "Agriculture is causing serious environmental problems. Agriculture is the largest single nonpoint source of water pollutants, including sediment, salts, fertilizers, pesticides, and manures" (National Research Council Board on Agriculture, 1989, 89).

In regard to fertilizers, the complaint involves dissolved fertilizer chemicals getting into the water supply. See Chapters 4, 11, 14, and 22 for more discussion on water resources and pollution. This chapter presents facts on fertilizers, manures, and human effluent.

Fertilizer use on farms in the United States, in terms of nitrogen, phosphorus, and potassium, is presented in Table 12-2 for the years 1950–1989. The greatest tonnage used was in 1980, with a decrease in 1985 and a further decrease in 1989. Reasons for decreasing use of chemical fertilizers are likely due to (see Figure 12-1):

■ The U.S. government policy of paying farmers to plant grass and trees on highly erodible soils. For example, crop acreage decreased eleven percent from 1982–1989.

Table 12-2. Chemical Fertilizer Use on Farms in the United States, 1950–1989*

Year	Total Fertilizer Use on Farms (millions of tons of nitrogen + phosphorus + potassium)
1950	4.1
1955	6.1
1960	7.5
1965	11.0
1970	16.1
1975	17.6
1980	23.1
1985	21.7
1989	19.6

Source: U.S. Department of Agriculture, Economic Research Service. **Agricultural resources: Inputs, outlook, and situation report. Washington, D.C.*

- The environmentally conscious effort of farmers to use fertilizers by less polluting techniques.

- The rising cost of fertilizers.

Fertilizers and Environment

All fertilizers sold in the United States are regulated by the respective states. There is general agreement among all states on how a bag of fertilizer must be labeled when offered for sale. An example is depicted in Figure 12-2. The first number in a complete fertilizer is always the percent total nitrogen, the second the available phosphorus, and the third number is water-soluble potassium. Because of tradition among all fifty states, however, these percentages of nutrients are listed in Figure 12-2 as:

- Fifteen percent total **nitrogen** (N).

- Forty percent available phosphorus reported as chemical equivalent **phosphorus pentoxide** (P_2O_5).

- Five percent water-soluble potassium reported as chemical equivalent **potassium oxide** (K_2O).

  The respective state fertilizer control officials have a national organization known as The Association of American Plant Food Control Officials. It is

Figure 12-1. Only the potentially most polluting fertilizer—nitrogen—continues to increase in use. Consumption of potassium (K_2O) and phosphorus (P_2O_5) have been decreasing since 1984. (Courtesy United States Department of Agriculture and Tennessee Valley Authority.)

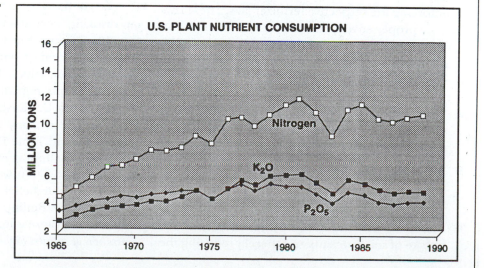

Figure 12-2. Although there is no national fertilizer law, each state regulates its own fertilizer and lime laws. There is, however, general agreement among the states on labeling. For example, this bag of fertilizer is a model. The numbers mean: 15% total nitrogen (N), 40% available phosphorus (P_2O_5), and 5% water-soluble potassium (K_2O).

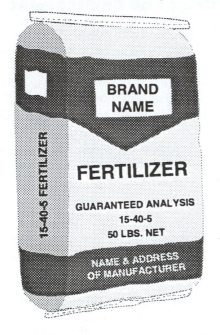

comprised of all scientists in the fifty states in charge of enforcing the respective fertilizer laws. These laws include proper environmental use of fertilizers. The Environmental Affairs Committee reported in 1990 that appropriate fertilizer application can improve the quality of the environment by:

- Increasing biomass per unit area which aids in stabilizing the soil and therefore reducing sediment, the number one pollutant.

- Increasing crop yields per acre, thereby reducing the necessity of producing crops on soils unsuited for cultivation.

- Increasing the percentage composition and total uptake of soil nutrients and thereby reducing the amount of soluble plant nutrients that percolate to pollute groundwater.

- Reducing the need to clear more protective forest land for crop production.

- Promoting more and better soil and plant tissue testing to reduce the application of fertilizers that may pollute the environment.

- Evaluating research data more thoroughly before fertilizer laws are changed.

- Supporting efforts to reduce pollution of fertilizers by reducing spills and leaks from storage facilities. For example, the Indiana Department of Environmental Management reported fifty-four fertilizer spills in 1988 and sixty-four in 1989. Serious efforts are being made in Indiana and in many other states to contain all fertilizers so they will not pollute the environment.

- Working with fertilizer dealers to reduce nitrate pollution of groundwater.

Nitrogen Fertilizer Pollution

Nitrogen fertilizer use in the United States continues to increase but the uses of phosphorus and potassium fertilizers are decreasing. For sake of the environment, this is sad because of greater nitrogen pollution hazard.

Nitrate nitrogen is a health hazard when it gets into our drinking water at a Health Advisory Level greater than 10 milligrams per liter of nitrate nitrogen (NO_3-N). Nitrates come from nitrogen fertilizers, animal manures, septic tank drain fields, municipal sewage treatment plants, rainfall and snowfall, and composts.

In 1985 the U.S. Environmental Protection Agency tested waters in many wells in the fifty states (see Figure 12-3 and Table 12-3).

Nitrate pollution of drinking water can be reduced by adsorption with an exchange resin. Contrary to popular belief, *nitrates cannot be reduced by boiling the water.* Boiling actually *increases* nitrate concentration because water boils away, leaving the *concentration of nitrates higher.*

Figure 12-3.   Agricultural counties with nitrate nitrogen (NO$_3$-N) in groundwater. (**Source:** United States Department of Agriculture. "The Magnitude and Costs of Groundwater Contamination from Agricultural Chemicals: A National Perspective." Staff Report AGES870318, 1987. Economic Research Service. Washington, D.C. Taken from: National Research Council. "Alternative Agriculture," 1989, page 110. National Academy Press, Washington, D.C. Field survey data from United States Geological Survey.)

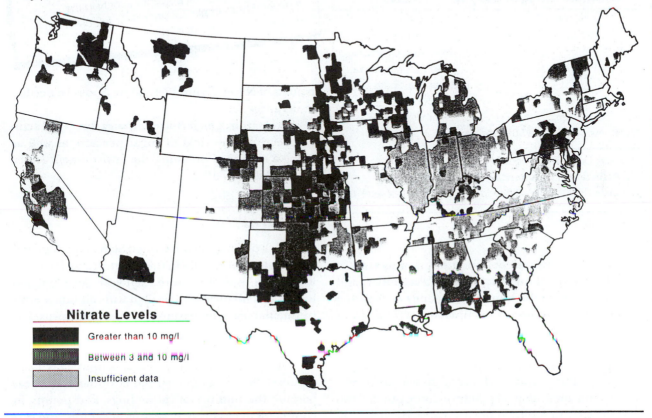

Nitrate Levels

▓	Greater than 10 mg/l
▒	Between 3 and 10 mg/l
░	Insufficient data

Avoiding nitrate pollution of well waters can be achieved by these mandates:

- Do not apply any nitrogen fertilizers in the fall.

- Do not leave the soil bare in winter, instead sow winter cover crops or leave crop residues on the soil surface.

- Apply nitrogen fertilizers only when a crop is growing.

- Apply animal manures only when the soil is not frozen and never more than ten tons per acre.

Potential pollution by nitrogen and perhaps other fertilizers is so great that the National Fertilizer Development Center at Muscle Shoals, Alabama, changed its name in late 1989 to the National Fertilizer and *Environmental* Research Center. This center established ten model demonstration sites throughout the United States during 1990, in cooperation with fertilizer dealers. These models will show other fertilizer dealers state-of-the-science technologies on environmental correctness. Ten more such demonstration sites will be established.

Table 12-3. Well Waters with Greater than 10 Milligrams per Liter of Nitrate Nitrogen in 1985[*]

Percentage of Wells Tested with More than 10 Milligrams per Liter of Nitrate Nitrogen (NO_3-N)	States
>20%	Kansas, Rhode Island
10-19%	Arizona, California, New York, Oklahoma
5-9%	Colorado, Minnesota, Nebraska, South Dakota, Texas
5%	All other of the 50 states

Note: Average for the 50 states — 6.4%

[*]**Source:** U.S. Environmental Protection Agency, as reported in: **Association of American Plant Food Control Officials** (1990). Official Report No. 43, West Lafayette, IN.

Note: Recommended applications are written like 50 pounds of nitrogen (N) fertilizer per acre. To determine how much of a specific nitrogen fertilizer is used, divide 50 by the percentage of N in the fertilizer. For example, for urea (45% N), divide 50 by 0.45 = 111.11 pounds of urea.

A study of nitrate pollution of groundwaters in Maryland was conducted by Weil, Weismuller, and Turner. They studied an area of sandy soils with a shallow groundwater. A comparison was made of nitrate nitrogen concentrations in the groundwaters following the application of 250 pounds per acre of nitrogen (N) fertilizer compared with thirty tons per acre of poultry manure. The crop grown was corn.

Winter application of poultry manure resulted two months later in a nitrate nitrogen level of about 100 milligrams per liter of the shallow groundwater — ten times the recommended level. Throughout the year, groundwater under nitrogen fertilizer and poultry manure applications were always in the range *above* 10 milligrams per liter of nitrate nitrogen. A significant observation was that under a nearby forest, the nitrate nitrogen level of groundwater was always less than 1 milligram per liter. No explanation was given for this unexpected result. Clearly, more research is needed.

But each soil, site, and crop is a different environment. For example, on a fine sandy loam soil that receives twenty-five inches of annual precipitation in Texas, the nonpolluting and economic amount of nitrogen fertilizer for rainfed grain sorghum is thirty pounds of N (see Figure 12-4). More N may be needed under irrigation.

Excess nitrogen fertilizer on sugarbeets will actually decrease the yield of sugar per acre as well as being a hazard of polluting the environment (refer back to Figure 7-10).

Manures

Livestock on farms in the United States and dogs and cats in cities, towns, and in the country produce about two billion tons of manure each year. An estimated half of this manure is deposited without much environmental harm on pastures, ranges, and lawns. It is the other half that is a potential hazard as well as a potential benefit to increase soil productivity.

Over the past twenty years, manure has been accumulating in larger amounts on smaller areas because the number of cattle, hogs, and poultry in confinement has been increasing. Feedlots, and hog and poultry houses now confine animals by the thousands of head. Odors and hazards of environmentally safe disposal of manures have increased. Safe disposal may now require thousands of acres for a large animal enterprise. Whether manures are environmental hazards or benefits depends on how they are managed.

Comparing the relative fertilizing value per ton of manure from various kinds of livestock (solid handling system) are in this order: chicken, turkey, sheep, beef cattle, horse, and dairy cattle. On the basis of potential pollution of ammonium nitrogen, the relative order is: chicken, turkey, and dairy cattle equal swine, sheep, and horse (see Table 12-4).

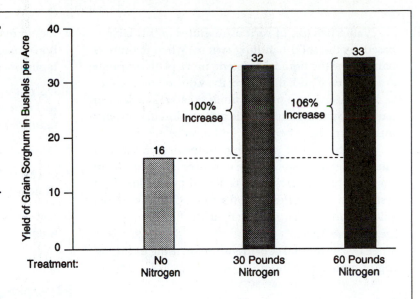

Figure 12-4. Grain sorghum yields are doubled by thirty pounds per acre of nitrogen fertilizer but are increased by only one bushel per acre by a second thirty pounds. The second thirty pounds wastes money and may pollute the environment. (Texas) Notes: (1) Average annual rainfall is twenty-five inches. (2) Soil is a Miles fine sandy loam. (3) Dryland conditions (not irrigated). (**Source:** United States Department of Agriculture.)

Table 12-4. Approximate Fertilizing Value of Animal Manures at Time of Application — Solid Handling System [*]

Kind of Livestock	Bedding or No Bedding	Dry Matter (percent)	Total Nitrogen Percentage	Ammonium	Phosphorus (P_2O_5) (Pounds per Ton)	Potassium (K_2O)	Dollars per Ton
Beef Cattle	No Bedding	15	11	4	7	10	4.70
Dairy Cattle	Bedding/	21	9	5	4	10	3.70
	No Bedding	18	9	4	4	10	3.65
Swine	Bedding/	18	8	5	7	7	4.15
	No Bedding	18	10	6	9	8	5.20
Sheep	Bedding/	28	14	5	9	25	7.40
	No Bedding	28	18	5	11	26	8.30
Chickens	Bedding/	75	56	36	45	34	26.30
	No Bedding	45	33	26	48	34	24.50
Turkeys	Bedding/	29	20	13	16	13	9.50
	No Bedding	22	27	17	20	17	12.30
Horses	Bedding	46	14	4	4	14	4.20

[*] **Source:** Sutton, A.L., Nelson, D.W., & Jones, D.D. (1983) **Utilization of animal manure as fertilizer.** West Lafayette, IN: Purdue University.

Value per ton of various manures in this table is based on the solid handling system. When manure is handled by the liquid pit system, more plant nutrients are conserved and, therefore, the value of the manure as a fertilizer is two to four times greater. The lagoon handling system results in the greatest loss in nutrients on a volume basis because of dilution by more water.

The hazard of polluting the environment with ammonium nitrogen in the manures is greatest from the solid handling system, followed by the liquid pit system and last by the lagoon system. Offensive odors will probably be the greatest from the liquid pit system (see Figure 12-5).

Environmental considerations in using or disposing of manure include:

- Incorporate into the soil by injection or plowing as soon as possible after application.

- To reduce odors, spread frequently and early in the day, especially in warm, damp weather.

- When the soil surface is frozen, apply manures only on level soils.

- Liquid manure pits generate methane gas and carbon dioxide—both lethal to animals and humans in high concentrations (see Chapters 13 and 20).

- Do not apply more than five tons of chicken or turkey manure or more than twenty tons of cattle manure per acre in any one year.

- Weed seeds are spread in manures.

- Excess soil salt can reduce plant growth as a result of heavy manure applications.

- Nitrates from manures may be toxic in drinking water.

Human Effluent

The Problem

In general, the term **effluent** refers to a flowing discharge of some noxious substance. **Human effluent** refers to liquid and flowing solid waste generated by humans.

Figure 12-5. A liquid pit system preserves more plant nutrients in manure than other methods of manure handling. (Courtesy United States Department of Agriculture, Soil Conservation Service.)

In 1988 the population of the United States was almost 250 million. From Table 12-5 it can be seen that 69.9 million people, twenty-eight percent of the population, were not served by *any* municipal sewage treatment plant. Are we a developed nation?

Septic tanks release four billion gallons of human effluent wastewaters into the U.S. soil and water environment each day. Municipal treatment plants pour out an additional 30.8 billion gallons. This totals 34.8 billion gallons of human-effluent—contaminated wastewater each day that must find environmentally safe disposal. New York state generates about ten percent of this total.

Municipal sewage treatment plants that serve twenty-six percent of the U.S. population treat sewage wastewaters with more than secondary treatment, as seen in Table 12-5. This means that seventy-four percent of our population is served by plants that release sewage wastewaters into our environment that receive less than secondary treatment. Also, the table indicates that one percent (about twenty-five million people) are served by municipal treatment plants that release *untreated* wastewaters into soils or surface waters. Clearly, we have a problem with pathogens and nitrates.

Wastewaters released from municipal sewage treatment plants are about ninety-five percent water and five percent solids. These solids, when separated from the water are known as **sewage sludge**.

In the United States there were eleven million tons of sewage sludge generated in 1990. The hazards of this sludge to spread disease is great but cannot be *completely* eliminated.

For example, certain parasitic worm ova (eggs) will survive in soil for seven years. The survival time for bacteria in soil averages about 1 year. Most viruses can survive less than six months in the soil. Protozoa usually die within about ten days. Table 12-6 indicates the common diseases in human effluent caused by parasitic worms, bacteria, viruses, and protozoa.

Testing for Pathogens

For routine testing of municipal wastewater sludge, fecal coliform and fecal streptococci bacteria are commonly used as indicators of the potential presence of pathogens in wastewater sludges and septages. These bacteria are abundant in human feces and therefore are always present in untreated sewage

Table 12-5. Number of People Served by Municipal Treatment Systems, by Level of Treatment 1960–1988[*]

Level of Treatment	1960	1978	1982	1986	1988	Percentage of U.S. Population in 1988
	(in millions of people served/not served)					
Not Served	70.0	66.0	62.0	67.8	69.9	28
Untreated Wastewater Discharge	**	**	**	1.6	1.4	1
Less Than Secondary Treatment	36.0	237.0	37.0	28.8	26.5	11
Secondary Treatment	**	56.0	63.0	72.3	78.0	31
More Than Secondary Treatment	4.0	49.0	53.0	54.9	65.7	26

[*]**Source:** U.S. Environmental Protection Agency (1990). **Twentieth annual report of the council of environmental quality**. Washington, D.C. p. 454.
[**]No Data Available

Table 12-6. Common Diseases Caused by Pathogens in Human Effluent

Pathogen	Common Disease
Parasitic Worms	Hookworm
	Pinworm
	Sheep liver fluke
	Cat liver fluke
	Beef tapeworm
	Pork tapeworm
Bacteria	Salmonella food poisoning
	Typhoid fever
	Bacillary dysentery
	Cholera
	Gastroenteritis
Viruses	Polio
	Pneumonia
	Common cold
	Severe diarrhea
	Respiratory infections
Protozoa	Gastroenteritis
	Diarrhea
	Dysentery
	Weight loss

- The effectiveness of the treatment process.

- The hardiness of the organism relative to other organisms of that type.

- The likelihood that it was present in the raw sludge.

- The availability and reliability of the testing procedures.

- Cost.

Testing requirements should be based on a knowledge of how the operating conditions of the sludge treatment process affect pathogen survival. For example, heating sludge to particular temperatures (e.g., 113° to 122° F) for a sufficient period of time will destroy all viruses and bacteria, but may not adequately reduce helminth ova. In this case, fecal indicator tests could be used to confirm the level of reduction of bacteria and viruses. However, we would need to test helminth ova directly.

Controlling Pathogens

Pathogens in human effluent can be controlled by heat, radiation, disinfectants, and by **desiccation** (drying) (see Table 12-7).

Official definitions of processes to *significantly reduce* pathogens in sewage sludge are given in Table 12-8. Official definitions of processes to *further reduce* pathogens in sewage sludge are listed in Table 12-9.

Nutrients in Sewage Sludge

The average plant-growth-enhancing (fertilizing) composition of sewage sludge is listed below:

- Organic carbon (C) 31%—Carbon is very valuable as a soil amendment, especially on soils low in organic matter such as surface mine spoils and soils disturbed by construction activities.

- Total nitrogen (N) 3.9%—This nitrogen will become available to plants when it is decomposed by bacteria within two to three years.

sludges. They are easily and inexpensively found in testing. Although fecal coliforms and fecal streptococci themselves are usually not harmful to humans, their numbers indicate the presence of fecal waste which may contain pathogens.

When more specific information is needed on the levels of pathogens in sludge, it is generally considered acceptable to test for one representative of each of the three more common types of organisms of concern — bacteria, viruses, and helminth ova (worm eggs). Deciding which organism to test for depends on several factors:

Table 12-7.   General Approaches to Controlling Pathogens in Wastewaters and Sewage Sludge *

Approach	Effectiveness	Process Examples
Kill pathogens with high temperatures (temperatures may be generated by chemical, biological, or physical processes)	Depends on time and temperature. Sufficient temperatures maintained for sufficiently long time periods can destroy bacteria, viruses, protozoan cysts, and helminth ova. Helminth ova are the most resistant to high temperatures.	• Composting (uses biological processes to generate heat) • Heat drying and heat treatment (use physical processes to generate heat; e.g., hot gases, heat exchangers) • Pasteurization (physical heat; e.g., hot gases, heat exchangers) • Aerobic digestion (biological heat) • Anaerobic digestion (biological heat)
Kill pathogens with radiation	Depends on dose. Sufficient doses can destroy bacteria, viruses, protozoan cysts, and helminth ova. Viruses are most resistant to radiation.	• Gamma and high-energy electron beam radiation
Kill pathogens using chemical disinfectants	Substantially reduces bacteria, viruses, and vector attraction. Probably reduces protozoan cysts. Does not effectively reduce helminth ova unless combined with heat.	• Superchlorination • Lime stabilization
Inhibit pathogen growth by reducing the sludge's volatile organic content (the microbial food source)	Reduces viruses and bacteria. Reduces vector attraction as long as the sludge remains dry. Probably effective in destroying protozoan cysts. Does not effectively reduce helminth ova unless combined with other processes such as high temperature.	• Aerobic digestion • Anaerobic digestion • Composting
Inhibit pathogen growth by removing moisture from the sludge	Reduces viruses and bacteria. Reduces vector attraction as long as the sludge remains dry. Probably effective in destroying protozoan cysts. Does not effectively reduce helminth ova unless combined with other processes such as high temperature.	• Air drying

*Source: U.S. Environmental Protection Agency (1989). **EPA/625/10, 89/006.** Washington, D.C.

■ Ammonium nitrogen (NH_4-N) 0.65% and nitrate nitrogen (NO_3-N) 0.05%—These two forms of nitrogen are available to plants immediately.

■ Total phosphorus (P) 2.5%—This phosphorus becomes available to plants within two to three years.

Table 12-8. Official Definition of Processes to Significantly Reduce Pathogens in Wastewaters and Sewage Sludge*

Aerobic Digestion	The process is conducted by agitating sludge with air or oxygen to maintain aerobic conditions at residence times ranging from sixty days at 59° F to forty days at 68° F with a volatile solids reduction of at least thirty-eight percent (see Figure 12-7).
Air Drying	Liquid sludge is allowed to drain and/or dry on underdrained sand beds, or on paved or unpaved basins in which the sludge depth is a maximum of nine inches. A minimum of three months is needed, for two months of which temperatures average on a daily basis above 32°F
Anaerobic Digestion	The process is conducted in the absence of air at residence times ranging from sixty days at 68° F to fifteen days at 95° F to 131° F, with a volatile solids reduction of at least thirty-eight percent.
Composting	Using the within-vessel, static aerated pile, or windrow composting methods, the solid waste is maintained at minimum operating conditions of 104°F for five days. For four hours during this period the temperature exceeds 131° F.
Lime Stabilization	Sufficient "quick" lime ($CaO/Ca(OH)_2$) is added to produce a pH of 12 after two hours of contact. (Agricultural limestone [$CaCO_3$] is not suitable because its pH does not exceed 8.4.)
Other Methods	Other methods or operating conditions may be acceptable if pathogens and vector attraction of the waste (volatile solids) are reduced to an extent equivalent to the reduction achieved by any of the above methods.

*Source: U.S. Environmental Protection Agency (1989). **EPA/625/10, 89/006**. Washington, D.C.

- Total sulfur (S) 1.1%— Like phosphorus and total nitrogen, most of the sulfur is in the organic form and availability to plants depends on bacterial decomposition.

- Total potassium (K) 0.4%—Although too low in com-position for balanced plant nutrition, within about two years this potassium becomes available to plants. Chemical potassium fertilizer must be added before sludge is a balanced fertilizer.

- Total calcium (Ca) 4.9%—This composition is usually sufficient to make a balanced fertilizer.

- Total magnesium (Mg) 0.5%—Balanced nutrition for most plants is about ten times as much calcium as magnesium. At this ratio the calcium and magnesium composition in sewage sludge is just about right (10 Ca: 1 Mg).

- Iron (Fe) 1.3%—A large part of this iron is in the organic form and is usually adequate for plant growth but not present in toxic amounts.

- Zinc (Zn) 0.28%—Adequate but not toxic.

- Copper (Cu) 0.12%—Adequate but not toxic.

- Manganese (Mn) 0.04%—Adequate but not toxic.

- Boron (B) 0.008%—Adequate but not toxic.

- Cobalt (Co) 0.0005%—Adequate but not toxic.

Table 12-9. Official Definition of Processes to Further Reduce Pathogens in Wastewaters and Sewage Sludge

Composting	Using either the within-vessel or the static aerated pile composting method, the solid waste is maintained at operating conditions of 131° F or greater for three days. Using the windrow composting method, the solid waste attains a temperature of 131° F or greater for at least fifteen days during the composting period. Also, during the high temperature period, there will be a minimum of five turnings of the windrow.
Heat Drying	Dewatered sludge cake is dried by direct or indirect contact with hot gases, and moisture content is reduced to ten percent or lower. Sludge particles reach temperatures well in excess of 176° F, or the wet bulb temperature of the gas stream in contact with the sludge at the point where it leaves the dryer is in excess of 176° F.
Heat Treatment	Liquid sludge is heated to temperatures of 356° F for thirty minutes.
Thermophilic Aerobic Digestion	Liquid sludge is agitated with air or oxygen to maintain aerobic conditions at residence times of ten days at 131° F to 140° F, with a volatile solids reduction of at least thirty-eight percent.
Beta ray Irradiation	Sludge is irradiated with beta rays from an accelerator at dosages of at least 1.0 megarad at room temperature (approximately 68° F).
Gamma Ray Irradiation	Sludge is irradiated with gamma rays from certain isotopes, such as $_{60}$Cobalt and $_{137}$Cesium, at dosages of at least 1.0 megarad at room temperature ca. 68° F).
Pasteurization	Sludge is maintained for at least thirty minutes at a minimum temperature of 158° F.
Other Methods	Other methods or operating conditions may be acceptable if pathogens are reduced to an extent equivalent to the reduction achieved by any of the above add-on methods.

- Molybdenum (Mo) 0.003%—Adequate but not toxic.

Hazardous Composition of Sewage Sludge

Elements not wanted in sewage sludge and occasionally hazardous to plants, animals, or humans are listed below.

- Sodium (Na) 0.6%—High sodium in soils indicate excessive salt content. In addition sodium disperses clay and humus particles which reduces the soil's capacity for adequate air and water storage.

- Aluminum (Al) 1.2%—In strongly acid soils aluminum may be toxic to plants but not to animals or humans.

- Lead (Pb) 0.14%—Nickel 0.03% — Cadmium (Cd) 0.01%—Chromium (Cr) 0.26% — Arsenic (As) 0.004%—and Mercury (Hg) 0.07% — These metals are potentially very hazardous in the environment but the amounts supplied by sewage sludge are usually not critical. Sometimes the cadmium level may be an environmental hazard to animals and humans.

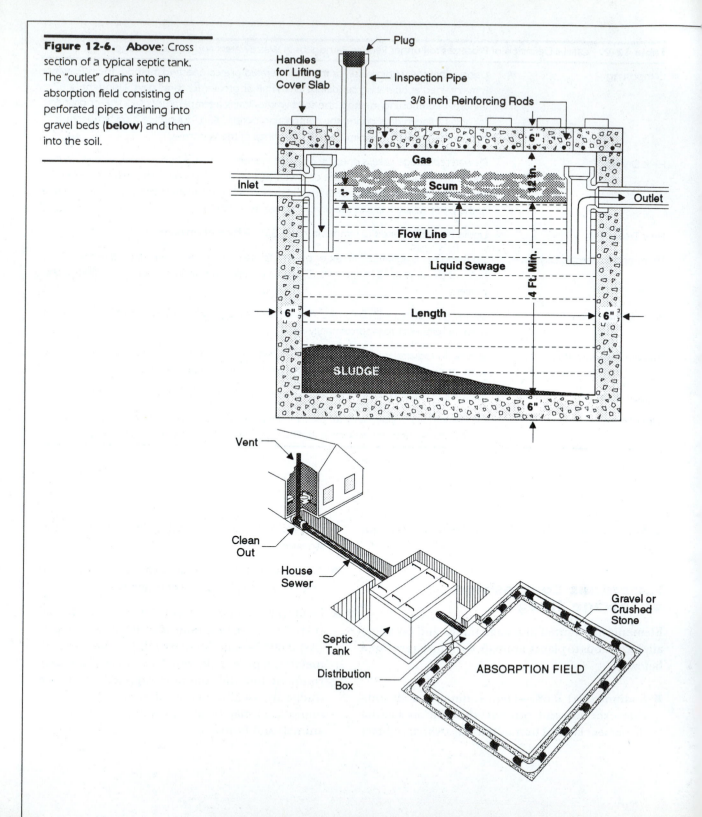

Figure 12-6. **Above**: Cross section of a typical septic tank. The "outlet" drains into an absorption field consisting of perforated pipes draining into gravel beds (**below**) and then into the soil.

Environmentally, the safest place to apply sewage sludge is on forestland. To reduce the hazard of toxic metal uptake by plants used for feed and food, the following guidelines are suggested:

- Maintain soil pH at 6.5 or higher. Most heavy metals are more soluble at lower pHs.

- Apply no more than five tons of sewage sludge per acre per year (dry-weight basis). On semiarid soils, reduce this to three tons per acre per year because of the salt hazard.

- To be "safe" from toxic cadmium, do not apply any sewage sludge on such food crops as beets, Swiss chard, spinach, and lettuce. These crops accumulate cadmium more than most plants.

Septage

Septage is: (1) Sludge produced in a septic tank. (2) That which is pumped from a septic tank in cleaning. (3) Effluent in a cesspool. A cesspool is a hole in the ground usually lined or filled with loose rock into which household effluent is piped. The liquid seeps into the surrounding soil, and the solids decompose anaerobically.

In 1988 there were about twenty-two million homes in the United States with septic tanks that served about seventy million people. Each person generates about sixty gallons of septage a day to dispose of 1/3 pound of solid human effluent. In 1988 it was estimated that four billion gallons of septage was generated each day in the United States.

Liquid septage is drained from a septic tank into a drain field where it soaks into the soil. Solids in the septic tank are pumped out about every three years and disposed of on land in one of these techniques (see Figure 12-6):

- Sprayed on cropland, pastureland, or forestland through an irrigation system.

- Spread on land through the furrow irrigation system.

Figure 12-7. This "honey wagon" is actually spreading livestock wastes from a dairy on a pasture in Flordia but it can also be used to spread sewage pumped from a septic tank. (Courtesy United States Department of Agriculture, Soil Conservation Service.)

- Injected into the soil from a tank truck and a chisel.

- Sprayed on the soil surface from the back of a tank truck, as shown in Figure 12-7.

- Buried in a sanitary landfill.

Disease-causing organisms are forever a human hazard in septage as well as in sewage wastewaters and sludges from municipal treating plants. Diseases carried by human effluents such as septage and sewage sludge include food poisoning, typhoid fever, dysentery, cholera, polio, pneumonia, hepatitis, common colds, anemia, nervousness, insomnia, digestive disturbances, and hookworm (see Table 12-6).

Summary

Plants are made up of chemical elements. Only sixteen elements are generally considered essential for plant growth. Those sixteen elements must be provided to every plant for it to survive. The elements come from the air, water, and soil. Oxygen from air and water, and carbon from the air make up the majority of every plant's weight. Hydrogen is also supplied from water. The other elements must come from the soil.

Sometimes a soil will have all of the necessary elements for plant growth, and in adequate amounts. But that situation is rare, and it seldom lasts long naturally when the soil is in agricultural production. Thus, the other elements must be supplied from somewhere else. The elements that must be supplied most often and in the largest quantities are nitrogen, phosphorus, and potassium. These are the "N," "P," and "K" listed on bags of complete fertilizer.

Fertilizer use is absolutely essential to the production of food in the United States. Without some levels of chemical fertilization of U.S. cropland, we simply could not produce the amount or quality of food that we do. At the same time, environmental concerns about the improper use of fertilizers are well-founded. Problems have developed when we have applied too much fertilizer, or too often, or at the wrong time, or in the wrong way. The most serious pollution from fertilization comes from excess nitrogen dissolving in water, moving with the water, and entering the surface water supply or the groundwater.

The amount of livestock manure generated in small areas has become a more serious problem in recent years. The problem is not the total number of livestock, it is how we produce them. Intensive confinement operations put many animals in a small space. In the past, cattle grazed on open pastures or ranges. The manure produced was never concentrated. In a feedlot that situation changes. Today, poultry houses may contain many thousands of chickens in a single large room. A single feedlot may confine thousands of cattle.

With the more intensive practices have come intensive manure buildups. Manures are valuable for fertilizers and they contribute organic matter to the soil. But, whenever too much animal manure is dis-posed of on too little land, the result can be a buildup of nitrates in surface water or groundwater.

Human effluent is also a major environmental problem. How to collect, treat, and safely dispose of human wastes of all kinds is one of our greatest problems. Sewage is one of the most pressing of those problem areas. In this country, municipal sewage treatment facilities handle much of the problem. Septic systems for individual homes are also an important part of the solution. Yet, both treatment plants and septic systems have potential problems, too. Careful monitoring of the effluent as it is being returned to the environment is important.

DISCUSSION QUESTIONS

1. To what do "N," "P," and "K" refer in chemical fertilizers?

2. What are the potential positive effects of chemical fertilizers on the environment?

3. What are the potential negative effects of chemical fertilizers on the environment?

4. How can we decrease the problem of nitrate contamination of groundwater?

5. Why has the problem of animal manure disposal become so much greater in recent years?

6. What techniques can be used to safely use or dispose of animal manures?

7. What is the magnitude of the human effluent disposal problem in the United States?

8. Are fecal coliforms and streptococci in sewage sludge normally harmful to humans? Why do we test for them? Why do we become so concerned when we find them in high concentrations in our water supply?

9. What pathogens are commonly found in human effluent? How are they most easily controlled?

10. What are the potential benefits and disadvantages of using sewage sludge as a soil amendment?

11. How often should septic tanks be pumped out? What should be done with the solids that are removed?

ADDITIONAL ACTIVITIES

1. Build a scale-model septic system to illustrate the relationship between the parts of the system.

2. Visit a poultry house, hog confinement facility, or beef feedlot. Find out how much manure is generated each year. How is the manure handled, treated, and disposed of?

3. Secure some sewage sludge from your waste treatment facility. Test it for bacteria and chemical composition. Is it being treated adequately before disposal?

4. Visit a fertilizer dealership or a garden center. Make a list of all of the different fertilizers that are available. Include the nutrient make-up (grade) of the fertilizers as well as the form in which it is available (liquid, granular, etc.).

CHAPTER 13
Solid Wastes

Courtesy United States Environmental Protection Agency.

Terms to Look for and Learn

Composting	Ocean Dumping Act
Incinerating	Recycling
Landfill	Solid Wastes

Learning Objectives

After reading this chapter and participating in the activities, you should be able to:

- Explain the seriousness of the solid waste disposal problem facing this country.

- List and describe the major categories of solid wastes in this country.

- Discuss landfill site selection, construction, and maintenance.

Overview

A 1978 Supreme Court ruled that according to the U.S. constitution, solid wastes were a "good" in interstate commerce and thus could not be restricted by individual state regulations. On January 11, 1991, U.S. District Judge Matthew Perry ruled that South Carolina could not close its borders against solid waste from North Carolina and eighteen other states. As of 1991, states with surplus solid waste were hauling and dumping it as much as several hundred miles into other states. Solid waste disposal is surely one of our most current and pressing environmental problems. Is there a solution?

Solid wastes consist of paper products, food and yard wastes, aluminum and iron, glass, wood, plastics, rubber and leather, textiles, and similar items that have been discarded. These wastes are disposed of in these ways:

- Recycled

- Incinerated

- Thrown away, often along the side of a road or highway

- Dumped in the ocean

- Buried in a landfill

- Composted

Managing Solid Wastes

Solid Waste Generated and Recovered (Recycled)

We are a very careless and wasteful society. We throw away an average of 3.67 pounds of solid waste per person each day. This is an increase of thirty-eight percent since 1960. Recycling is the process of recovering solid waste for reuse. On the brighter side we recover (recycle) 0.4 pound per person per

Table 13-1. Municipal Solid Waste Generated and Recovered, 1960–1990 *

Year	Gross Discarded Pounds per Day per Person	Daily U.S. Total (million tons)	Materials Recovered Pounds per Day per Person	Net Discarded per Day per Person	Daily U.S. Total Net (million tons)
1960	2.65	87.5	0.18	5.8	81.7
1965	2.88	102.3	0.17	6.2	95.9
1970	3.22	120.5	0.21	8.0	112.1
1975	3.18	125.3	0.23	9.1	115.5
1980	3.43	142.6	0.32	13.4	126.5
1985	3.49	152.5	0.35	15.3	129.7
1990	3.67	167.4	0.40	18.4	135.7
Percentage Increase/Decrease, 1960–1990	+38%	+91%	+122%	+217%	+66%

*Source: U.S. Environmental Protection Agency, Office of Solid Waste and Emergency Response.

day, an increase of 122 percent from 1960. But again on the darker side, our net discards (generated minus recycled wastes) have increased sixty-six percent since 1960. We can do better—and must do better because we are running out of environmentally safe places to stash the trash (see Table 13-1 and Figure 13-1).

Kinds of solid waste recovered (recycled) by percentage of that discarded are in this relative order: paper, food and yard waste, metals, glass, wood, plastics, and rubber and leather, textiles. Of the eight kinds of solid waste recycled, paper leads the list but only at a 36.8 percent recovery rate in

1990, a slight increase over the thirty percent of paper products that were recycled in 1960. Food and yard wastes are next in percentage recycled—28.2 percent in 1990, a decrease of 11.2 percent from 1960 (see Table 13-2).

Incinerated

An estimated two percent of all solid waste is destroyed by burning (**incinerating**). This may vary from a backyard open fire to a huge furnace operated by a municipality. In some instances the heat gener-

Figure 13-1. In 1990 the average person in the United States threw away 3.67 pounds of solid waste per day but recycled only 0.4 pounds. Where did the net 3.27 pounds per day go? Here is some of it. Is this a civilized way to stash the trash? (Courtesy United States Environmental Protection Agency.)

Table 13-2. Kinds of Solid Waste Discarded and Percentage Recovered by Kind and Year, 1960–1990*

Year	Paper	Food and Yard Waste	Metals	Glass	Wood	Plastics	Rubber and Leather	Textiles
1960	30.0	39.4	12.8	7.8	3.7	0.5	2.1	2.1
1965	33.5	35.4	11.1	8.8	3.6	1.5	2.3	2.0
1970	32.4	32.0	12.0	11.1	3.6	2.7	2.7	1.8
1975	29.6	33.2	11.5	11.4	3.8	3.8	3.2	1.9
1980	32.5	29.7	10.1	11.0	3.8	5.9	3.2	2.0
1985	35.5	29.4	9.0	8.9	3.9	2.5	2.0	2.0
1990	36.8	28.2	8.9	8.3	3.6	2.3	2.0	2.0
Percentage Increase/Decrease, 1960–1990	+6.8%	-11.2%	-3.9%	+0.5%	-0.1%	+1.8%	-0.1%	-0.1%

*SOURCE: United States Environmental Protection Agency, Office of Solid Waste and Emergency Response.

ated is used to make electricity. Incineration may seem to be an ideal environmental way to dispose of unwanted solid waste but these problems arise:

- Smoke generated can pollute the ambient air.

- If chlorinated pesticides or polyvinyl chloride (PVC) plastics are burned, the fumes may be toxic.

- If the solid waste contains metals or glass, a person may become injured in handling it, and ashes may be difficult to dispose of.

- The fire may escape and be difficult to control.

- There may be city ordinances against open burning.

Thrown Away

When solid wastes are thrown into a gully or hidden in the woods, rats, mice, opossums, woodchucks, raccoons, and snakes may find food and shelter among the trash. Furthermore, oils, acids, chemical fertilizers, and pesticides can spill or leach out and contaminate the environment (see Figure 13-2).

Dumped in the Ocean

The **Ocean Dumping Act** authorizes the U.S. Environmental Protection Agency and the Corps of Engineers to regulate ocean dumping by means of a permit system. A 1988 amendment prohibits dumping of sewage sludge, industrial waste, and infectious medical wastes into the ocean. It also establishes a scheme for phasing out currently permitted dumping of sludge.

To enforce the law, the U.S. Coast Guard monitors barges using on-board inspections and electronic surveillance. Barges are fitted with "black boxes" that measure the vessels' draft (depth) in the water and transmit the data to a Coast Guard vessel. A sudden change in a barge's draft prior to reaching a designated site could indicate illegal ocean dumping. Although only a few permits have been issued, permittees are tracked carefully because of the potential for pollution and the public's reaction to ocean dumping.

Buried in a Landfill

A **landfill** is a trench dug in the soil to bury solid waste. Civil engineers claim it is their province to design a

Figure 13-2. Thrown away! A homeless man (at arrow) seems to be "at home" in this trash. The scene could be in any large city in the world but actually is in Djakarta, the capitol of Indonesia. (Courtesy Food and Agriculture Organization of the United Nations.)

satisfactory landfill, but public health officials and geologists also claim the expertise to design landfills. In addition, soil scientists claim knowledge of where and how to establish an environmentally safe landfill. However, no one wants a landfill in his or her back or front yard (see Figure 13-3). In fact, one of the easiest ways to arouse a public battle in the 1990s is to propose a new landfill almost anywhere. People who discard solid waste every day want somewhere to put it—a landfill. But nobody wants a new landfill near them.

Site selection for a landfill should consider these guidelines:

- Avoid deep sands with shallow water tables. This reduces the hazard of toxic wastes seeping into drinking water.

Figure 13-3. There is a better way to dispose of waste materials. Compacting and baling all solid wastes into bales weighing about three thousand pounds makes it easy to bury them in a sanitary landfill. The landfill will not settle unevenly and can be more quickly used for a park or recreation area. The active life of a landfill will also be extended by perhaps fifty percent (San Diego, California). (Courtesy United States Environmental Protection Agency.)

- In humid areas, percolation depth of hazardous chemicals is greater than in arid regions. For this reason a deeper soil should be selected in humid climates to filter the toxins before they reach the groundwater.

- Avoid stony or fractured limestone soils.

- Select a soil with a pH of 6.5 or above. At these pHs, toxic metals such as cadmium, mercury, lead, chromium and copper are less soluble and therefore are less likely to leach into the groundwater.

- Avoid wetlands.

- Establish a waterproof trench with a compacted clay or plastic film bottom and sides (see Figure 13-4).

- Compact the solid waste to reduce the space needed.

- Prohibit the dumping of bulky items such as old refrigerators, stoves, and automobile tires. These can be recycled. Tires can be burned in special ovens or in areas where the smoke is the least environmental hazard.

- Daily input in the landfill should be compacted with a crawler-type tractor.

- Cover each day's input with soil to reduce rodent and insect habitation.

- When the landfill is full, a final covering of at least two feet of productive soil should be added and seeded to an adapted perennial grass or planted to trees (see Table 13-3).

If properly located, compacted, covered, and stabilized by perennial vegetation, a filled and completed landfill could be managed as a "green space" for a city or county park. Under no circumstances should it be used as a building site because of the hazard of methane gas formation from anaerobic decomposition of organic matter in the solid waste (see Figure 13-5). Landfills are filling rapidly and new sites within economic hauling distance are becoming scarce (see Figure 13-6).

Figure 13-4.   Cross section of a landfill with a clay liner to reduce percolation loss. Plastic sheets can substitute as a liner to retain liquids so they will not contaminate groundwater. (**Source:** United States Enviromental Protection Agency. "Guide to Technical Resources for the Design of Land Disposal Facilities," EPA/625/6-88/018, 1988.)

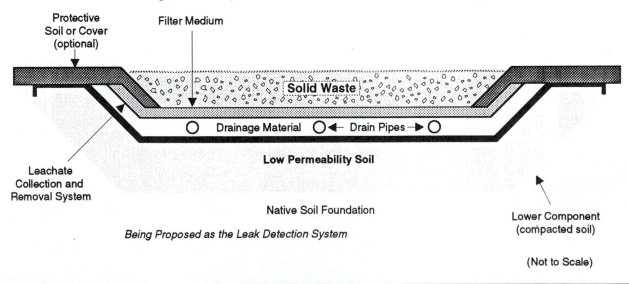

Table 13-3. Functional Suitability of Various Soil Textural Classes for Use in Covering a Sanitary Landfill *

Function	Clean Gravel	Clayey/ Silty Gravel	Clean Sand	Clayey/ Silty Sand	Silt	Clay
Prevent rodents from burrowing	G	F-G	G	P	P	P
Keep flies from emerging	P	F	P	G	G	E**
Minimize moisture entering fill	P	F-G	P	G-E	G-E	E**
Minimize landfill gas venting through cover	P	F-G	P	G-E	G-E	E**
Provide pleasing appearance and control blowing paper	E	E	E	E	E	E
Support vegetation	P	G	P-F	E	G-E	F-G
Be permeable for venting of decomposition gas***	E	P	G	P	P	P

Coding: E, excellent; G, good; F, fair; P, poor.

*SOURCE: Brunner, D.R., Hubbard, S.J., Keller, D.K., and Newton, J.L. (1971). **Closing open dumps**. Washington, D.C.: Environmental Protection Agency, Solid Waste Management Office, p. 15.
**Except when cracks extend through the entire cover.
***Only if well drained.

Composted

The **composting** of urban wastes and the use of the finished compost as an organic fertilizer and mulch has been generally successful in many countries of Europe and in Japan. Composting refers to the process of holding organic materials in an environment favorable to their rapid, partial decomposition. Will

the system be acceptable in the United States? Transfer of technology from one country to another is extremely hazardous because of differences in cultural backgrounds.

Reasons for general success in the practice of composting urban wastes in Europe and Japan, but with only partial success in the United States, may be summarized as follow.

- Farms in Europe and Japan are smaller than in the United States.
- Labor for farming is cheaper and more plentiful than that in the United States.
- Hand labor is more traditional in Europe and Japan.
- Cost of chemical fertilizer, including the cost of application, is cheaper in the United States.
- Traditional attitudes in Europe and Japan favor organic sources vs chemical sources of plant nutrients.

Compost made from solid wastes in a typical United States urban area was evaluated over a nine-year period under field conditions for its ability to increase yields of potatoes, rye, and oats. Conclusions are as follow.

- Fresh compost reduced yields the first year because of competition from microorganisms for available essential elements, especially for nitrogen. The second year after application the crop yields were increased slightly by compost.
- Compost that had aged in the open for a period of six months to a year was responsible for increasing crop yields almost twelve percent over those that had received equivalent plant nutrients in chemical fertilizer.

The equivalent amounts of compost and chemical fertilizer applied to the soil were forty tons and fifty pounds, respectively, per acre. The compost improved the ease and cost of preparing the soil for a desirable

Figure 13-5. Solid wastes here at Virginia Beach, Virginia were covered with soil in 1971. The resulting large mound was stabilized by trees and grasses and made into a park. Figure 13-5A shows the work of covering the landfill and Figure 13-5B shows the entrance to the park as it appears today. The final photo in Figure 13-5C shows part of the park as seen from the highway. (Courtesy United States Environmental Protection Agency.)

A

B

C

seed/plant bed, increased the quality of soil structure, reduced soil erosion, and made more soil water available to plants. The general conclusions were as follow.

■ Compost is not as economical as chemical fertilizers in enhancing plant growth.

■ Compost is not as effective as animal manures in improving the physical condition and fertility of soils.

■ Compost contains particles of glass and plastic that are objectionable, especially in home gardens where hand manipulation of soil is common.

Figure 13-6. Solid waste is rapidly filling the landfills in the densely populated East. Some solid waste companies are trucking the trash into southern and midwestern states. (**Source:** National Soild Wastes Management Association.)

grades of paper for recycling prior to final disposal and are further encouraged to recycle yard trash and other mechanically treated solid waste into compost for agricultural and other acceptable uses.

No more than one-half the (twenty-five percent reduction) goal may be met with yard trash, white goods (refrigerators, etc.) construction and demolition debris, and tires.

No person shall knowingly dispose of the following special wastes in landfills: (1) lead-acid batteries; (2) used oil; (3) white goods; (4) yard trash, after January 1, 1993. Yard trash that is separated at the source may be accepted at a solid waste disposal area where the area provides and maintains separate yard trash composting facilities (*Research Perspectives*, 1990).

For Further Information

Table 13-4 gives information for each of the fifty states and in Washington, D.C. Agencies to write or telephone to get more information on solid waste management are included.

Model Solid Waste Law

North Carolina passed a model solid waste law in 1989. Goals stated in this law are as follow:

It is the goal of the State to promote methods of solid waste management that supplement disposal in landfills. In order of preference these are: 1) Waste volume reduction at the source, 2) Recycling and reuse, 3) Composting, 4) Incineration to produce energy, 5) Incineration to reduce volume, 6) Disposal in landfills.

It is the goal of this State that at least twenty-five percent of the total waste be recycled by January 1, 1993.

Units of local government are encouraged to separate all plastics, metal, and all

Summary

As much as each of us might want to keep trash from other states out of our own state, that is illegal. Solid wastes are considered an interstate good from the perspective of the federal law. As such, no state may prevent another's solid wastes from being transported across its boundaries and along its highways, waterways, or railroads. That does not mean, however, that a locally operated landfill must accept the solid wastes. Dumping it is up to local officials and landfill operators.

Solid wastes consist primarily of paper products, food and yard wastes, aluminum and iron, glass, wood, plastics, rubber and leather, and textiles. They must be either recycled, incinerated, discarded, dumped in the ocean, buried in a landfill, or composted.

Table 13-4. For Additional Information on Solid Waste, Write or Telephone the Agency in your State

Alabama
Department of Environmental Management
Solid Waste Division
1715 Congressman Wm. Dickinson Dr.
Montgomery, AL 36130
(205) 271-7700

Alaska
Department of Environmental Conservation
Solid Waste Program
P.O. Box O
Juneau, AK 99811-1800
(907) 465-2671

Arizona
Department of Environmental Administration
Quality-O.W.P.
Waste Planning Section, 4th Floor
Phoenix, AZ 85004
(602) 257-2317

Arkansas
Department of Pollution Control and Ecology
Solid Waste Division
8001 National Drive
Little Rock, AR 72219
(501) 562-7444

California
Recycling Division
Department of Conservation
819 19th Street
Sacramento, CA 95814
(916) 323-3743

Colorado
Department of Health
4210 E. 11th Avenue
Denver, CO 80220
(303) 320-4830

Connecticut
Recyling Program
Department of Enviromental Protection
Hartford, CT 06106
(203) 566-8722

Delaware
Department of National Resources and Environmental
Control
89 Kings Highway
P.O. Box 1401
Dover, DE 19903
(302) 736-4794

District of Columbia
Public Space and Maintenance Administration
4701 Shepard Parkway, SW
Washington, DC 20032
(202) 767-8512

Florida
Department of Environmental Regulation
2600 Blairstone Road
Tallahassee, FL 32201
(904) 488-0300

Georgia
Department of Community Affairs
40 Marietta St., NW, 8th Floor
Atlanta, GA 30303
(404) 656-3898

Hawaii
Litter Control Office
Department of Health
205 Koula Street
Honolulu, HI 96813
(808) 548-3400

Idaho
Department of Environmental Quality
Hazardous Materials Bureau
450 W. State Street
Boise, ID 83720
(208) 334-5879

Illinois
Illinois EPA
Land Pollution Control Division
2200 Churchill Road
P.O. Box 19276
Springfield, IL 62706
(217) 782-6761

Table 13-4 (Continued). For Additional Information on Solid Waste, Write or Telephone the Agency in your State

Indiana
Office of Solid and Hazardous Waste Management
Department of Environmental Management
105 S. Meridian Street
Indianapolis, IN 46225
(317) 232-8883

Iowa
Department of Natural Resources
Waste Management Division
Wallace State Office Building
Des Moines, IA 50319
(515) 281-8176

Kansas
Bureau of Waste Management
Department of Health and Environment
Topeka, KS 66620
(913) 296-1594

Kentucky
Resources Management Branch
Division of Waste Management
18 Reilly Road
Frankfort, KY 40601
(502) 564-6716

Louisiana
Department of Environmental Quality
P.O. Box 44307
Baton Rouge, LA 70804
(504) 342-1216

Maine
Office of Waste Reduction and Recycling
Department of Economic and Community Development
State House Station #130
Augusta, ME 04333
(207) 289-2111

Maryland
Department of Environment
Hazardous and Soild Waste Administration
2500 Broening Highway
Building 40
Baltimore, MD 21224
(301) 631-3343

Massachusetts
Division of Solid Waste Management
D.E.O.E.
1 Winter Street, 4th Floor
Boston, MA 02108
(617) 292-5962

Michigan
Waste Management Division
Department of Natural Resources
P.O. Box 30028
Lansing, MI 48909
(517) 373-0540

Minnesota
Pollution Control Agency
520 Lafayette Road
St. Paul, MN 55155
(612) 296-6300

Mississippi
Non-Hazardous Waste Section
Bureau of Pollution Control
Department of Natural Resources
P.O. Box 10385
Jackson, MS 39209
(601) 961-5047

Missouri
Department of Natural Resources
P.O. Box 176
Jefferson City, MO 65102
(314) 751-3176

Montana
Solid Waste Program
Department of Health and Environmental Science
Cogswell Bldg., Room B201
Helena, MT 59620
(406) 444-2821

Nebraska
Litter Reduction and Recycing Programs
Department of Environmental Control
P.O. Box 98922
Lincoln, NE 68509
(402) 471-4210

Table 13-4 (Continued). For Additional Information on Solid Waste, Write or Telephone the Agency in your State

Nevada
Energy Extension Service
Office of Community Service
1100 S. Williams Street
Carson City, NV 89710
(702) 885-4420

New Hampshire
Waste Management Division
Department of Enviromental Services
6 Haxen Drive
Concord, NH 03301
(603) 271-2900

New Jersey
Office of Recycling
Department of Enviromental Protection
CN 414
401 E. State Street
Trenton, NJ 08625
(609) 292-0331

New Mexico
Solid Waste Section
Environmental Improvement Division
1190 St. Francis Drive
Santa Fe, NM 87503
(505) 457-2780

New York
Bureau of Waste Reduction and Recycling
Department of Enviromental Conservation
50 Wolf Road, Room 208
Albany, NY 12233
(518) 457-7337

North Carolina
Solid Waste Management Branch
Department of Human Resources
P.O. Box 2091
Raleigh, NC 27602
(919) 733-0692

North Dakota
Division of Waste Management
Department of Health
1200 Missouri Avenue, Room 302
Box 5520
Bismark, ND 58502-5520
(701) 224-2366

Ohio
Division of Litter Prevention and Recycling
Ohio EPA
Fountain Square Building, E-1
Columbus, OH 43224
(614) 265-7061

Oklahoma
Solid Waste Division
Department of Health
1000 N.E. 10th Street
Oklahoma City, OK 73152
(405) 271-7159

Oregon
Department of Environmental Quality
811 S.W. Sixth
Portland, OR 97204
(503) 229-5913

Pennsylvania
Waste Reduction and Recycling Section
Division of Waste Minimization and Planning
Department of Environmental Resources
P.O. Box 2063
Harrisburg, PA 17120
(717) 787-7382

Rhode Island
Office of Environmental Coordination
Department of Environmental Management
83 Park Street
Providence, RI 02903
(401) 277-3434

Table 13-4 (Continued). For Additional Information on Solid Waste, Write or Telephone the Agency in your State

South Carolina
Department of Health and Environmental Control
2600 Bull Street
Columbia, SC 29201
(803) 734-5200

South Dakota
Energy Office
217-1/2 West Missouri
Pierre, SD 57501
(605) 773-3603

Tennessee
Department of Public Health
Division of Solid Waste Management
Customs House, 4th Floor
701 Broadway
Nashville, TN 37219-5403
(615) 741-3424

Texas
Division of Solid Waste Management
Department of Health
1100 W. 49th Street
Austin, TX 78756
(512) 458-7271

Utah
Bureau of Solid and Hazardous Waste
Department of Environmental Health
P.O. Box 16690
Salt Lake City, UT 84116-0690
(801) 538-6170

Vermont
Agency of Natural Resources
103 S. Main St., West Bldg.
Waterburg, VT 05676
(802) 244-6916

Virginia
Department of Waste Management
Division of Litter Control and Recycling
11th Floor, Monroe Building
101 N. 14th Street
Richmond, Va 23219
1-800-Keepit

West Virginia
Department of Natural Resources
Conservation, Education, and Litter Control
1800 Washington Street E.
Charleston, WV 25305
(304) 348-3370

Washington
Department of Ecology
Mail Stop PV-11
Olympia, WA 95804
1-800-Recycle

Wisconsin
Department of Natural Resources
P.O. Box 7921
Madison, WI 53707
(608) 266-5741

Wyoming
Solid Waste Management Program
Department of Environmental Quality
Herschler Building
122 W. 25th Street
Cheyenne, WY 82002
(307) 777-7752

For more information on solid waste management call the U.S. Environmental Protection Agency Solid Waste Hotline at 1-800-424-9346. In Washington, D.C., call (202) 382-3000. Or write: U.S. Environmental Protection Agency, 401 M Street SW, Washington, D.C. 20460.

America is perhaps the most wasteful of all societies. We generate huge amounts of solid wastes and reuse only a small part of them each year. Of all the recycling we do, paper is the thing we most often recycle. There is certainly no shortage of trees for making paper, yet we recycle a larger percentage (by ten-fold) of our paper than of metals which are nonrenewable resources.

Incineration holds some promise as a means of disposal, but there are problems. The same is true for composting. Ocean dumping is a technique that has been used widely in the past. Fortunately, ocean dumping is coming under better control in this country. The same is not necessarily true in other countries.

Landfills are really the only satisfactory disposal method available today for the United States. But landfills are not welcome neighbors. As a result, we will be unable to develop adequate landfills for very much longer. Replacing our existing landfills as they fill up is going to be a very difficult political problem.

DISCUSSION QUESTIONS

1. Why is solid waste disposal becoming such a serious problem in the 1990s?

2. How is ocean dumping of solid waste in U.S. waters monitored?

3. Landfills seem to be an effective way to dispose of solid wastes. Why do we seem to have so many problems with landfills in the United States?

4. What are the alternatives we have for disposing of our solid wastes?

5. What is composting, in regards to solid waste disposal? Why does composting seem to have less promise in the United States than in Europe and Japan as a means of solid waste disposal?

6. What kinds of materials make up our solid wastes in the United States?

ADDITIONAL ACTIVITIES

1. Contact the office listed for your state in Table 13-4. Request information on the amount and nature of solid wastes being generated in your state. Ask for information about where and how the materials are being disposed of.

2. Take a class field trip to your local landfill. Find out how much solid waste is coming there, where it is coming from, and how it is being handled.

3. Inventory the trash in your home for a full week. Make a list of the kinds of materials being thrown away. What conclusions can you draw from this exercise about your family's lifestyle. Can you suggest ways to reduce the volume of trash based on your findings?

4. Plot a graph to examine the thirty-year trends in the types of municipal solid waste in the United States. Can you draw any conclusions from the data?

5. As a class or group project, pick up all of the trash along a stretch of highway. Separate the trash into kinds as indicated in Table 13-2. How do the percentages of each kind compare with the others. Write a letter to the editor of your local newspaper and report your findings. Try to answer this question: Why do people throw trash away along a highway?

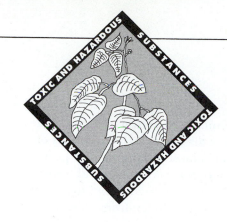

CHAPTER 14

Toxic and Hazardous Substances

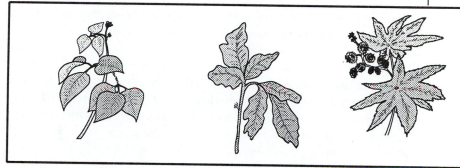

Terms to Look for and Learn

Allergen

Biomagnification

Castor Bean

Contact Dermatitis

Dumbcane

Foxglove

Hazardous Waste

Jimson Weed

Juglone

Lily of the Valley

Nightshade

Philodendron

Poison Ivy/Poison Oak

Pokeweed

Rhododendron

Toxin

Water Hemlock

Wisteria

Learning Objectives

After reading this chapter and participating in the activities, you should be able to:

- List and discuss common poisonous plants.

- Explain the processes by which common poisonous plants affect sensitive people.

- Explain the symptoms caused by the common poisonous plants.

- Explain the treatments recommended for problems caused by common poisonous plants.

- List and discuss common toxic substances in sewage sludge and septage.

- Discuss the common household hazardous wastes.

237

Overview

The word toxic (Latin *toxicum*; Greek *toxikon*) means poisonous and toxin is used to refer to a toxic substance. Hazardous means involving chance, risky, or dangerous. These two words are often used interchangeably in environmental literature. Certainly a **toxin** is hazardous and some hazardous substances are toxic.

This chapter will focus on poisonous plants, toxic chemical laws, toxins in the Great Lakes, toxic substances in sludge and septage, and household hazardous waste. The discussion of hazardous industrial, agricultural, and military wastes is beyond the scope of this chapter.

Poisonous Plants

Plants poisonous to other plants but not to people include black walnut and butternut. Plants poisonous to skin of humans are dominated by poison ivy and western poison oak. Also discussed will be the principal plants that produce toxic reactions of humans when swallowed.

Poison ivy, poison oak, and other plants discussed here that are poisonous if eaten generally should not be used as houseplants, especially where there are children. Cats and other household pets will sometimes chew on such plants and become poisoned, too. But, because their foliage (leaves) is so attractive, several of the poisonous plants discussed are commonly used as household plants. People who use such plants should be aware of the danger, symptoms, and treatment for poisoning from the plants.

Walnut and Butternut

Black walnut and butternut belong to the same genus of plants: *Juglans*. Both have a sap toxic to many other plants, even to their own seedlings. This toxic substance has been named **juglone**.

Roots of walnut and butternut trees contain toxic juglone, as well as the leaves, bark, and hulls of the fruit (a nut). The following plants in close proximity to walnut or butternut roots soon die with symptoms resembling a wilt caused by bacteria or fungi: tomato, potato, alfalfa, apple, rhododendron, white pine, red pine, white birch, beans, and cucumbers.

For some unexplained reason, the following plants seem to grow better under black walnut and butternut trees: Kentucky bluegrass, orchardgrass, red clover, wild grape, and black raspberry.

Poison Ivy (Toxicodendron [Rhus] species) and Poison Oak

Poison ivy and **Poison oak** are responsible for more cases of allergic contact dermatitis than all of the other substances combined (see Figures 14-1 and 14-2). One or more contacts with these plants is required before the individual becomes sensitized. There is no relationship either in the ability to become sensitized nor in the severity of the response to other allergies the individual may have, such as food allergies, hay fever, or asthma. Once any area of the body is sensitized, the whole body becomes sensitive to the plant; however, only those areas of the skin that actually come into contact with the plant develop a rash.

Some people never develop sensitivity to poison ivy; others may develop sensitivity after several years of being immune. If the plant resin gets on the hands, it may be distributed to other areas of the body such as the face, arms, and genitals. Dogs and cats are immune to the toxin but may carry the resin on their fur and then on to humans. The smoke from burning plants is harmless unless the smoke carries unburned plant particles. The pollen from poison ivy and poison oak is not allergenic, nor is honey made from the nectar toxic.

Following exposure to poison ivy or poison oak by a sensitized individual, usually twelve to forty-eight

Figure 14-1. Poison ivy.

Figure 14-2. Western poison oak.

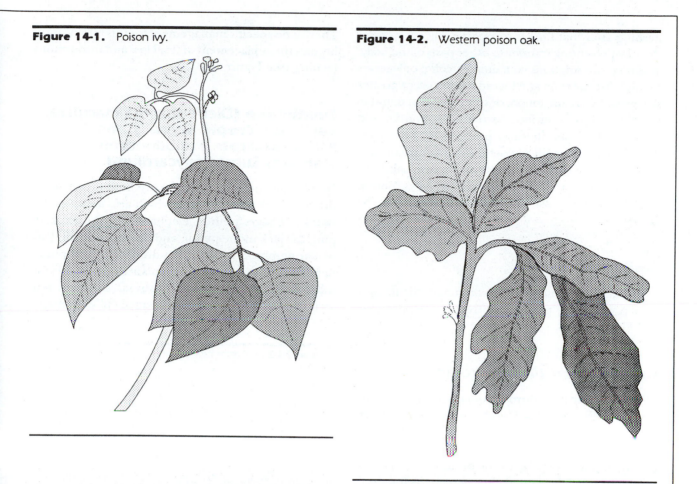

hours are required before there is a visible response on the skin. Initially, there is redness and swelling. During the subsequent twenty-four hours blisters may develop. The blister fluid does not contain the **allergen** (the substance that induces the allergy) and cannot spread the rash to others or to other parts of the body. If you get a poison ivy or poison oak rash, scratching it will not cause the rash to spread unless more resin is present. But, scratching it can result in secondary infections. So, the old advice, "don't scratch the rash" is right. But, it is right for a different reason than the one usually given.

Itching is a constant and prominent feature. Crusting and scaling begin in a few days. In the absence of complications caused by scratching or inappropriate medication, the dermatitis rarely persists more than ten days.

The severity of the response varies greatly. Some victims are disabled while others have only mild reddening. In addition to this individual variation, the response will be determined by the total area of exposure and the amount of plant resin involved. Poison ivy and poison oak reactions seem to be more intense in adults than in children.

Cold, wet dressings prepared from ice water and a cotton cloth provide relief during the acute state. Lotions such as calamine are widely used, but lotions containing antihistamines, local anesthetics, or metal salts should be avoided.

It takes approximately ten minutes for the allergen to penetrate the skin. The only purpose served by washing after that time is to remove any excess resin thereby preventing transfer to other parts of the body. Washing with solutions containing alcohol only serves to solubilize the resin and spread it faster over a greater skin area. No cream, lotion, or spray has been found to exert an effective surface barrier to poison ivy and poison oak. Desensitization procedures are equally disappointing. A medical doctor can give injections to reduce the severity of poison ivy and poison oak.

If patients are sensitive to poison ivy or poison oak, they will be cross-sensitive to other plants in the family.

Anacardiaceae

This plant family includes poison sumac, the mango, poison wood, the cashew nut tree, the pepper tree, and Florida holly.

Castor Bean (Ricinus communis)

The **castor bean** grows throughout the West Indies, and as a naturalized weed in Florida, along the Gulf Coast states to Texas, and along the Atlantic Coast to New Jersey; in southern California, Hawaii, and Guam. It is grown commercially for castor oil in the Gulf Coast states and is widely planted elsewhere for its foliage.

Castor bean is an annual, growing to fifteen feet, higher in the tropics. The large-lobed leaves are up to three feet across. Spiny seed pods form in clusters along spikes. The pods contain small, plump seeds. They are usually mottled with black or brown and have a pleasant taste but *should not be eaten because they are poisonous.*

The seed contains *ricin*, a protein that acts to prevent cell reproduction in the walls of the intestines. Small quantities of the relatively harmless cathartic, castor oil, are also present in the seeds.

The onset of poisoning develops only after a latent period of many hours after eating the seeds. It is characterized by nausea, vomiting, and diarrhea. Other effects that become apparent are due to breakdown of the function of the intestine.

The castor bean is an extremely toxic plant; the ingestion of two to six well-chewed seeds can be fatal.

There is no specific management for this intoxication beyond the replacement of fluid loss and intravenous feeding (see Figure 14-3).

Dumbcane (Dieffenbachia maculata, cultivar "Tropic Snow") and Philodendron (Philodendron scandens Subsp. oxycardium)

Dumbcane and philodendron are both members of the botanical family *Araceae* which also includes the skunk cabbage, pothos, elephant ear, the cuckoo-pint, the jack-in-the-pulpit, and the water arum. Particular emphasis is placed on dumbcane and philodendron because these two plants are responsible for more cases of symptomatic distress (although not serious poisoning) in preschool-aged children in the

Figure 14-3.   Castor bean.

United States than all other plants combined. Most plants in the family *Araceae* produce similar toxic effects.

Dumbcane is a tall, erect, unbranched plant with large oblong leaves splotched with white or ivory markings. There are a number of horticultural varieties. It is cultivated outdoors in the West Indies, southern Florida, and Hawaii. Dumbcane is a common, decorative potted plant in offices, waiting rooms and lobbies, and is becoming increasingly popular as a houseplant.

The **philodendrons** are climbing vines. The leaves are often large and may vary in shape or notchings on a single plant.

The irritant effect of these plants is due to sharp needles of water-insoluble *calcium oxalate*. Chewing on the leaf produces an almost immediate intense pain which may be followed by swelling inside the mouth and the formation of blisters. The speech, in severe cases, becomes almost unintelligible, hence the name dumbcane. These plants can also produce injury to the eye and contain an irritant producing **contact dermatitis** (skin rash).

The pain and swelling inside the mouth will recede slowly without treatment. Cool liquids held in the mouth may bring some relief (see Figures 14-4 and 14-5)

Foxglove (Digitalis purpurea)

The **foxglove** is a European plant that is grown as an ornamental flower in most of the United States and is now naturalized in the western states.

It is a biennial or perennial herb, easily recognized when in flower. The flowers grow along the top of the central stalk (raceme). They are tubular, about three inches long and droop downwards. They are purple or pink (rarely white) and are usually spotted along the inside bottom of the tube.

The entire plant contains the digitalis glycosides along with irritant saponins. The saponins are responsible for an initial irritation in the mouth, nausea, vomiting, and abdominal pain with cramping and diarrhea. The digitalis glycosides are responsible for the major toxic effects on the heart.

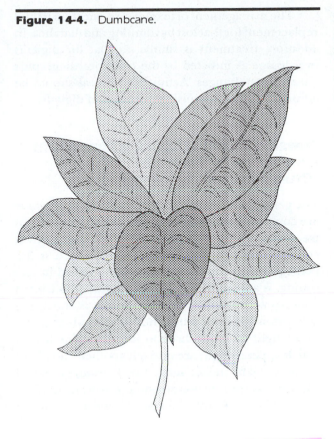

Figure 14-4. Dumbcane.

Figure 14-5. Variations in leaves of philodendron.

The management of foxglove intoxication is fluid replacement for that lost by vomiting and diarrhea. In addition, treatment is similar to that for digitalis overdosage as indicted by the electrocardiographic (heartbeat) changes. Activated charcoal should be administered to diminish absorption of digitalis.

Jimson Weed (Datura stramonium) and Angel's Trumpet (Brugmansia suaveolens)

The genus *Datura* and the closely related *Brugmansia* are composed of weeds, bushes, or small ornamental trees with distinctive trumpet-shaped flowers which contain similar toxins. **Jimson weed**, which is the plant of this group most frequently involved in poisoning, is a weed that grows throughout the United States (except Alaska) and Canada. It is an annual with a stout green or purple hollow stem which grows to a height of about three to four feet. The leaves, which appear singly, are coarsely toothed and about three to eight inches long. Erect, trumpet-shaped white or pale purple flowers emerge at the forks of the stems. The seed pods are spiny capsules, about two inches in diameter, filled with numerous small, kidney-shaped, hard, brownish-black seeds.

The whole plant, including the nectar, has been involved in accidental poisoning. The plant contains the *belladonna* alkaloids, which include atropine and scopolamine.

Touching one eye after handling the plant will result in dilation of that eye only. Ingestion results in thirst and such extreme drying of the mouth that there is difficulty in speech and swallowing. The skin also becomes dry and the body temperature is elevated. Occasionally a rash develops, particularly on the neck. The heart rate may increase. Large quantities eaten produce blurring of vision and dilation of the pupils. There may be excitement, hallucinations, and delirium. Coma is seen only after massive ingestions or following convulsions due to elevated body temperature.

Specific treatment is not necessary except for the management of highly elevated body temperatures in small children. The other symptoms may be alleviated by the cautious administration of physostigmine. The dilation of the pupils may persist for some days. The resulting discomfort in bright light may be alleviated by dark glasses or having the victim remain indoors in a darkened room for a few days.

Lily of the Valley (Convallaria majalis)

Lily of the Valley is a Eurasian plant that is widely cultivated throughout the north temperate zones of the United States. It is a small perennial with two oblong-oval leaves. It has an arched flower stalk (raceme) bearing small, fragrant, bell-shaped, white flowers on the down side. The plant occasionally forms orange-red, fleshy berries. It spreads by underground roots to form thick beds.

The whole plant, and water from vases in which the flowers have been placed, contains digitalis-like glycosides and irritant saponins. The saponins are responsible for an initial irritation of the mouth, nausea, vomiting, abdominal pain and cramping, and diarrhea. The digitalis-like glycosides are responsible for the toxic effects on the heart.

The management of lily of the valley intoxication is fluid replacement for that lost by vomiting and diarrhea and otherwise as for digitalis overdosage, as indicated by electrocardiographic changes.

Nightshade (Solanum species)

Solanum is a very large genus of plants with approximately seventeen hundred species. These include the bittersweet, the Jerusalem cherry, the Carolina horse nettle, and the common white potato.

The nightshades are a closely related group of species, generally distinguishable only by a botanist. *Solanum americanum* appears throughout the United States, but primarily in the eastern portion, from Nova Scotia to Florida and westward to North Dakota and Texas. The European nightshade (*Solanum nigrum*) has been introduced and is now a weed on both coasts of the United States, Alaska, Hawaii, and Guam.

The **nightshades** are annual weeds. They tend to be sprawling plants, usually not more than one to three feet high. The small, five-toothed, white flowers form singly or in clusters. The berries are black when mature.

Human poisoning from most species of *Solanum* is generally attributed to eating the immature fruit which contain the *solanine glycoalkaloids*. Solanine has a low toxicity for adults, but there have been fatal intoxications when eaten by children. Solanine poisoning is often confused with bacterial gastroenteritis. There is a scratchy feeling in the throat, nausea, an elevated body temperature, and diarrhea. Symptoms usually appear only after a latent period of several hours after ingestion.

The management of solanine poisoning consists of general symptomatic care for the gastroenteritis. In addition, replacement of body fluids is important, particularly in children, to compensate for the losses due to the vomiting and diarrhea.

Pokeweed (Phytolacca americana)

Pokeweed grows in damp fields, moist woods, and roadside ditches from Maine to Minnesota and southward to Texas, along the Gulf of Mexico into Florida. The plant arises from a huge rootstock, up to six inches in diameter, from which stout, red purplish branched stems appear, usually to a height of five to six feet. The leaves appear singly and are from four to twelve inches long and have smooth edges. The flowers are greenish white to purplish, small, appearing on a vertical stalk in racemes. The flattened, purple-black berries are attached to the stalk by short stems and give the plant additional common names such as inkberry, red ink plant, and pigeon berry.

The poisonous principle, *phytolaccatoxin* and related *triterpenes*, is mostly concentrated in the leaves and roots. The berries are relatively nontoxic. The young sprouts and stems may be eaten after boiling and *discarding the cooking water*. The cooked product can be purchased in cans commercially. Intoxications generally arise from eating the uncooked leaves in salads or using the roots mistakenly for horseradish or parsnips.

After ingestion, there is a delay of about two to three hours prior to the onset of symptoms. There are then abdominal cramps, profuse sweating, and persistent vomiting which is later accompanied by diarrhea. The intoxication may continue for up to forty-eight hours. The severity of the abdominal pain may require medication. Fluids lost from the protracted vomiting and diarrhea must be replaced, particularly in small children (see Figure 14-6).

Rhododendron (Rhododendron species)

There are about eight hundred species of **rhododendrons**. The genus *Rhododendron* includes the former genus *Azalea*. The rhododendrons are cultivated in most of the United States with the exception of the north central states and subtropical Florida. One (*R. maximum*) is the state flower of West Virginia and another (*R. macrophyllum*) the state flower of

Figure 14-6. Pokeweed.

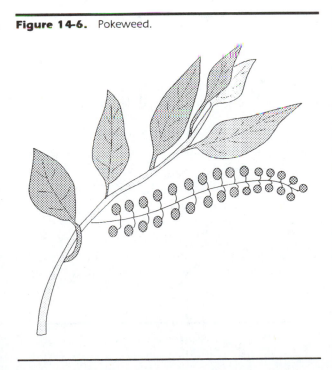

Washington. Azaleas are popular potted plants at grocers and florists during the Easter season. Both azaleas and rhododendrons are widely used as shrubbery for home, school, office, and business landscaping.

The rhododendrons are evergreen, semievergreen, or deciduous shrubs, depending on temperature. The leaves are simple and alternate. The flowers are of various colors, usually white, pink, or red, appear in clusters, and are bell-shaped or funnelform.

Rhododendrons have caused serious intoxications in children who have chewed on the leaves. *Poisoning has also occurred from eating honey made from their nectar.* The poisonous principles are the *diterpinoid grayanotoxins.* There is a transitory burning in the mouth on ingestion. This is followed several hours later by increased salivation, vomiting, diarrhea, and a prickling sensation in the skin. The victim may complain of headache, muscle weakness, and visual disturbances. In serious intoxications, there is a slowing of the heart rate and a severe fall in blood pressure.

The victim may require replacement of fluid losses, respiratory support, atropine for the bradycardia, and ephedrine for hypotension if fluids and body positioning are inadequate to maintain the blood pressure. Electrocardiographic monitoring may be required. The intoxication, regardless of its severity, rarely persists for more than twenty-four hours (see Figure 14-7).

Water Hemlock (Cicuta maculata)

Except for Hawaii, species of **water hemlock** grow in the entire United States, Mexico, and Canada. It is the most toxic native plant in Alaska.

The various species of water hemlock have a similar appearance. They are members of the carrot family. They may grow to a height of six to eight feet. In the underground portion of the mature plant there is a bundle of chambered, short tuberous roots.

Figure 14-7. Rhododendron terminal bud, leaves, and flower.

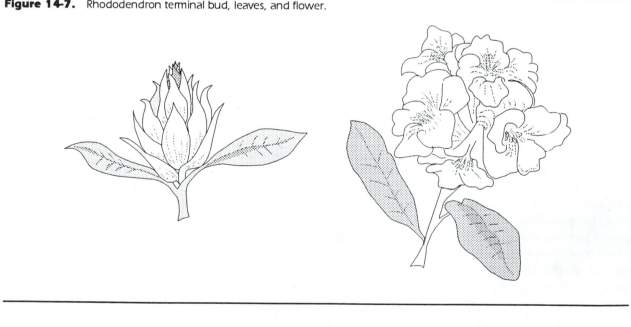

The stem may be mottled or striped with purple. The stems are hollow except at the nodes where the leaves emerge. The yellow oily sap which emerges from a cut stem has a distinct raw parsnip odor. The leaves are compound, up to two feet long. The individual leaflets are narrow, up to four inches long. The leaf edges are toothed. The leaflet veins run to the notches between the teeth tips. The small, whitish, heavily scented flowers grow in umbrellalike clusters (umbels). These plants grow only in wet or swampy areas.

All parts of the plant, particularly the roots, contain the unsaturated aliphatic alcohol, cicutoxin. Between fifteen minutes to one hour after chewing on the plant the victim becomes nauseated, begins salivating copiously, and may vomit. This is followed with epilepticlike convulsions, in which the teeth are clenched, the muscles rigid, and the back arched. During the convulsions the victim is unconscious.

The victim's mouth should be cleared and the respiration assisted if necessary. Because the effects of the toxin may persist for some hours, resulting in additional seizures, a rapidly acting, intravenous anticonvulsant may be required.

Wisteria (Wisteria sinensis)

The **wisteria** is a cultivated vine, hardy in the North, but most common in the southeastern states where it has also become naturalized. It has masses of showy, sweetpea-like flowers in racemes. These are generally bluish purple, but there are pink and white varieties. The leaves are alternate and compound. The seeds are borne similar to thick peapods.

All parts of the plant contain an alkaloid of unknown structure. Statements in the older literature that the flowers are nontoxic are in error.

Ingestion causes nausea, abdominal pain, and repeated vomiting. Diarrhea is slight or absent. Serious intoxications (from chewing the bark) have resulted in sufficient fluid loss from vomiting to result in shock.

Replacement of fluid losses due to vomiting is an essential part of the management. The administration of antiemetic drugs may be useful. The victim is usually symptom-free within twenty-four hours.

Toxic Chemical Laws

Passed in 1976, the Toxic Substance Control Act authorized the U.S. Environmental Protection Agency to test new chemicals and to maintain an inventory of chemicals in commercial use. The law bans the manufacture of compounds that pose an unreasonable risk to human health or the environment, such as polychlorinated biphenyls.

The major toxic chemical laws administered by the U.S. Environmental Protection Agency are outlined in Table 14-1. In addition to these federal laws, each of the fifty states has environmental laws on hazardous substances that assess civil and criminal fines for violation.

Toxins in the Great Lakes

Ninety percent of the fresh surface water in North America and eighteen percent of the world's fresh surface water are in the Great Lakes. That amounts to sixty-five trillion gallons (5,439 cubic miles) of freshwater in the five lakes and their tributaries. The lakes and their connecting channels border eight states and two provinces, cover 94,250 square miles, and comprise one-third of the U.S. boundary with Canada. Their coastlines total 9,674 miles, and their watershed covers more than 201,000 square miles.

Toxic substances in waters of the Great Lakes accumulate in aquatic plants (phytoplankton) and animals (zooplankton). These are eaten by fish and the fish are eaten by herring gulls. At each step in the food chain, the concentrations of toxins increase. One toxin, polychlorinated biphenyls, thus concen-

Table 14-1. Major Toxic Chemical Laws Administered by the U.S. Environmental Protection Agency (EPA) *

Statute	Provisions
Toxic Substances Control Act	Requires that EPA be notified of any new chemical prior to its manufacture and authorizes EPA to regulate production, use, or disposal of a chemical
Federal Insecticide, Fungicide and Rodenticide Act	Authorizes EPA to register all pesticides and specify the terms and conditions of their use, and remove unreasonably hazardous pesticides from the marketplace
Federal Food, Drug, and Cosmetic Act	Authorizes EPA in cooperation with FDA to establish tolerance levels for pesticide residues on food and food products
Resource Conservation and Recovery Act	Authorizes EPA to identify hazardous wastes and regulate their generation, transportation, treatment, storage, and disposal
Comprehensive Environmental Response, Compensation and Liability Act	Requires EPA to designate hazardous substances that can present substantial danger and authorizes the cleanup of sites contaminated with such substances
Clean Air Act	Authorizes EPA to set emission standards to limit the release of hazardous air pollutants
Clean Water Act	Requires EPA to establish a list of toxic water pollutants and set standards
Safe Drinking Water Act	Requires EPA to set drinking water standards to protect public health from hazardous substances
Marine Protection Research and Sanctuaries Act	Regulates ocean dumping of toxic contaminants
Asbestos School Hazard Act	Authorizes EPA to provide loans and grants to schools with financial need for abatement of severe asbestos hazards
Asbestos Hazard Emergency Response Act	Requires EPA to establish a comprehensive regulatory framework for controlling asbestos hazards in schools
Emergency Planning and Community Right-to-Know Act	Requires states to develop programs for responding to hazardous chemical releases and requires industries to report on the presence and release of certain hazardous substances

*Source: U.S. Environmental Protection Agency.

trates almost fifty thousand times from smelt to herring gull eggs (see Figures 14-8 and 14-9). The effect of toxin levels increasing as you go higher up the food chain is called **biomagnification**.

The ill effects of biomagnification extend to humans, especially to infants of women who eat fish from the Great Lakes. Lower birth weight and slower physical and mental development of the children resulted when mothers ate Great Lakes fish.

Because of toxins in waters of the Great Lakes and in the fish, all bordering states of the United States and provinces of Canada, as of 1987, had issued warnings against eating Great Lakes fish. However, the advisories from the states and provinces were not always in harmony. This means that there is a serious scientific health problem but with different interpretations by independent scientists. This is a healthy sign when scientists disagree.

Figure 14-8. Toxic substances in the Great Lakes include Polychlorinated Biphenyls, Dioxin, Mercury, and pesticides. (**Source:** International Joint Commission, 1985.)

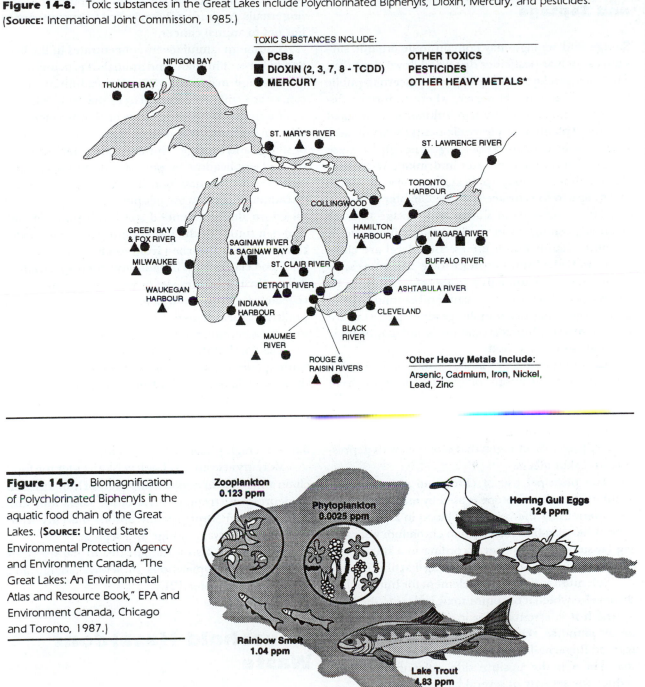

Figure 14-9. Biomagnification of Polychlorinated Biphenyls in the aquatic food chain of the Great Lakes. (**Source:** United States Environmental Protection Agency and Environment Canada, "The Great Lakes: An Environmental Atlas and Resource Book," EPA and Environment Canada, Chicago and Toronto, 1987.)

Toxic Substances in Sludge and Septage

Sludge and septage may contain toxic organic substances such as pesticides discarded down the toilet. These are not known to be of great concern to public health of humans. However, where industrial discharge is mixed with human effluent in municipal treatment plants, many trace elements (heavy metals) may exist in concentrations high enough to cause human illness. High nitrates in drinking water may also result from heavy applications of sewage sludge or septage to the surface soil (see Chapter 3).

Trace elements in sludge and septage that are essential to humans in small amounts but toxic in amounts slightly above the essential level are chromium and selenium. Two more trace elements essential to laboratory animals and therefore likely to be necessary for humans are arsenic and cadmium. Lead and mercury may occur in sludge and septage but are not essential for humans or animals and may be toxic in small amounts to both.

Chromium is a heavy metal existing in the Earth's crust in concentrations of one hundred to three hundred parts per million. In small amounts it is essential in glucose metabolism for human beings and for animals. In 1981, in large amounts chromium was declared a carcinogen that causes growth depression and skin ulcers.

The principal use of chromium is for making stainless steel. One isotope is used by medical doctors for determining the volume of blood in a person. In June, 1990, slag-filled soil high in chromium was causing lawsuits in New Jersey according to a number of articles in *The Wall Street Journal* during that time period.

Selenium is an essential element for humans but the level at which it becomes toxic is close to the level at which it is essential. Toxic effects include tooth decay, jaundice, skin disorders, chronic arthritis, deformed fingernails and toenails, and swellings under the skin. On the positive side, selenium tends to reduce the severity of several kinds of cancer.

Arsenic in land applications of sludge and septage may get into the human body and cause weakness, prostration, aching muscles, and discoloring of the fingernails. Arsenic has been linked to human cancer but not to animal cancer.

Cadmium cumulatively concentrates in the liver and kidneys of humans and animals. Permanent kidney damage may result. Whether cadmium causes cancer was still controversial as of the mid-1980s.

Cadmium is the most likely of all trace elements to get in the human food chain from land application of sludge and septage. However, wind deposition of silt and application of phosphorus fertilizer carry cadmium into human food. There is a special hazard of cadmium toxicity for people who smoke tobacco raised on sludge-amended soil. Absorption of cadmium in smoke by the human lungs is much greater than cadmium absorbed from food.

Food classes high in cadmium are respectively as follow: Irish potatoes, leafy vegetables, other root vegetables, cereals, oily fats (red meats, fish, and poultry). Trace elements that reduce human absorption of cadmium include iron, calcium, and zinc.

Lead is highly toxic to humans and animals, causing anemia, kidney damage, and nervous disorders. Lead damage to children is more severe, resulting in intellectual impairment, hyperactivity, and damage to the central nervous system. Less lead gets into the food chain when the soil has a pH less than 5.5 and a high phosphorus content.

Mercury at even very low levels can enter the food chain and result in nervous disorders of humans such as depression, tremors, irritation, salivation, and diarrhea. Plant roots readily absorb mercury from land-applied sewage sludge and septage.

Nitrates are also a human health hazard from excessive land applications of sewage sludge and septage (see Chapter 12).

Household Hazardous Waste

Household **hazardous waste** refers to substances that would be toxic to humans. Such substances include pesticides, paint thinners, drain openers, and grease and rust solvents. These are usually discarded in solid wastes and end up in the least hazardous place for them — a landfill. The most hazardous final resting place for these hazardous wastes is down the toilet and into the waste treatment plant or the septic tank and the drain field.

The U.S. Environmental Protection Agency financed a comparative study of hazardous waste in New Orleans, Louisiana, and Marin County, California. The study concluded that hazardous waste discarded per household at both locations totaled 5.8 pounds per year. This was about 0.4 percent of all household waste discarded. Batteries, electrical materials, cosmetics, automotive maintenance materials, and pesticides were the principal hazardous wastes at both locations.

Summary

Toxins are substances that produce direct effects on contact or when ingested. Plants such as poison ivy, poison oak, poison sumac, and others contain resins that are toxic on contact with the skin. When a sensitive person touches the plant, some of the resin rubs off onto the skin. The resin can also be spread to other parts of the body and to other persons. The allergen can also be transmitted by getting it on clothing, by pets, or even in the unburned plant parts from a brush fire.

Poison ivy and poison oak produce over half of all cases of allergic dermatitis reported, but not everyone is sensitive to the allergen in poison ivy or poison oak. On the other hand, just because you have never been affected by contact with such plants does not mean you will never become sensitive. The reaction to exposure ranges from a light rash to large, clear blisters. Treatment includes cold, wet dressings and lotions formulated especially for this purpose. Physicians can administer medicine as an injection to help lessen the severity.

The castor bean is toxic if ingested. It interferes with the function of the intestines producing nausea, vomiting, and diarrhea. The only treatment is fluid replacement.

Leaves of the dumbcane and philodendron are poisonous when chewed or swallowed. Symptoms include pain, swelling, and later blisters inside the mouth. They may affect the victim's speech. They can cause irritation to the eye as well as contact dermatitis. Other plants that produce toxic effects from chewing or swallowing the leaves or other plant parts include foxglove, Jimson weed, angel's trumpet, lily of the valley, nightshade, pokeweed, rhododendron, water hemlock, and wisteria.

Unfortunately, all of those plants grow well in most of the United States. In fact, many are used as landscape plantings or as potted plants in homes. One of the authors has dozens of rhododendrons in his yard as ornamental shrubbery and foundation plantings. He also has several varieties of philodendron as decorative plants indoors.

All of our surface water reservoirs contain some toxins. In this country, one of the most important waterborne toxin problems is in the Great Lakes. It is not that other water bodies are necessarily cleaner. The problem is that the Great Lakes are so big and border an area with such a huge population. Many scientists recommend against eating fish caught in the Great Lakes. But, what is true there, is also true in many other lakes and rivers in the United States, too. Fortunately, the ongoing efforts of environmentalists and environmental scientists are beginning to improve this situation.

When we discard toxic substances in our sewage and trash, they eventually find their way into the ecosystem. Batteries of all kinds discarded in the trash release heavy metals into landfills. If the landfill leaks, the heavy metals eventually can enter the groundwater. The amounts of such hazardous wastes discarded by individuals is relatively small. As a result, you will probably do relatively little damage to the environment as an individual. At the same time, we should all do our part to minimize such environmental degradation. Even though household toxic and hazardous waste amounts are small, you should still take into account the potential environmental effect of dumping them into the trash, sewer, or septic system.

DISCUSSION QUESTIONS

1. What causes poison ivy and poison oak reactions on people? How can the affected areas spread well beyond the point of contact with the plant?

2. After a person contacts poison ivy or poison oak, what can be done to minimize the spread of the affected area? What should not be done?

3. What treatments are recommended for poison oak or poison ivy reactions?

4. What are the symptoms of dumbcane or philodendron poisoning?

5. What treatments are recommended for dumbcane or philodendron poisoning?

6. What are the symptoms and treatments for foxglove poisoning?

7. What are the symptoms of Jimson weed poisoning?

8. What are the symptoms of nightshade poisoning?

9. What common poisonous plant is eaten by many people? If an intoxication occurs, what are the symptoms and treatment?

10. What are rhododendrons? How do they cause poisoning? What are the symptoms and treatment?

11. What are the symptoms and treatment for water hemlock poisoning?

12. What toxic substances are present in sewage sludge and septage? What are their most common effects?

ADDITIONAL ACTIVITIES

1. Examine the plants in and around your home to determine if any of them are among those described in this chapter. Odds are very good that you will find at least one if you make a careful search.

2. Contact your local landfill operator. Find out his or her policies regarding dumping of motor oil, automotive batteries, and other hazardous waste.

3. Ask a nurse or doctor to come to your class to discuss hazardous and toxic materials around the home.

CHAPTER 15
Human Population

Terms to Look for and Learn

Arithmetic Increase

Birth Rate

Boomsters (Cornucopians)

Death Rate

Doomsters (Malthusians)

Exponential Increase

Growth Rate

Linear Function

Overpopulation

Population Density

Thomas Malthus

Zero Population Growth

Learning Objectives

After reading this chapter and participating in the activities, you should be able to:

- Explain why the human population is expanding at such a phenomenal rate.

- Describe the historical trends in human population.

- Relate the population "explosion" to ecological concepts described earlier in this book.

- Discuss the implications of that growth rate on the ecosystem from the perspectives of the biologist.

- Discuss the effects of the human population from the alternative perspective of economics.

Overview

Conceivably, one of the most important points that has been made in this book was brought out in the first chapter. There is no single branch of science that can claim to be an all-encompassing environmental science. It may be that ecology can legitimately claim to provide a framework for understanding the other sciences that deal with the environment. Even if that is the case, neither ecology, nor agronomy, nor geology, nor any of the other branches of environmental science can claim to have all of the answers. That truism is certainly valid in the case of human population.

The human population is larger than it has ever been. And it is expanding at such a rapid rate that it is expected to double within the next forty years. There are two very dissimilar views of the future of humanity that will result from a growing population. The perspectives are based on the conflicting theories of the sciences of biology and economics.

Respected biologists look at the dynamics of population growth and carrying capacity and see one version of the future. This is nowhere more graphically demonstrated than in *The Population Explosion*, by Erlich and Erlich. On the other hand, respected economists can look at the same growing population and see a bright future, filled with innovation, creativity, and improved living conditions based on natural and man-made resources yet undiscovered. The epitome of that view is detailed in *The Ultimate Resource*, by J.L. Simon.

Which view is right? Probably both in some ways and in other ways, probably neither. That is the very reason for studying environmental science the way it is presented in this book. There are alternative explanations for virtually any phenomenon. It is only by examining those alternative explanations that the truth can be approached. The remainder of this chapter will examine the human population. First, we will learn about the dynamics of the population explosion. Then, we will examine population from those two differing perspectives: biological and economic.

Dynamics of Population Growth

In the human population, the rate of net increase is generally referred to as the growth rate. Two major factors that determine the growth rate in any population are birth rate and death rate.

Birth rate is defined as the number of live births per one thousand population per year. In 1989, the average worldwide human birth rate was twenty-eight per one thousand (0.028). **Death rate** is the exact opposite. It is defined as the number of deaths per one thousand population per year. In 1989, the human death rate was about ten per one thousand (0.010) (Erlich & Erlich, 1990). **Growth rate** is defined as the percent net increase in the population. It is calculated by subtracting the death rate from the birth rate and converting the result to a percentage, as shown in Figure 15-1.

That would seem to be a relatively small figure, but when applied to a total human population in 1989 of 5.2 billion, 1.8 percent takes on more significance. That represents a total number of live births of 144 million compared to just fifty-one million deaths. Thus, for the year 1989, the human population expanded by a net of ninety-three million, or almost 0.1 billion to an estimated 5.3 billion in 1990.

Arithmetic Increase vs Exponential Increase

A discussion of arithmetic growth rates and exponential growth rates is necessary to understand the changes in the human population. The difference between the two is minor at first, but it becomes quite astounding when long periods of time are considered.

To illustrate the process of **arithmetic increase**, let us assume a linear rate of increase of one percent per year. If the total population of a species starts at

Figure 15-1. The worldwide human population growth rate in 1989.

$$\text{Birth Rate} = 28/1000 = 0.028 = 2.8\%$$
$$\text{Death Rate} = 10/1000 = 0.010 = 1.0\%$$
$$\text{Growth Rate} = (28 - 10) / 1000 = 0.018$$
$$= 1.8\%$$

Figure 15-2. Arithmetic increase compared to exponential increase. By examining the two graphs, you can easily see why populations grow so rapidly over the years.

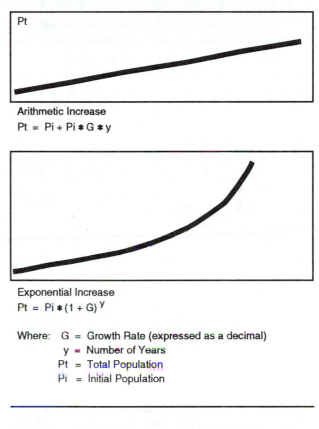

Arithmetic Increase
$$Pt = Pi + Pi * G * y$$

Exponential Increase
$$Pt = Pi * (1 + G)^{y}$$

Where: G = Growth Rate (expressed as a decimal)
         y = Number of Years
        Pt = Total Population
        Pi = Initial Population

one million, then it would grow to 1,010,000 at the end of one year. In other words, the net increase would be ten thousand for that year (1,000,000 x .01 = 10,000). In an arithmetic expansion, the annual increase would remain at ten thousand per year. At the end of year two the total population would be 1,020,000, and at the end of year three it would be 1,030,000, etc. The increase would be linear, as shown in Figure 15-2. The formula for estimating that population would be a linear function of the number of years, as also shown in Figure 15-2.

After a century of increasing at an arithmetic rate, that population would have increased by 100 x .01, or one hundred percent, reaching a total of two million. If the rate of increase were two percent, after one hundred years, the population would be three million. At a rate of increase of three percent, after one hundred years, the population would be four million.

But, that is not what happens! Each year, the annual growth rate is applied to the current population which has increased over the previous year by the growth rate. The population thus describes an upward curving line. It is an **exponential increase** of the number of years (see Figure 15-2).

As we have seen, the total population of a given species and in a given ecosystem increases in one year by its growth rate—i.e., (births – deaths)/1000/year and is expressed as a percentage/year. After a century of increasing at an exponential rate of one percent per year, that population would have increased by 1.01^{100}, or 270 percent, reaching a total of 2,704,814. If the rate of increase were two percent, after one hundred years the population would be 7,244,646. At a rate of increase of three percent, after

one hundred years the population would be 19,218,632. Compare these numbers to the equivalent ones resulting from the arithmetic expansion illustrated above and the difference between the two should be quite glaring.

History of the Human Population

As humans initially evolved, the population was obviously very small. At a time when people gathered plants and berries, hunted small animals, and lived in the open or in caves, the birth rate was probably very low. A female probably gave birth only a few times in her reproductive life. The life span was likely only about forty years and the death rate was relatively high. Fortunately for us, the birth rate slightly exceeded the death rate, and so those numbers very gradually increased.

For early human ancestors, the population growth rate must have been extremely small over the eons. One obvious reason is that early prehumans faced a number of very effective predators. Thus, the effects of exponential growth rates, as explained by the logistic curve, for human population growth, was partially offset by predation. Of course the growth rate was also affected by disease, accidents, and early forms of warfare. In a very real sense, warfare is also a form of predation, although rarely for food gathering. The limiting factor on the human growth rate was almost certainly the lack of a readily available year-round supply of food energy resources from the ecosystem.

Regardless of the causes, for millions of years of human evolution, the growth rate of our ancestors was nearly zero. *Homo sapiens* developed as a species perhaps 300,000 years ago. Eventually came the invention of more lethal weapons and the discovery of more effective methods of collecting and storing foodstuffs. Modern humans established a more reliable, year-round food source. Then they began to increase their birth rate and decrease their death rate. Yet the net growth rate still remained small compared to today's. Our ancestors produced perhaps as many as ten million humans in the world by the time organized agriculture began.

The deliberate cultivation of crops and the domestication of livestock began about 6000 BC. The development of agriculture was the thing that triggered the first real population explosion that humankind had ever known (see Table 15-1).

When it became possible for one person to produce food for more than his or her own family, civilization became possible. The development of other trades followed along with the rise of organized communities and commerce. It is hard for us to imagine the lifestyle of prehistoric hunters and farmers as comfortable. Yet, their lives were much more secure and comfortable than those of the hunter/gatherers of previous generations. This more comfortable life made possible a higher birth rate and a

Table 15-1. World Population over 8,000 Years *

Year	Human Population (in billions)
6000 B.C.	0.01
AD 1	0.2
AD 1500	0.5
AD 1750	0.8
AD 1800	1.0
AD 1900	1.6
AD 1950	2.5
AD 1960	3.0
AD 1970	3.7
AD 1980	4.4
AD 1990	5.3
AD 2000 (**projected**)	6.5
AD 2050 (**projected**)	11.0

*SOURCES: Camp, W.G. & Daugherty, T.B. (1991). **Managing Our Natural Resources**. Albany, NY: Delmar Publishers, Inc. Simon, J.L. (1981). **The Ultimate Resource**. Princeton, NJ: Princeton University Press.

lower death rate. Thus, the population growth rate increased somewhat. A gradual upward curve in the total human population began about eight thousand years ago. The gradual increase continued through about the beginning of the seventeenth century.

The agricultural revolution was followed by an industrial revolution which increased human productivity in both manufacturing and agriculture. That released a still larger proportion of the population to pursue other occupations than subsistence farming. That made possible the growth of huge cities and still more elaborate systems of manufacture and commerce. Again, peoples' lives became more secure and more comfortable. As a result, the human population growth rate increased again.

By the time of the birth of Christ, the world's human population was about 300 million. That was a major increase from the estimated ten million at the beginning of the new stone age (about 6000 BC) As late as AD 1800, there were fewer than one billion people. By 1900, that had grown to 1.6 billion. In 1979 it reached 4.3 billion and by the turn of the century, world population should exceed six billion (see Table 15-1 and Figure 15-3).

The combination of the effect of the exponential growth rate and the increase in the growth rate itself produced a population explosion. That explosion continues to gain momentum.

Population Growth Momentum

In order for a country to achieve a stable population, it is necessary to reach what is commonly referred to as **zero population growth**. That would require the balancing of the death rate and the birth rate. But, the death rate is a function of two factors: the average age of the population and the average life span in that country. If those two things are known, then the birth rate needed for zero population growth can be estimated.

In the United States, the number of children produced by each couple over their entire reproductive lifespan would need to be about 2.1 to bring about a stable population. To achieve the same stable population in India, the average number of children per couple would need to be about 2.4 because of the shorter average life span.

Perhaps you noticed that the worldwide death rate described earlier seemed low. If the average lifespan of humans is less than seventy years, how could only one percent of the population die in 1989? It is because of the population momentum resulting from the upward swing of the population curve. That

Figure 15-3. World population over 8,000 years. (Courtesy William G. Camp & Thomas B. Daugherty, **Managing Our Natural Resources**, © 1991, Delmar Publishers, Inc.)

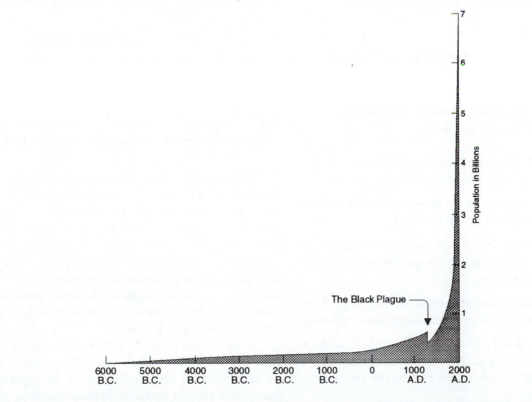

phenomenon means that the average age of the population in the rapidly growing nations is very low.

In fact, in 1989, over forty percent of the people in the developing nations were under fifteen years of age (Erlich & Erlich, 1990). At the same time, in the developed nations with an older population, the life span is becoming longer because of improved living conditions and health care. Thus, in both the developing and the developed nations, the death rates are well below what would be predicted simply by dividing the total human population by the average life span.

As long as the population curve is increasing, the effect is a momentum in population growth. The population momentum works against efforts at promoting zero population growth. The net effect of this growth momentum is that, even if the number of children produced per couple were lowered immediately to that required for replacement, it would take fifty to sixty years to achieve zero population growth. It would take that long for death rates to catch up with birth rates in a population with such a young average age.

Doomsters vs Boomsters

The biological and ecological questions that are raised by the amazing human population growth figures are obvious. They are things like:

- How large can the human population become before the ecosystem can no longer support it?

- What will happen to the ecosystem as the human population goes beyond its limit?

- What limiting factor will produce a significant deterioration of human life quality and force the growth rate downward?

- How bad must the quality of life on a worldwide basis become, before the population explosion will stop?

- Can humankind make a collective, conscious decision to limit its population growth before such a limiting factor forces the end to the population explosion?

Such questions go back to the seventeenth century theories of **Thomas Malthus** (1766–1834), as we will see later in this chapter. Those who forecast a dark future for humanity based on population growth are often connected with Malthusian theories. Such scientists and activists are often referred to as **doomsters** because of the gloomy outlook their warnings foresee.

Yet, not everybody agrees that the long-term effect of human population growth is necessarily bad. In fact, it is possible to make a very powerful argument that a much larger human population would be good. Some experts believe that more people could mean even further improvements in the quality of life on Earth. The questions these scientists ask are quite different from the questions of the Malthusians in the previous paragraph. They ask things like:

- What new ways of food production can be devised to feed the people?

- What new sources of energy can be developed that are nonpolluting?

- How can future resources substitute for present ones in improving the quality of life on the planet?

- What new technologies will we need to invent to solve these problems?

- How can we devise more equitable ways to share the world's wealth so that everyone lives better?

These questions are clearly more optimistic than those of the doomsters. Indeed this point of view forecasts a bright future for a human population much larger than the present one. Because of their inherent optimism, this group of theorists is often referred to as **boomsters** or "cornucopians." They see hope for a booming future for humanity. They forecast a time in which wealth will be more widespread than at present and in which it will be possible to work against poverty and starvation.

In the next section, we will examine the propositions of first the doomsters and then the boomsters. In each section, the arguments of the respective group will be put forward without criticism. As a student-scholar in a democratic society, it will be your responsibility to examine the opposing views and search for truth as you determine it.

"Human Population Growth Is Bad"—The Biological Perspective

Human population growth may very well be the most pressing problem facing the environment, according to most biologists and ecologists. For the student with a clear understanding of the concept of the logistic curve and of exponential population growth rates, the truth of that statement should be obvious. In the absence of predation, essentially any organism could be expected to increase in numbers exponentially, according to the logistic theory. For all practical purposes, humans are one of the very few organisms on Earth without a predator to help control its population. Thus, the carrying capacity of the ecosystem for humans is the only natural control of human numbers.

Even with an improvement in the way humans generate and use energy—even with cleaner and more efficient industries and cleaner cities, the Earth must have a finite carrying capacity for humans. About two centuries ago, Thomas Malthus examined the rate of human population growth. He began a series of mathematical speculations that predicted the doom of humanity because of population growth —Malthusian speculations persist to this day. Malthusian theories predicted that we would soon exceed the upper limit in the number of humans that the planet could support.

Had it not been for phenomenal increases in agricultural productivity over the past century and improvements in processing and transportation, his predictions almost certainly would have been proven right before now. The rate of population growth is higher now than ever before in human history. Indeed, the human growth rate worldwide is still increasing. Fortunately, there are many indications that the rate of increase is beginning to slow somewhat.

Limiting Factors

As explained in Chapter 2, there must always be some limiting factor on the total population size of any species within a given ecosystem. The limiting factor can be any one of many different requirements. In plants, it could be the number of freeze-free days, the amount of zinc in the soil, the frequency and severity of droughts, or many other factors. For most animals, the limiting factor is usually one of three things: food energy, water, and space. Let us examine those three factors as they relate to humans.

Food Resources

The total worldwide capacity for the production of food energy far exceeds the number of calories that would be needed to sustain a human population considerably greater that the six billion expected by AD 2000.

There are several problems that surface in regard to worldwide food production. The 1985 world food production was calculated to have been adequate to feed approximately six billion people on a purely vegetarian diet, or four billion people with some animal products (fifteen percent of calories) in their diets, or about 2.5 billion people with a full diet consisting of thirty-five percent calories from animal products. Over a billion people are overfed today while more than another billion are continuously hungry (Erlich & Erlich, 1990).

Thus, much of the food problem is distribution. The richest one-third of humanity have more food than is needed and the poorest one-third have too little. Not by coincidence, it is in the richest countries that most of the food is produced. It is there that mechanical and scientific agriculture are practiced. It is in the poorest countries that the least food is produced because their agriculture is subsistence-level and animal or human powered (see Figure 15-4).

Another problem is a result of the Second Law of Thermodynamics (see Chapter 1). Each time energy undergoes a transformation, the amount of available energy decreases. The conversion of food energy stored in grain to fish results in approximately a fifty percent loss in available food energy. In other words,

Figure 15-4. Compare the technologies in use by this subsistence farmer in Africa and this commerical farmer in Indiana. (Photos courtesy Roy Donahue [Africa] and A.R. Stevens [Indiana].)

it takes at least two pounds of processed and concentrated fish food to produce a pound of live fish.

To produce a pound of live hog requires at least four pounds of feed. This is true, even in the most efficient farming operations because the animals must always expend energy to metabolize their food. **Homeothermic** animals must use food energy to maintain their body temperatures. All animals must use some energy to produce body parts that cannot be used for human food. And energy is consumed by animals in simply moving from one place to another.

A third problem in terms of food production is that as the human population continues to grow, the capacity of the food producing regions of the Earth is being lowered by erosion and plant nutrient depletion. Soil degradation is a major world problem. It threatens the ability of the planet to maintain even the current level of food energy production in the long run.

Water Resources

Many geologists believe that water will be the first limiting factor in food production for humans. That may well be because we are currently relying for much of the production of our food on groundwater that cannot be replaced in our lifetimes. As an example, in the central and southern Great Plains of the United States, the great Ogallala aquifer supplies much of the irrigation water for some of the most productive farming in the world. The Ogallala aquifer was formed as a result of the last Great Ice Age between ten thousand and twenty thousand years ago. Only about 1/2 inch of water is added from surface infiltration each year. Yet, we are pulling groundwater from that aquifer at the rate of four to six feet per year, primarily for irrigation. Even if the pumping of water from wells were stopped immediately (which cannot happen) it would take centuries for nature to replace the lost water into the Ogallala aquifer.

As another example, the aquifer underlying the Everglades region of southern Florida has been used so heavily in the past fifty years that the water table is dropping dangerously. As the fresh groundwater is removed, it is being replaced by seepage of salt water from the ocean. Thus, the fresh aquifer is becoming too saline to support plants. Even thousands of years could not correct that problem. The falling water tables in the Southwestern United States is yet a third example of this growing problem.

At the same time that the aquifers are being lowered dangerously in Southern California and other places, the Mississippi River pours massive amounts of fresh water back into the Gulf of Mexico every day. The rain forests in Northwestern Washington state produce huge amounts of runoff every year. Irrigation practices, industrial applications, and urban uses of water are terribly wasteful. In fact, actual water consumption could be drastically curtailed without seriously affecting our economy or our lifestyle. Thus, the total quantity of fresh water produced each year from precipitation is not the problem—distribution and misuse by greedy people appear to be the problems.

There is a series of political questions that must be answered in order to solve the impending severe water shortages. Should it be the cities like Los Angeles and Miami or the agriculture in the surrounding areas that get the water as it becomes more scarce? Will agricultural uses of water continue to be subsidized by tax monies to keep food prices artificially low? Or will the real costs of irrigation water be passed to the farmer, forcing food prices substantially upward?

Even though agriculture must and will become more efficient in the use of water, food production in many parts of the world will soon become limited because of water shortages. Thus, it appears that the real limitation imposed by water shortages will be expressed as higher food production costs to the farmer and higher food costs to the consumer.

Space (Density)

The third potential problem is that of space. With some animals such as birds of prey, territory require-

ments mean serious limitations within their habitat. The California condor appears to be destined for extinction largely because it cannot adapt its requirements for space. That does not appear to be a problem in humans who have been successful in adapting to crowded living conditions. Many people even prefer to live in high-rise apartments.

Population density is generally measured in terms of persons per square mile (or square kilometer). Worldwide, population densities range from about fifty-five per square mile for the whole continent of Africa to an estimated 14,218 per square mile in the city of Hong Kong (see Table 15-2).

Thus, it is clear that density alone is not the limiting factor. What must be remembered is that for every extremely dense area, such as Hong Kong, there must be a much larger area with low enough density to allow for excess food production to export to the people centers. Again, the real limitation imposed by density is that of food.

Overpopulation

When we consider the great range in population densities, two questions arise. What is overpopula-

---

Table 15-2. Human Population Densities*

Area	Population (per square mile)
Africa	55
Habitable Africa	117
Europe (excluding the former USSR)	261
Japan	857
The Netherlands	1,031
Taiwan	1,604
Hong Kong	14,218

*SOURCE: Erlich, P.R., & Erlich, A.H. (1990). **The population explosion**. New York: Simon and Schuster.

tion? At what human population level will the Earth be overpopulated?

An area suffers from **overpopulation** "when its population can't be maintained without rapidly depleting nonrenewable resources (or converting renewable resources into nonrenewable ones) and without degrading the capacity of the environment to support the population. In short, if the long-term carrying capacity of the area is clearly being degraded by its current human occupants, that area is overpopulated." (Erlich & Erlich, 1990).

The pollution of coastal marine waters, the acidification of surface waters, the lowering water tables, the erosion of topsoils, and the loss of genetic diversity all mean that the global ecosystem is being affected by human activity. If one assumes that those effects are negative, then the global ecosystem is being degraded. Therefore, by the Erlichs' definition, clearly most parts of the world, and indeed the world as a whole are "already vastly overpopulated."

Multiple Impacts

According to Erlich and Erlich, there are basically three major sources of the impact (I) of humans on the environment:

- The number of people who are living within the system. This is represented by the population (P).

- The rate at which those people consume the renewable and nonrenewable resources. This is a difficult construct to measure directly, so the indirect measure of consumption is assumed to be wealth. The wealthy naturally consume more than the poor. This is represented by measures of affluence (A).

- The disruptive effects of the technologies used by those people to produce the goods they consume. This can be estimated by examining the environmental disruptiveness of the manufacturing, processing, transporting, and marketing technologies. It is represented by a technological index (T).

In this perspective, the total impact of humans is seen as a simple **function** of the three sources of impact:

$$I = P \times A \times T$$

Even though this may be an oversimplification, the model does provide a way to examine the relationships among people, their activities, and the environment. As population increases, assuming constant levels of consumption and technology, the impact on the environment increases. As we have seen, the human population is expected to double over the next forty years. Thus, the impact of human activities on the ecosystem should also double during that time.

The dynamics of population growth make it very unlikely that the total number of people on Earth will level off any time soon. Because of population momentum, even if the worldwide population growth rate were to fall today to the rate required for replacement, the total number of humans on Earth would continue to increase for a long time. The realities of the world's politics and economics make a decrease of consumption in the rich parts of the world very unlikely. The rich like being rich and the poor aspire to be rich. Thus, the only source of environmental impact that can be changed immediately is the impact of technology. Sadly, even if the impact of technology on the environment were halved over the next forty years, by that time, the growth in human numbers would offset the technological improvement.

Thus, from an ecological perspective, there are only two alternatives. Either human population must be brought under control as soon as possible or we can expect a massive increase in the death rate in the future. That resulting "die-offs" will result from disease and starvation is inevitable. The impact of human activity on the environment must degrade the carrying capacity of the Earth for humans just as it did on Isle Royale for the moose (see Chapter 2).

Population Control (From this Perspective)

According to this view, the only hope for the future of humanity is to bring this population explosion under control and to do so soon. That will most certainly mean ending religious prohibitions against birth control. It definitely means governmental encouragement and perhaps mandates for birth control in those countries with growing populations. It probably means mandatory sterilization for large numbers of persons who cannot afford to raise children.

"Human Population Growth Is Good"—The Economic Development Perspective

Exactly the opposite view of the effects of population growth are held by some economists. The most notable among those holding this view during recent years has been Julian Simon, whose 1981 book, *The Ultimate Resource*, made a very strong argument that population growth is good. The remainder of this section will explain the position and arguments of this group of scientists, who have been referred to as "boomsters" in contrast to the doomsters, led by the Erlichs, and whose views have been covered in the preceding section.

Are Resources Really "Limited?"

Food

Over the past thousand years, the per capita food situation, on a worldwide basis, has been improving steadily. There is a higher production of food today than there has ever been. The limit of food production is not in sight, even with current technology. Each year, in the United States, land is set aside from food production to prevent huge surpluses that would glut the market and push food prices downward. The most important government programs in this country are those designed to suppress food production. Agriculturally productive land is not in short supply. Much more potential cropland is available than can be profitably used now. At the same time, food prices are at the lowest point in human history, and they are still declining. For example, in 1970, thirteen percent of disposable income in the United States was spent on food. In 1985, that figure was ten percent.

Agricultural land is not really a fixed resource in any practical sense of the word, either. Given current farming practices, much more usable land could be available. But if food were to become scarce, prices would rise and that would make more intensive production practices economically attractive. The inevitable result would be substantial additions to the food production land base from land that would currently be considered marginal. Beyond that, research and improvements in hydroponics (water-based farming) and aquaculture (aquatic animal and plant farming) may have made the old concept of a limited food supply outmoded (see Figure 15-5).

Natural Resources (Minerals and Fossil Fuels)

The predictions of shortages in world mineral production have been based on technological forecasting techniques. Technological forecasts are based on a procedure that has a number of fundamental flaws. The procedure involves:

- Estimating currently known sources of the mineral, such as copper, lead, zinc, etc.

- Assessing current means of extraction for those known deposits.

- Estimating the amounts of the mineral that can be profitably removed given current (or perhaps future) prices.

- Estimating current worldwide use of the mineral at current prices.

- Dividing the removable supply by the annual use to produce an estimate of the number of years of supply left.

Figure 15-5. Research and improved practices in aquaculture and hydroponics could make the "doomsters" theories about limits on food production outmoded. Figure 15-5A shows a bench of lettuce being grown hydroponically at Pulaski County High School, in Virginia. (Photo courtesy Bill Camp.) Figure 15-5B shows four different systems for aquacultural production: pond, cage, and two different kinds of tanks. (Photo courtesy Dr. George Libey, Virginia Polytechnic Institute and State University, Blacksburg, VA.)

A

B

That is the procedure on which much of the argument of the doomsters is made. The problems with technological forecasting are multiple (Simon, 1981):

■ It is clearly impossible to estimate the availability of the mineral. The first problem is agreeing on a definition of the mineral. For instance, take the case of copper. If the copper salts dissolved in sea water were counted, the estimates of available supplies would be increased many fold. There are vast differences between the amount of known and proven quantities available for the minerals and fossil fuels used by humans today and the best estimates of how much is actually present in the Earth's crust. That results in the paradox that, as we use more of our resources, still more is available. As an example, between 1950 and 1970, as the world used great quantities of tin, zinc, lead, and copper, the known reserves of those minerals increased, in spite of the fact that each year the world demand for the minerals had increased (see Table 15-3).

Table 15-3. Changes in the Known World Reserves (between 1950 and 1970) of Selected Minerals, in Terms of Years of Estimated Use at Current Usage Rates*

Tungsten	-30%
Tin	+10%
Manganese	+27%
Zinc	+61%
Lead	+115%
Copper	+179%
Bauxite (aluminum ore)	+279%
Oil	+507%
Chromite	+675%
Iron	+1,221%
Potash	+2,360%
Phosphates	+4,430%

*Source: Simon, J.L. (1981). **The ultimate resource**. Princeton, NJ: Princeton University Press.

- Estimates of supply ignore recycling. As long as it is less expensive to extract a mineral from the Earth's crust, there will be little recycling. As soon as the easily extractable ores are depleted, recycling will become more economical as a means of production and will supplement or replace mining.

- Minerals and fossil fuels have a very high elasticity of supply. That is an economics concept that says, in effect, as prices rise, supplies that will be made available will rise even faster. Conversely, as prices fall, available supplies fall even faster. An example of the effect of this principle was the worldwide production of crude oil during the 1990–91 Persian Gulf War. As Iraqi oil was removed from the world oil supply, prices shot up temporarily. As a result, the other oil-producing states quickly increased oil production. As a result, the world prices of crude oil fell to below the prewar levels even before Iraqi oil again became available.

- Estimates of current reserves are highly speculative. Estimates of the amount of oil in the ground are inexact at best. Estimates of undiscovered reserves are pure guesswork.

- In actuality, we do not know the extent of our natural resources. In spite of the large numbers of estimates being made and published, nobody has ever made a complete inventory of the natural resources of the planet. Nobody has ever needed to do so because there has never been a shortage of supply of natural resources.

- Estimates of future supply rely on current technologies of extraction and recovery. Nobody can estimate the effects of yet-unimagined technologies on the recoverable supply of future natural resources.

- Finally, estimates of available supplies ignore substitutability of one resource for another. This is such a critical concept that it deserves more explanation.

Substitutability

Simon explained that the value of a natural resource is not in the resource itself; rather, a resource is valuable only when it provides a service. The concept of utility was discussed in Chapter 3, and what Simon called "service" is the same as what this book calls "utility."

That means that we do not value oil, as such. We value the utility that we get from using oil. We use oil to produce lubricants, plastics, and other products. But, our primary interest in oil is as a source of energy. Yet, there are many sources of energy that can be substituted for oil.

In spite of accidents at Three Mile Island in 1979 and Chernobyl in 1986, nuclear generation of power is critical in providing energy for the world. For instance, in 1991, France got about seventy-five percent of its electricity from its fifty-five nuclear power-generating plants. That is not the case in the United States, where two decades of attempts to build nuclear plants have resulted in few successful completions and the loss of billions of dollars invested in plants that never opened. The failure in this country to achieve heavier reliance on nuclear energy has resulted largely from environmental concerns and antinuclear activists who oppose nuclear energy on principle.

Regardless of the problems with nuclear power, the fact remains that it is technologically feasible to substitute nuclear-generated energy for oil or coal-generated electricity. Other sources of energy as substitutes for fossil fuels include solar power, geothermal, biomass, wave, and many others that we already know about. Future sources of energy are impossible to predict. There is only one thing of which we can be certain: when the cost of producing energy from alternate sources becomes less than that from using fossil fuels, the alternate sources will be used. If the costs of using the fossil fuels included the costs of "scrubbing" the emissions to make them environmentally neutral, then alternate sources will become economically feasible sooner.

Energy is just one example. An article in a recent edition of the *New York Times Magazine* reported that in the mid-1800s, a British economist looked at this country's growing dependence on coal and forecast that "the conclusion is inevitable… that our present progressive condition is a thing of limited duration." (Tierney, 1990). The article concluded that the perceived coal shortage and a growing shortage of whale oil for home lamps in the same time frame led to the development of oil wells. Crude oil produced a quite satisfactory substitute for the earlier forms of energy. Tierney further reminded us that in 1905 Theodore Roosevelt warned his countrymen of an impending "timber famine," and that in 1926 the Federal Oil Commission Board of the United States warned that the nation had only a seven-year supply of petroleum left.

The Paradoxes

More people does not mean more hunger. Hunger results from a maldistribution of food, not a shortage of it. As the population density increases, the transportation infrastructure improves (roads, shipping channels, railways, etc.) So, in reality, more people means less famine. We hear about the terrible famines in some parts of the world. But famines are not new and they are not worse than they have been before. The news media simply allow us to be more aware of them than our ancestors were. Beyond that, the famines in Ethiopia, Sudan, and Somalia during the 1980s and 1990s resulted from military and political decisions, not from actual lack of food. In all three cases, the opposing sides in the conflicts used food as a political tool and used starvation as a means of influencing events. Food lines in Eastern Europe and the nations of the former Soviet Union resulted from inefficient distribution and monetary systems, not from lack of production.

Using more resources does not mean less resources remaining. It means a more thorough search for existing resources which are effectively limited only by economics. It means more innovation to develop substitutes. It means more reliance on renewable energy sources like solar power and biomass conversion. And it means more recycling of previously used resources.

Population Control (From this Perspective)

Population growth must slow, but mandatory birth control measures are not necessary. Rapid birth rates exist only in the developing countries. In those settings, the infant death rate requires parents to overproduce children to ensure they will have offspring who will reach adulthood. That is necessary in those countries for two reasons: children supply a cheap source of labor at an early age and adult children are necessary to care for aging parents who are too old to work. It also results from ignorance about effective birth control measures, religious restrictions, and an inability to afford contraceptives.

In the developed countries, population growth is at or near zero. The actual number of births in the United

States, for instance, is below that required for replacement, in the long term. The increase in the U.S. population at this point is a relic of the age of the population explosion and of immigration. The average completed family in this country would need to be about 2.1 children to maintain a stable population. It stood below that at about 1.9 in 1990 (Erlich & Erlich, 1990).

The solution is to assist the developing countries to improve their standards of living and quality of life so that families there will voluntarily lower their reproduction rates. Any solution must include providing education on birth control and contraceptives to assist in bringing the growth rate under control. Crash programs are unnecessary because there is no population crisis in the foreseeable future.

A Case Study

The two academics whose work formed the basis for much of this chapter are Paul Erlich, Professor of Ecology at Stanford University and Julian Simon, Professor of Economics at the University of Maryland. In his book, Dr. Simon issued a challenge to anyone (especially Erlich) to bet on the future of their theories. His challenge was for anyone to select a set of minerals, grains, fossil fuels, or other commodities. The person would agree to pay to Simon the present price for a thousand dollars-worth of the commodity at some future date at which time Simon would then purchase back the commodity at that future date's actual price. In effect, Simon was betting that the price of the commodity would fall. His reasoning was that if the resource were actually becoming more scarce, as the doomsters believed, its price should rise. Simon's theory was that resources were actually becoming less scarce and that, therefore, their prices should fall.

The December 2, 1990 issue of the *New York Times Magazine* reported that the bet was made in 1980— Erlich betting that the prices of a set of minerals would go up over the next 10 years and Simon betting they would fall. Dr. Erlich selected the minerals and set the duration of the bet. Tierney wrote:

The bet was settled this fall without ceremony. Erlich did not even bother to write a letter. He simply mailed Simon a sheet of calculations about metal prices—along with a check for $576.07. Simon wrote back a thank you note, adding that he would be willing to raise the wager to as much as $20,000, pinned to any other resources and any other year in the future... The prices fell for the same cornucopian reasons they had fallen in previous decades—entrepreneurship and continuing technological improvements.

Summary

There is no doubt that the world human population is larger today than it has ever been. There is also no doubt that it is growing at a greater rate than ever before. The effects of exponential expansion are such that a population increases at an increasing rate until some limiting factor forces the population growth rate downward.

Population growth is a function of the difference between live births and deaths in a given population. The growth rate (expressed as a percentage) multiplied by the population equals the net population growth for a year. Each year, the growth rate is multiplied by the current population. As a result, with a constant growth rate, the actual population increases at an increasing rate. That is what is meant by the exponential growth.

The history of the human population was one of privation and relative weakness for millions of years. Prehumans were slow and weak. They were preyed upon by other prehumans and carnivorous animals. They were subject to food shortages brought on by droughts, storms, and climate changes. As a result, the total number of humans remained small until the development of agriculture. The industrial revolution provided a further impetus to the human population growth rate.

The rapidly growing population unquestionably has major implications for the environment. Yet, there are two very different schools of thought about what those implications mean for the future of humanity.

"Malthusians" see a growing population and apply biological principles to it. Based on the direct extension of the concept of the logistic curve, that means problems for the future of humanity. They see a time when the carrying capacity of the planet for humans will be exceeded. Some believe that that time has already come. When that happens, they believe, the ecosystem will deteriorate with the result that massive die-offs of people will be the only possible result. Some even argue that the ecosystem will be so degraded by mankind that it will become incapable of supporting human life and that the end of *Homo sapiens* will be the inevitable result.

Based on the observations of biologists and ecologists over the past century, that conclusion is perfectly logical. The solution, they argue, is immediate and mandatory birth control to bring the human population under control — if it is not too late already. It is little wonder that another nickname for people who believe those theories is "doomsters."

From the perspective of the biologist, the ultimate resource is food, and exponential population growth is the obvious problem. The carrying capacity of the planet (in terms of food resources) must, by definition, someday be exceeded. You may remember the moose on Isle Royale (see Chapter 2) eventually exceeded the carrying capacity of their ecosystem. They experienced large-scale starvations to bring the population well below the maximum it had reached.

There is a very different school of thought which is advanced by a group who are sometimes called "cornucopians." They believe that people do not operate on the same principles as fruit flies and moose. They argue that people can identify and solve problems that other forms of life cannot address. Using this logic, they believe that the term "carrying capacity" is not directly applicable to people. People need things so they invent them. They look at a finite planet and see recycling and substitution and renewable sources of energy that make natural resources essentially unlimited. It is not surprising that another name for this group is "boomsters."

From the perspective of some economists, the ultimate resource is people. They argue that economic development requires expanding markets and economic growth. They point out that economic growth makes possible further mechanization and scientific development. Yet, that argument ignores the obvious truism that the carrying capacity of the Earth must have a limit. It also ignores that massive damage to the Earth's topsoil, air, surface and marine water, and biological diversity that has already been done. It ignores the fact that we cannot continue to degrade the ecosystem as we have in recent history without seriously diminishing the carrying capacity of the Earth for human population.

Which group is right? No one knows. Both sides believe that they are right and that the other group is misinterpreting reality. It is probable that both groups are right in some ways and wrong in other ways. As was pointed out in Chapter 1, there is more than one way of looking at the environment. In this case, *some* ecologists and *some* economists certainly differ on the interpretation of the same facts.

DISCUSSION QUESTIONS

1. What are the differences between arithmetic and exponential growth?

2. Why did the human population remain so small for so long, only to surge in the past few hundred years?

3. What are the views of the doomsters in terms of human population?

4. What are the views of the boomsters in terms of human population?

ADDITIONAL ACTIVITIES

1. With an electronic calculator, use the formulas for arithmetic and exponential expansions to compare the total value of one million dollars over fifty years at eight percent interest with the two different procedures.

2. Examine the U.S. Census for the past one hundred years for your state and county. Make a graph to show the population changes over that period of time.

3. Organize a class debate over the future of humanity that is likely to result from continued population growth.

4. Review and report to the class on an article about either human population growth or economic development.

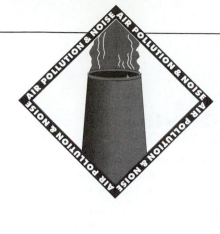

CHAPTER 16
Air Pollution and Noise

Courtesy Bill Camp.

Terms to Look for and Learn

Acid Rain
Carbon Dioxide
Carbon Monoxide
Chlorofluorocarbons
Emergency Episode
Emissions
Lead

Methane
Nitrogen Oxide
Noise
Photosynthesis
Respiration
Sulfur Dioxide
Suspended Particulates

Learning Objectives

After reading this chapter and participating in the activities, you should be able to:

■ Describe the makeup of the planet's atmosphere.

■ Outline anthropogenic atmospheric emissions and their sources.

■ Discuss some of the effects of air pollution.

■ Discuss trends in anthropogenic air pollution.

■ Discuss air quality emergency episode actions.

■ Discuss noise as a form of air pollution.

Overview

"One air—one world," may be the greatest truth ever spoken in the 1900s. The solid Earth's surface is only twenty–nine percent solid land and seventy–one percent water. Above this spinning and rotating ball nearly eight thousand miles in diameter are swirling clouds that extend five miles high above the two poles, eight miles high in temperate climates and ten to twelve miles above the Earth in the tropics. The outer reaches of the atmosphere extend as high as twenty–two thousand miles above the planet's surface.

Whereas, land can be bought and sold (in fee simple) and oceans claimed many miles from land, the air—the atmosphere—belongs equally to all of us. No one has the right to own the air even above property owned in fee simple (undisputed right).

The term "pristine" (Latin *pristinus* [former]) was used at one time to mean original, pure, or untouched by humans. Nowhere can we find pristine air because air everywhere has been influenced *anthropogenically* (by people). In blunt language, all air everywhere is polluted—it is only a question of what kind of pollution and how much.

All air contains slightly greater than seventy–eight percent nitrogen, slightly less than twenty–one percent oxygen, less than one percent argon, and traces of other gases. These minor gases include neon, helium, krypton, and water vapor, as well as carbon dioxide, ozone, sulfur dioxide, nitrogen oxides, carbon monoxide, methane, and some odors.

Solid suspended particulates (particles) always in the air include: salt, silt, carbon ash, pollen grains, bacteria, fungal spores, and viruses. Of special interest to environmental science are the emissions and concentrations of these gases that will be discussed in this order: carbon dioxide, carbon monoxide, nitrogen oxides, sulfur dioxide, chlorofluorocarbons, methane, lead, and suspended particulates.

Carbon Dioxide

Carbon dioxide is a colorless, odorless, nonpoisonous, noncombustible, naturally-occurring gas with a faint acid taste. It can be used to put out fires. Carbon dioxide is usually marketed under pressure as a liquid in steel drums. In this form it is injected into soft drinks to improve their taste and to produce the effervescent "fizz." Another familiar commercial product is dry ice, which is frozen carbon dioxide (see Table 16-1). In the

Table 16-1. Global Annual Emissions of Carbon Dioxide from all Sources Compared with Average Annual Concentration of Carbon Dioxide in the Atmosphere in Hawaii *

Year	Emissions (million tons of carbon)	Average Annual Concentration of Carbon Dioxide in the Ambient Atmosphere ** (parts per million by volume)
1960	2,823	317
1970	4,500	326
1980	5,811	338
1987	6,215	349
% Increase 1960-1987	+120%	+10%

* **Source:** U.S. Environmental Protection Agency.
** **Note:** Data from Mauna Loa, Hawaii.
Note: The only other stations for which long-term carbon dioxide data are available are the South Pole, American Samoa, and Point Barrow, Alaska. Data from all stations are similar to those from Hawaii.

atmosphere dry ice changes to a gas without first passing through a liquid form. Gloves must be used in handling dry ice to avoid freezing the hand. Humans cannot breathe air containing more than ten percent carbon dioxide without losing consciousness.

The Carbon Cycle

The process of **photosynthesis** was described in detail in Chapter 1. The first simplified step in this process is:

$$12 H_2O + 6 CO_2 + energy \xrightarrow{chlorophyll} C_6H_{12}O_6 + 6 H_2O + 6 O_2$$

Water + Carbon + Sun-light → Glucose + Water + Oxygen Gas
Dioxide

But, the reverse is also occurring. Plants use some of this glucose as energy to "run the machinery" of their cells. This is known as **respiration**—the reverse process of photosynthesis. Plant and animal respiration can be shown by simplified equation this way:

$$C_6H_{12}O_6 + 6 O_2 \dashrightarrow 6 CO_2 + 12 H_2O + Energy$$

Glucose Oxygen → Carbon Water for
Dioxide Biological Work

Thus, plants use carbon dioxide from the air to manufacture glucose which supplies energy for plant, human, and animal life. Some carbon (C) in carbon dioxide is *stored* in woody plants, mostly as cellulose, $C_6H_{10}O_5$. When the wood is decomposed or burned,

carbon dioxide is released (emitted) into the atmosphere for use again in photosynthesis by other plants (see Table 16-2).

Atmospheric CO_2 Levels

As long as the rates of respiration and photosynthesis are balanced, the amount of CO_2 entering the atmosphere from the first would be equal to the amount being removed by the second, but that is not happening. In fact, humans are using chemical energy stored in ages past. Every time we burn a gallon of gasoline or a lump of coal, we release carbon that was removed from the atmosphere millions of years ago. It should

Table 16-2. Estimated Relative Net Emissions of Carbon Dioxide in the World from Anthropogenic (human-caused) Sources *

Source of Carbon Dioxide	Relative Net Input** of Carbon Dioxide to the Ambient Air Environment ***
Fossil fuel burning	100 **
New tropical forest clearing	10
Soil carbon loss in cultivation	3
Desertification (only partly anthropogenic)	1
Soil carbon loss from burning	0.4
Urbanization of farmland	0.2
Tropical shifting cultivation	0

*SOURCE: U.S. Environmental Protection Agency.
**The worldwide emission of CO_2 from fossil fuel burning is set arbitrarily to 100 here. To interpret the numbers, new tropical forest burning produces ten percent as much CO_2, desertification, one percent, etc.
***NOTE: Contribution to the environment after allowance for uptake of carbon dioxide during photosynthesis.

be easy to see why carbon dioxide is being emitted (released) into the atmosphere much faster than it is being removed. People are causing this release by:

- Fossil fuel burning (carbon compounds in gasoline and coal).

- Tree cutting and burning faster than regrowth, as in the tropics and elsewhere in urban developments.

- Soil organic matter decomposition faster than it is being replaced by manures and crop residues.

- Desertification by the drying and decomposing of organic residues faster than plants replace them by new growth.

- Soil carbon loss by burning crop residues.

- Urbanization by the making of "desert-island" and "heat island" cities out of farmland, woodland, and grassland.

But even with the increased emissions of CO_2, atmospheric levels of the gas are not rising as fast as should be expected. Why is the carbon dioxide level in the atmosphere not climbing any faster than it is? There are three basic reasons: (1) net increase in biomass, (2) carbon deposits in landfills, and (3) CO_2 dissolving in surface water, especially oceans.

Higher CO_2 levels in the atmosphere promote plant growth. After all, CO_2 is the most basic of all "fertilizers" because it is used directly in food production. As the atmosphere becomes richer in CO_2, photosynthesis increases. The result is that worldwide biomass levels are increasing somewhat.

Part of the CO_2 is being buried in landfills. Remember that respiration and photosynthesis should balance each other in nature. But human consumption is not a part of the natural system. When a ton of paper or other wood products is buried in a landfill, the carbon in it is not all returned to the atmosphere for many years.

Finally, as atmospheric CO_2 levels rise, more is dissolved in the surface waters of the planet. Most of that is dissolved in the oceans where it can be held for

millennia. Most of that is locked up in the form of calcium carbonate ($CaCO_3$). Part is used by aquatic plants in photosynthesis.

Still, there is no doubt that the worldwide concentration of atmospheric carbon dioxide is increasing. There is doubt as to what it means. Does it help to cause global warming? Does it cause green plants to grow faster? Is there a human and animal health hazard? Based on present trends, the authors conclude that:

- Global warming, based on long-time temperature recordings, so far has not been clearly shown to be an actual problem. For a much more detailed discussion of this controversial issue, see Chapter 20.

- Crop plants will grow faster in concentrations up to 3-1/2 times present levels of carbon dioxide in the ambient atmosphere. Trees and shrubs grow better with up to 10 times present levels of carbon dioxide. Higher levels of carbon dioxide also help plants to tolerate some otherwise toxic effects of pollutants.

- Doubling present levels of carbon dioxide may have some adverse effects on humans and some animals. Apparently it reduces calcium utilization.

Many people today are exposed to twice the normal concentration of carbon dioxide in big cities, in subways, and in submarines. They seem to experience little adverse effect from the exposure.

Carbon Monoxide

Carbon monoxide is highly poisonous to humans and animals. It is a colorless, odorless, and tasteless gas, which burns in air. When inhaled by humans and animals, it reduces the ability of blood to carry oxygen. It is a lethal gas; death may follow in a few minutes after breathing air containing two percent to four percent carbon monoxide. Cigarette smokers are especially at risk by a factor of four over nonsmokers.

Emissions of carbon monoxide in the United States have *decreased* twenty percent from 1975 to

1985. During the same period, atmospheric concentrations have *decreased* by forty–two percent. Such decreases of so poison a gas is an indication of an improvement in the human environment in the United States (see Table 16-3).

Why is the carbon monoxide concentration in the ambient atmosphere falling more than twice as fast as emissions into the atmosphere? Something must be working right in our environment!

What is helping our environment to reduce the percentage of carbon monoxide in ambient air are several groups of soil bacteria that oxidize harmful carbon monoxide (CO) to harmless carbon dioxide (CO_2). These bacteria are naturally more plentiful in productive soil rich in organic matter. When soil is sterilized with heat, steam, antibiotics or a ten percent salt solution, all of these "good" bacteria are killed.

There are three general groups of bacteria capable of oxidizing CO to CO_2:

■ Two species of anaerobic (no oxygen) methane-producing (CH_4) bacteria.

■ Two species of bacteria that oxidize CO in pure cultures to CO_2.

■ One species that oxidize CO to CO_2 in the presence of a sulfur radical (SO_3).

Note: Coauthor and Soil Scientist Roy Donahue has collected extensive laboratory data on CO to CO_2 oxidation. His research indicated that these bacteria in very productive soil are capable of oxidizing more than two hundred tons of carbon monoxide per square mile per year. This is more than that which is emitted into the atmosphere from all sources in the United States.

The concentration of carbon monoxide in ambient air in 10 selected cities in the United States is given in Table 16-4. Los Angeles with twenty–three parts per million of carbon monoxide is more than double the concentration in other cities. Five cities have a concentration of ten parts per million, three have eight parts per million and one has a carbon monoxide concentration of seven parts per million. Please note that seven parts per million is the average concentration in 1985 listed in Table 16-3 for the entire United States.

Table 16-3. National Annual Carbon Monoxide Emissions and Their Concentrations in the Ambient Atmosphere *

Year	Carbon Monoxide Emissions per Year (million tons)	Carbon Monoxide Concentrations (parts per million)
1975	89	12
1980	85	9
1985	71	7
% Decrease, 1975–1985	-20%	-42%

* **Source:** U.S. Environmental Protection Agency.

Table 16-4. Carbon Monoxide Concentrations in Ambient Air in Ten Selected Cities *

City	Carbon Monoxide (in parts per million)
Los Angeles, California	23
Anaheim, California	10
Baltimore, Maryland	10
Hartford, Connecticut	10
Reno, Nevada	10
San Diego, California	10
Houston, Texas	8
Jersey City, New Jersey	8
Philadelphia, Pennsylvania	8
Chicago, Illinois	7

* **Source:** U.S. Environmental Protection Agency, 1988.

Nitrogen Oxides

Nitrogen oxide concentrations in ambient air in the United States have decreased by fourteen percent from 1975 to 1985. This is true even though the total emissions have increased by ten percent during the same period (see Table 16-5).

Sources of nitrogen oxide emissions are approximately in this relative order:

Motor vehicle exhausts	51%
Burning soft coal	46%
Nitric acid plants	1%
Solid waste disposal	1%
Miscellaneous	1%

Nitrogen oxides combine with water to form nitric acid. The result of this process in the atmosphere is acidic precipitation (**acid rain**). In soil moisture and surface water, it produces an acidified environment (see Figure 16-1).

Sulfur Dioxide

Sulfur dioxide is a colorless, nonflammable gas with a very strong suffocating odor. It is also an important contributor to acid rain, athough to a lower extent than nitrogen oxides. Sulfur dioxide emissions in the United States decreased fourteen percent from 1975 to 1985, but concentrations in ambient air decreased forty percent during the same decade. Sources of sulfur dioxide are (see Table 16-6):

Burning of high-sulfur coal	82%
Sulfuric acid plants	14%
Transportation	3%
Miscellaneous	1%

Table 16-5. National Annual Nitrogen Oxides Emissions and Their Concentrations in the Ambient Atmosphere *

Year	Nitrogen Oxides Emissions Per Year (million tons)	Nitrogen Oxides Concentrations (parts per billion)
1975	20	29
1980	22	28
1985	22	25
% Increase/ Decrease, 1975-1985	+10%	-14%

*Source: U.S. Environmental Protection Agency.

Figure 16-1. Automobile exhaust is the origin of more than half of all sources of nitrogen oxides in ambient air in the United States. (Photo courtesy Bill Camp.)

The relative sensitivity of selected woody plants to sulfur dioxide fumes is presented in Table 16-7. Note especially that white birch and white pine are sensitive (easily killed) by sulfur dioxide but that white flowering dogwood is tolerant (resistant) (see Figures 16-2, 16-3, and 16-4).

Table 16-6. National Annual Sulfur Dioxide Emissions and Their Concentrations in the Ambient Atmosphere *

Year	Sulfur Dioxide Emissions per Year (million tons)	Sulfur Dioxide Concentrations (parts per billion)
1975	28	15
1980	26	11
1985	24	9
% Increase/ Decrease, 1975-1985	-14%	-40%

*Source: U.S. Environmental Protection Agency.

Chlorofluorocarbons

Chlorofluorocarbons (CFC-11, 12) are anthropogenic (people-made) gases. CFCs have been accused of depleting the ozone layer which protects human skin from cancer. They are synthetic chemicals used in refrigerators, air conditioners, spray cans, and in the manufacture of plastic foam.

The global annual releases (emissions) of chlorofluorocarbon 11 has decreased ten percent and CFC 12 has decreased nine percent from 1975 to 1985. During the same decade, the concentrations in the atmosphere have increased eighty–six percent and ninety–two percent, respectively. For the same period, the ozone concentration has decreased seventeen percent. From these data it is difficult to predict if the world efforts to reduce CFCs have been a failure or a success (see Table 16-8). Can the incidence of skin cancer be predicted? The authors say more relevant research is needed before any firm conclusions can be made. For a discussion of the mechanisms by which ozone absorbs ultraviolet light and how CFCs are thought to react with ozone to destroy it, see Chapter 20.

Table 16-7. The Relative Sensitivity of Selected Woody Plants to Sulfur Dioxide in North America *

Relative Sensitivity to Sulfur Dioxide	
Sensitive	**Tolerant**
Alder	Cedar, western red
Aspen, large-toothed, trembling	Dogwood, white flowering (Figure 16-3)
Ash	Fir
red	Silver
green	White
Birch, white (Figure 16-2)	Forsythia
Blueberry	Ginkgo
Elm, Chinese	Hawthorn
Hazel	Juniper
Larch, western	Basswood
Locust, black (Figure 16-3)	Maple
Maple	Norway
Rocky Mountain	Silver
Manitoba	Oak, pin, northern red
Mock-orange	Pine, pinyon
Mountain ash	Poplar, Carolina
Mulberry	Spruce, blue
Pine, white (Figure 16-4)	White cedar
Willow, black	

*Source: Davis, D.D., & Wilhour, R.G. (1976). **Susceptibility of woody plants to sulfur dioxide and photochemical oxidants.** Washington, D.C.: U.S. Environmental Protection Agency - 600/3-76-102.

Methane

Methane is a colorless, odorless, flammable, gaseous hydrocarbon (CH_4) present in natural gas. Natural gas used as a fuel has other gases mixed with it so it can be detected by odor by humans. Methane gas may

Figure 16-2. If a clump of white birch trees in the front yard have sickly looking leaves similar to the leaf at (A), the problem may be air pollution by sulfur dioxide. The leaves of white birch should be healthy, as at (B). (Courtesy United States Department of Agriculture.)

A   B

Figure 16-3. White flowering dogwood (A) is resistant to sulfur dioxide; whereas, black locust (B) is sensitive to this gas. (**Source:** United States Department of Agriculture, Forest Service.)

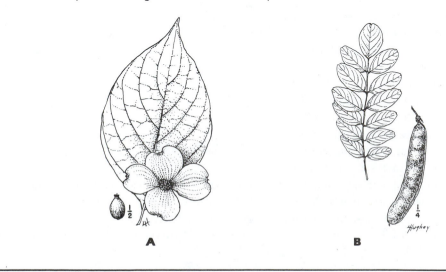

A   B

Figure 16-4. White pine is sensitive to sulfur dioxide. (A) Normal white pine tree. (B) White pine with normal air for several years, then sulfur dioxide for a few weeks. (Courtesy United States Department of Agriculture.)

A   B

Table 16-8. Global Annual Emissions of Chlorofluorocarbons -11 and -12 and Their Concentrations in the Ambient Atmosphere Compared with Ozone Concentrations *

Year	Global Annual Emissions of Chlorofluorocarbons CFC-11 (in million pounds)	CFC-12	Global Concentrations of Chlorofluorocarbons CFC-11 (in parts per trillion)	CFC-12	Ozone Concentrations (in parts per billion)
1975	684	889	120	200	153
1980	552	732	179	307	145
1985	618	810	223	384	127
% Increase/ Decrease, 1975-1985	-10%	-9%	+86%	+92%	-17%

*Source: U.S. Environmental Protection Agency.

Note: In the U.S. and France, one trillion is 1 followed by 12 zeroes; in Great Britain and Germany, it is 1 followed by 18 zeroes.

explode when mixed with air or oxygen. It is formed in nature by anaerobic decomposition of vegetable matter as in marshes and paddy rice fields; hence, the name marsh gas. It is also produced synthetically by heating carbon monoxide and hydrogen. Methane is used as a fuel, a source of carbon black, and as a source of hydrogen for making nitrogenous (NH_4) fertilizers.

Methane is one of the sewer gases and is extremely toxic to humans and animals. The other major sewer gases are hydrogen sulfide, ammonia, and carbon dioxide. This combination of odors is variously described as that of rotten eggs, skunklike, or a mixture of garlic and coffee.

Global concentrations of methane gas increased by twelve percent for the decade 1975–1985. This is hardly an indication of a more favorable environment (see Table 16-9).

Lead

Lead is a human health hazard. It is a bluish-gray metal whose salts are poisonous to humans and animals. Tetraethyl lead has been used in gasolines to increase their octane rating and antiknock qualities.

Lead has been used extensively in paints. In many forms, lead can be absorbed through the skin and may cause mental symptoms and early death.

Lead use is rapidly being phased out in gasolines and in paints, as is indicated in Table 16-10. U.S. lead emissions have decreased eighty–six percent during 1975–1985, and atmospheric lead concentrations have decreased seventy–seven percent during this period.

Suspended Particulates

Suspended particulates are mostly solid particles of dust, soot, ash, and smoke that originate from dry soil particles (mostly silt), incinerators, open trash fires, wood fires, and nearly all manufacturing processes. Allergenic agents such as pollen may also be included (United States Environmental Protection Agency).

Suspended particulates means, in simple language, "dirt in the air." The dirtiest cities, as judged by the relative concentration of suspended particulates are Riverside, Los Angeles, Chicago and Philadelphia, Anaheim and Baltimore, San Diego, Jersey City, Houston, and Hartford (see Tables 16-11 and 16-12).

Table 16-9. Global Annual Concentrations of Methane in Ambient Atmosphere *

Year	Global Methane Concentrations (in parts per billion)
1975	1,525
1980	1,639
1985	1,711
% Increase/Decrease	
1975–1985	+12%

*****Source:** U.S. Environmental Protection Agency.

Table 16-10. National Annual Lead Emissions and Their Concentrations in the Ambient Atmosphere *

Year	Lead Emissions per Year (in thousand tons)	Lead Concentrations (in milligrams per cubic meter)
1975	162	1.0
1980	78	0.6
1985	23	0.23
% Increase/ Decrease,		
1975–1985	-86%	-77%

*****Source:** U.S. Environmental Protection Agency.

Total suspended particulates emissions have decreased in the United States by twenty–seven percent from 1975 to 1985. Concentrations, however, increased five percent from 1975–1980, but decreased twenty–six percent from 1980–1985.

Suspended particulates in the air act as allergens to certain people, but not to all people alike. Particulates may be discharged from cotton mills, wool mills, broiler processing plants, shoe factories, jute mills, flour mills, and fur processing plants, to name just a few sources. Pollen from such cultivated and wild plants as corn, ragweed, and certain trees is suspended in the atmosphere. A whole range of allergic reactions to pollen are commonly known collectively as hay fever.

Ragweed pollen is one of the common sources of allergenic agents. There are so many people that are allergic to ragweed that the Canadian Government Travel Bureau of Ottawa, Canada, publishes a factual and appealing bulletin, "Canada Havens from Hay Fever." The bulletin includes full-page drawings of three species of ragweeds: common ragweed (see Figure 16-5), giant ragweed, and perennial (western) ragweed. The bulletin also includes a map of the Provinces of Canada with the average ragweed pollen air-index for 207 locations.

Pollen grains from the following trees produce allergenic toxicity in approximately this descending order: junipers, birches, sycamore, beech, oaks, elms, mulberry, walnut, hickories, poplars, and mesquite.

Global Warming

There are many discussions and some disagreement on the question: "Is the Earth getting hotter?"

Official and reliable records on annual global surface temperatures are reported by the Carbon Dioxide Analysis Center at Oak Ridge National Laboratory, Oak Ridge, Tennessee, for the years 1900 to 1988. The records show that during only three years, 1915, 1943, and 1960, the annual global temperature

Table 16-11. The Dustiest Cities in the United States *

City	Suspended Particulate Matter (in micrograms per 10 square meters)
Riverside, California	95
Los Angeles, California	65
Chicago, Illinois	47
Philadelphia, Pennsylvania	47
Anaheim, California	43
Baltimore, Maryland	43
San Diego, California	40
Jersey City, New Jersey	36
Houston, Texas	32
Hartford, Connecticut	30

*Source: U.S. Environmental Protection Agency

Table 16-12. National Annual Total Suspended Particulates Emissions and Their Concentrations in the Ambient Atmosphere in the United States *

Year	Total Suspended Particulates Emissions (in million tons)	Total Suspended Particulate Concentrations per Year (in micrograms per cubic meter)
1975	11	62
1980	9	65
1985	8	48
% Increase/ Decrease, 1975-1985	-27%	+5%, 1975–1980 -26%, 1980–1985

*Source: U.S. Environmental Protection Agency.

Figure 16-5. Common ragweed—
Ambrosia artemisiifolia. A. Plant; B. head
of male flowers; C. "seed."

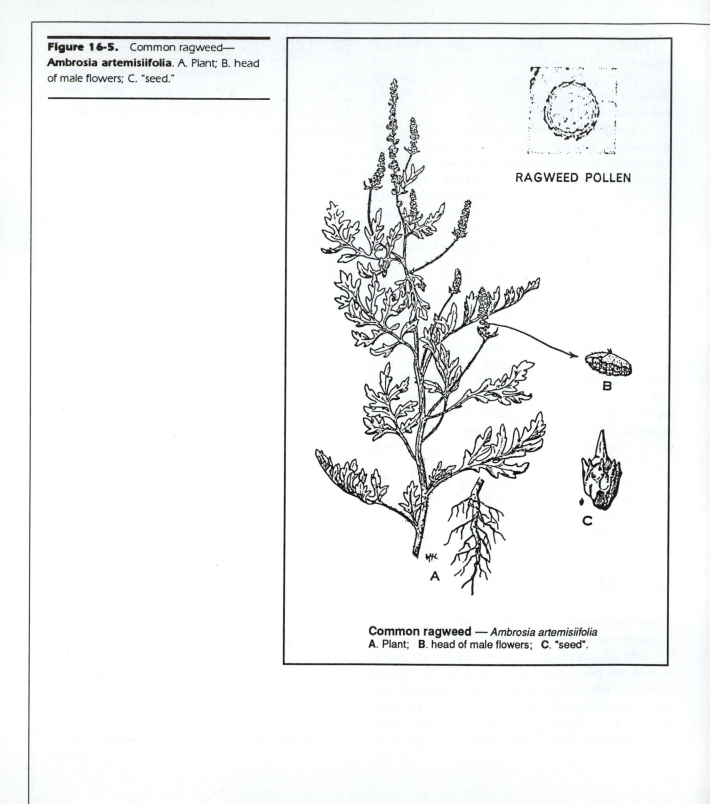

RAGWEED POLLEN

Common ragweed — *Ambrosia artemisiifolia*
A. Plant; B. head of male flowers; C. "seed".

was equal to the *average* for all years, 1900 to 1988. During thirty–four years the annual global temperature was *warmer* than average and for fifty–one years it was *colder* than average. During the last twelve-year period for which data are available, 1977–1988, the average annual increase in air temperature was 0.35°F. This compares with the twelve previous years, 1964–1976, when eight of the twelve years were slightly colder than average and four were slightly warmer.

Is this **global warming**? For a more thorough discussion on global temperatures and "global warming," see Chapter 20.

Acid Rain

The air pollutants accused of causing acid rain are nitrogen oxides, sulfur oxides, and carbon dioxide. Note that nitrogen oxides and carbon dioxide are also suspected of causing global warming.

The most acidified waters are in the northeastern and eastern upper Midwest. There are 29,368 lakes, more than 10 acres in area, within these regions. Of those, 3,660 have been declared acidified, or twelve percent of the lakes (see Table 16-13). See Chapters 8 and 9 for more details on this.

In a 1984 position statement, the Society of American Foresters stated:

- "Direct effects of acidic deposition on forest trees have not been detected under field conditions."

- "There is no conclusive evidence that acidic deposition has significantly increased the rate of forest soil acidification over the last two decades." (Society of American Foresters 1984).

No one doubts the research data; what soil scientists doubt is whether the acidified lakes have been caused by polluting acid rain, acid granite, acid soil, or a combination of both acid rain and acid rock and soil.

Table 16-13. Waters Troubled by Acid Rain *

Area	Total No. of Lakes and Streams Over Ten Acres	Lakes and Streams Acidified Number	Percent
Adirondacks, New York	450	171	38
New England			
Seaboard lowland lakes	848	68	8
Highland lakes	3,574	71	2
Appalachia			
Forested lakes	433	43	10
Forested streams	11,631	1,396	12
Atlantic Coastal Plain			
Northeastern lakes	187	21	11
Pine barrens streams	675	378	56
Other streams	7,452	745	10
Florida			
Northern highland lakes	522	329	63
Northern highland streams	669	187	28
Eastern Upper Midwest			
Low silica lakes	1,254	201	16
High silica lakes	1,673	50	3
Total	29,368	3,660	12

* **Source:** National Acid Precipitation Assessment Program.

Hazardous Air Pollution Episodes

In some of our larger cities, air quality levels often become very poor. Particularly in hot, still weather, air pollution levels can become severe. When this happens, a so-called **emergency episode** can occur. Assuming that emergency episode levels of air pollution are reached and the public is so informed through the news media, the following actions or precautions have been proposed by the U.S. Environmental Protection Agency:

- Curtail physical activity, both indoors and outdoors. The more active you are, the more breaths you take—and the more pollutants you breathe in.

- Stay indoors as much as possible and keep windows closed. Pollution levels are usually lower indoors than out, and buildings themselves act as filters of sorts, blocking or absorbing some pollutants.

- Avoid smoke-filled rooms. If you are a smoker, stop or cut down smoking.

- Do not use your fireplace.

- Do not use your incinerator.

- Do not drive if possible. If you must travel, use public transportation. If you must drive, form car pools, avoid busy streets and expressways. If you have a choice, take a bridge instead of a tunnel. If you must use a tunnel, keep car windows and ventilators closed.

- If you are on the street and a bus or truck emits a cloud of exhaust in front of you, try holding your breath until you pass through it.

- Do not wear contact lenses.

- In the winter months, use a humidifier or vaporizer to add moisture to your home. Moisture helps you to breathe easier.

- Cut down on water use. Electricity is needed to pump water and to run sewage treatment plants. Generating electricity requires the burning of coal and oil.

- Cut down on use of electricity. Keep nonessential lights off. Postpone running washing machine, dryer, dishwasher, and other nonessential appliances.

- Lower room temperature in your home if health considerations permit.

- Postpone indoor cleaning jobs that circulate dust, such as sweeping and vacuuming.

- Postpone outdoor jobs that raise dust, such as raking leaves, sweeping sidewalks, and excavating land.

- Do not use the phone unless it is essential. Telephone circuits can be overburdened in emergencies.

Special Precautions

For the elderly, chronically ill, heart and lung patients, bronchitis, asthma, and emphysema sufferers, postoperative patients, and newborn infants:

Stay indoors, keep windows closed, and follow the other appropriate suggestions listed above. In addition:

- If you have an air filtering system or air conditioner, turn it on.

- If you are on medication, take it at the first sign of worsening symptoms and call your physician.

Federal Agencies with Jurisdiction in Air Pollution Control

The federal agencies with jurisdiction by law or with special expertise on air quality are as follow:

- Department of Agriculture: Forest Service (effects on vegetation).

- Department of Health and Human Services: Public Health Service.

- Environmental Protection Agency: Air Pollution Control Office.

- Department of the Interior: Bureau of Mines (fossil and gaseous fuel combustion) and Bureau of Sport Fisheries and Wildlife.

- Department of Transportation: Assistant Secretary for Systems Development and Technology (auto emissions).

- Coast Guard (vessel emissions).

- Federal Aviation Administration (aircraft emissions).

Noise

Noise is unwanted sound, but opinions vary as to what is unwanted. Loud music is a sweet sound to some and a nuisance to others. A car without a muffler may please a "hot-rod" driver. A gentle breeze through the trees may be pleasant to someone needing solitude from noisy big-city life. Trying to sleep when jet airplanes fly low overhead is an experience no one needs.

Loud noises over a period of time can cause partial loss of hearing, general stress, or a decrease in muscular coordination. Poultry, birds, and most people cannot adjust to loud noises, but domestic animals can tolerate such noise readily. "Sounds-to-produce-fear" are heard on television prior to danger. Most people can be in the next room and just by listening to the music can determine when someone is about to be attacked or murdered.

Relative noise levels of several environments are as follow: Los Angeles apartment, New York apartment, general average U.S. urban living, small town, village, farm, and the Grand Canyon. According to noise level standards established by the U.S. Department of Housing and Urban Development, the noise levels of both Los Angeles and New York apartment living are acceptable to most people for no more than eight hours a day. Conclusion: If you can't tolerate the noise, move!

Noise Hazard to Human Hearing

It may be hard to believe, but power lawnmower noise damages more people's hearing than does commercial aircraft noise. The full relative hazard to hearing is: jackhammer, power lawnmowers, home shop tools, motorboats greater than 45 horsepower, trucks for personal use, commercial aircraft, motorcycles, chain saws, subways, highway buses, and snowmobiles (see Figure 16-6).

Summary

The atmosphere around the planet ranges in total thickness up to twenty–two thousand miles above the surface. Its effective thickness is much less than that,

Figure 16-6. The operator of a jackhammer is subjected to a noise level loud enough to impair hearing. (Courtesy United States Enviromental Protection Agency.)

A

Figure 16-7. These two pictures illustrate the progress made in the United States in recent decades. The smoke coming from these smokestacks in the 1960s (A) would not be permitted under the United States Clean Air Amendments of 1990. The emissions in the photograph from the 1990s (B) is more typical of industrial emissions today. (Figure 16-7A courtesy United States Environmental Protection Agency; Figure 16-7B courtesy Bill Camp.)

B

however. Clouds appear in the atmosphere up to about five miles high over the poles to as much as twelve miles high over the equator. Air is one of the few resources for which ownership has never been claimed by any country. The atmosphere consists mostly of nitrogen (seventy–eight percent) with a considerably smaller proportion of oxygen (twenty–one percent) and very small amounts of argon and of other trace gases. In addition, there are countless suspended particulates in the atmosphere at all times (see Figure 16-7).

One important trace gas is carbon dioxide (CO_2). It is a naturally-occuring gas which results from the oxidation of most carbon compounds, and in particular is a by-product of the process of respiration of plants, animals, and humans. Human production of CO_2 gas has increased dramatically over the recent

past because of burning of fossil fuels, clearing and burning of forests, and a number of other causes. Carbon dioxide has been described as a dangerous "greenhouse gas" in recent years, because some scientists believe it will cause global warming. At the same time, CO_2 is an essential gas, because it is necessary in photosynthesis and promotes plant growth and human food production (see Chapter 7).

Carbon monoxide (CO), on the other hand, is definitely a dangerous gas. Even in small concentrations, CO can be lethal to animals and humans. Fortunately, a number of organisms in the environment convert CO to the harmless CO_2 gas through oxidation.

Chlorofluorocarbons (CFCs) are anthropogenic gases which appear to be harmful to the environment, particularly to ozone gas in the upper atmosphere. As you will learn in Chapter 20 (Global Temperatures), CFCs react with and "destroy" (actually, transform) beneficial ozone gas. CFC emissions have been reduced somewhat, but they need to be eliminated as soon as possible.

Another dangerous anthropogenic emission is lead. Lead has been used in gasolines to make it more effective as a fuel. When leaded gasoline burns, the lead enters the atmosphere. Lead emissions and lead levels in the atmosphere have been reduced dramatically during the past several decades. This is a great success story of the environmental movement and should be shouted on the daily national news. Unfortunately, such good news seldom gets told.

Suspended particulates are mostly solid particles so small and lightweight that they can remain in the atmosphere for a long time. Fine dust, soot, pollen, and other particles are included. Human activity produces part of the suspended particulates in the atmosphere, but certainly not all. Events such as the eruption of Mount Pinatubo in the Philippines in 1991 produce massive amounts of atmospheric particulate matter.

Local concentrations of particulates and harmful gases in the atmosphere vary greatly over time. Concentrations become very high in cities like Los Angeles sometimes. When that happens, an air quality emergency episode may be declared by local officials. There are a number of precautions you should take if an air quality warning is issued in your area.

Two potential environmental problems are global warming and acid rain. Not all scientists agree that either problem is real. Even if they do occur, there is not consensus on what the long-term effects will be. For more details on global temperatures, see Chapter 20.

The final air pollution problem addressed in this chapter is noise. Unwanted sounds have a damaging effect on humans and animals. Loud noises produce hearing loss, nervousness, and other problems. Some of the most damaging sources of noise are very commonplace items: power lawnmowers, power machinery, motorboats, chainsaws, and others.

DISCUSSION QUESTIONS

1. What is the make-up of the atmosphere?

2. What are the emission gasses introduced into the atmosphere by human activity?

3. Have suspended particulate emissions and levels in the United States been going up or down?

4. What happens to most of the carbon dioxide (CO_2) that human activity puts into the atmosphere?

5. Are carbon monoxide emissions and levels increasing or decreasing in the United States? Why? What are the implications for human health?

6. Are nitrogen oxide emissions and concentrations going up or down?

7. Are sulfur dioxide emissions and levels rising or falling?

8. Are chlorofluorocarbon emissions and levels going up or down?

9. What should you do in case an air quality emergency episode is declared by local officials?

10. What are the effects of noise on animals and humans?

ADDITIONAL ACTIVITIES

1. Invite a meteorologist from your local television station to visit your class to discuss air pollution in your area.

2. Arrange for a joint project with the chemistry class in your school to analyze air samples from your community for CO_2, CO, NO_x, and particulate matter levels.

3. Contact the weather station serving your local area and find out the reported air quality index each day for the past year. Compute the weekly averages and plot the data on a graph. Can you

explain the changes in terms of the kind of weather at various times of the year?

4. Place a series of clean glass specimen plates for a microscope outside for twenty–four hours. Put one in an open area far away from anything that would cause a disturbance. Place one near a busy street, one in a wooded area, one in your classroom, and select several settings of your own. Examine the solids that settle onto the specimen plates under a microscope to determine what kinds of particulates you find. What can be done to reduce the concentration of particulates?

C H A P T E R 17

Oceans, Seas, and Estuaries

Photo courtesy Bill Camp.

Terms to Look for and Learn

Abyssal Plain	Latitude Gradient
Aphotic	Littoral
Backshore	Mean Sea Level
Bathymetry	Nearshore
Brackish	Neritic
Coastline	Photic
Continental Rise	Phytoplankton
Continental Shelf	Saline
Continental Slope	Shoreline
Estuary	Thermocline
Foreshore	Zooplankton

Learning Objectives

After reading this chapter and participating in the activities, you should be able to:

■ Discuss the bathymetry of the ocean bottoms.

■ Explain the effects of depth gradient on the ocean.

■ Explain the effects of latitude gradient on the oceans.

■ Explain the differences resulting from the coastal to open-water gradient on the ocean.

■ Discuss the marginal seas and compare them to the great ocean bodies.

■ Discuss estuaries and the role they play on the marine environment.

■ List the major kinds of marine organisms harvested for human use.

287

Overview

Oceans must be considered a major part of anything having to do with the environment. They store more energy than any other part of the ecosystem. Oceans hold the vast majority of the water on the planet. They generate the forces that drive the winds that make our climates. They provide much of the food for all of the organisms on Earth. Their borders outline the continents on which we live.

The level of human knowledge about the oceans is growing constantly. In the last few decades the sciences of oceanography and marine ecology have begun to provide a rapidly growing knowledge base about the open oceans. Around the borders of the oceans lie a series of ecosystems that are also of great importance—the estuaries. It is in the brackish waters of these coastal areas that many species live out parts or all of their lives. Moreover, the estuaries are a vital link in the human food chain. Finally, in addition to the vast oceans and their coastal estuaries, the planet's enclosed seas are of great importance to the ecosystem.

In this chapter we will examine the oceans, estuaries, and enclosed seas of the planet. We will examine some of their geology, ecology, and physics. We will look at the roles they play in providing food for the rapidly expanding human population.

Oceans

As we learned in Chapter 8, the oceans contain about 97.5 percent of all the planet's water. They cover nearly seventy–one percent of the Earth's 197,000,000 square miles of surface. They range in depth from the shallowest beach to over thirty–six thousand feet in the Marianas Trench. In the Northern Hemisphere, sixty–one percent of the surface is covered by water while in the Southern Hemisphere, eighty–one percent is water (Ketchum, 1983a). The Southern Hemisphere is sometimes called the "Water Hemisphere" (see Figure 17-1).

Bathymetry of the Oceans

The planetary ocean system is generally defined as all of the great open body of saline water that perma-

Figure 17–1. Oceans provide food, recreation, transportation, and other natural resources for humans. They act as a critical buffer for the climate. They provide the driving force for most of the planet's weather. Notice the fish swimming through this wave. (Photo courtesy Bill Camp.)

nently covers most of the Earth's surface. There are three great interconnected oceans: Pacific, Atlantic, and Indian. Both the Pacific and Atlantic Oceans are often thought of as consisting of northern and southern halves. Thus, we hear of the North Atlantic, North Pacific, South Atlantic, and South Pacific Oceans. The largest of the oceans is the Pacific and the smallest is the Indian (see Table 17-1). In terms of volume, the Pacific Ocean contains half of all of the water on the planet—166 out of a total 332 million cubic miles.

Extending outward from the land masses is a relatively shallow and flat region called the **continental shelf**. It is here that the rising and falling levels of the sea between the ice ages produced great changes in the **shoreline**. As will be discussed in more detail later, sea level changes in excess of 330 feet have occurred between the great ice ages and the warmest interglacial periods. The continental shelf rises above sea level and becomes coastal plain during such times. When it is under water, the continental shelf receives most of the sediment that reaches the ocean from soil erosion on land.

At a point averaging forty–three miles from the shoreline and ranging from just a few feet to over eight hundred miles, the continental shelf begins to drop off more rapidly. The point at which the slope begins to steepen is the continental shelf break. The bottom of the sea curves downward along the **continental slope** to depths in the order of ten thousand to thirteen thousand feet. The steepness of continental slope averages about 4°, but ranges from 1° to 25°.

Eventually, as the crust of the ocean nears the great sea bed, its slope becomes more gentle. That point is referred to as the **continental rise**. The ocean floor is relatively flat and smooth. This is referred to as the **abyssal plain**. In most parts of the oceans, the abyssal plains average between fifteen thousand and twenty thousand feet below the water surface. They are generally broad, level expanses covered by sedimentation of various kinds.

The oceans range in depth from less than an inch at the shoreline to over 6.8 miles. Across the abyssal plain there are great mountains and mountain ranges. The highest mountain ranges are those that reach the surface. Midocean islands are the result of undersea mountains rising above the surface of the sea.

There are also deep canyons and trenches. Canyons appear to be the result of erosion just like canyons on land. They even have some of the appearance of canyons on land. Trenches are formations generally near the point where the abyssal plain begins to give way to the continental rise—the margin of the sea floor. The origin of trenches is not as simple as that of submerged canyons. They may result from tectonic (plates in the Earth's crust) movements or volcanic activity. They are deep, steep-sided, and narrow. The greatest trenches reach amazing depths. The deepest known part of the ocean is the Challenger Deep of the Marianas Trench, at 36,161 feet.

The marine environment is often classified in terms of zones. Ocean zonation is generally based on three gradients: latitude, depth, and coastal to open water. Figure 17-2 illustrates the major aspects of the ocean bathymetry. The term **bathymetry** is used in oceanography just as the term topography is used in geography. It refers to the shape of the ocean floor—its slopes, canyons, flat plains, ridges, and so forth.

Table 17-1. Areas and Volumes of the Planet's Oceans *

Ocean	Area (thousand square miles)	Volume (million cubic miles)
Atlantic	37,798	84
Pacific	68,215	166
Indian	29,991	71
Arctic	3,281	2

*SOURCE: Speidel, D.H., & Agnew, A.F. (1988). The world water budget. In Speidel, D.H., Ruedisli, L.C., & Agnew, A.F. (Eds). **Perspectives on water—Uses and abuses**. New York: Oxford University Press.

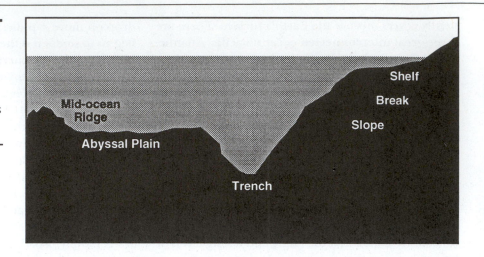

Figure 17-2. The ocean zones of a typical coastal region are illustrated (not to scale). The trench may be many thousands of feet deep and the mid-ocean ridge could be thousands of miles from the shore.

Starting at the shoreline, the margin of the oceans comprise a **littoral** zone of coastal waters. It is there that the land meets the shore and humans interact most closely with the sea.

Depth Gradient

Light Penetration

Ocean depths can be divided in several different ways. One of the most important is with regard to light penetration. For biological processes to take place, solar energy (and in a few cases other forms of energy) must be converted to food energy. For all practical purposes that means photosynthesis. Only in surface waters is there enough light to support photosynthesis. The zone in which this occurs is the **photic** (light) zone. The zone below which too little light penetrates to support photosynthesis is the **aphotic** (absence of light) zone (Barnes & Hughes, 1987).

The photic zone may be only a few inches in murky water, or it may be as deep as eight hundred feet in the open, clear tropical ocean. In coastal waters, the photic zone averages about 330 feet. In open water it averages about 1,640 feet deep (see Figure 17-2). It is in the photic zone that photosynthe-sis occurs, but almost all biological activity throughout the ocean depends on the food generated there.

Temperature Changes

In addition to light penetration, another important change that occurs with depth is temperature. Warm water is slightly less dense than cold water. As we have seen, light and heat penetrate to only a few hundred feet of the surface in even the clearest water. The surface water is thus warmed and becomes less dense. The warm water rests on top of the colder, denser water beneath it. At the point where warm surface water gives way to cold deeper water, a sharp drop in temperature occurs.

On land, air near the surface is warmed and rises. In the ocean, the same phenomenon occurs with water, but with opposite results. In the atmosphere, the warm air masses in one location, then pushes upward forcing the cold air masses above it downward at another location. In the ocean, the warm water rises and just stays there. The water temperature at the ocean surface in the tropics can be as high as 77° F (25° C) while the deep water is closer to 36° F (2° C). And at that depth there is nothing to cause it to mix.

The result of this is what is called the **thermocline**. In the highest latitudes (near the poles), the thermocline is small and occurs within about 330 feet of the water surface. In the low latitudes, the thermocline is a permanent feature and it begins and ends in the top two–thirds of a mile of water. In the mid-latitudes, a seasonal thermocline develops every summer in the top 330 to 500 feet in addition to the permanent thermocline.

The reason the thermocline is so important is that waters above and below it virtually never mix. The result is an ocean system that is unbelievably stable. Deep waters can absorb heat over the ages and keep it in place from one ice age to another. They can act as a gigantic climatic buffer (stabilizer). They can absorb massive quantities of atmospheric gases and hold them essentially forever. Nutrients in the form of all kinds of dissolved compounds can be locked up for time on a geologic scale.

An example of this effect is with the carbon dioxide generated by burning of fossil fuels. As atmospheric CO_2 levels increase, more is absorbed by the ocean surface. That leads to more rapid phytoplankton and algae production by photosynthesis in the photic zone. Eventually, much of the carbon is locked up in insoluble organic compounds and falls to the bottom of the ocean. There it becomes part of the ocean sediment where it will remain as long as the Earth remains a stable planet. The effect of this phenomenon in reducing global warming, for instance, has never been adequately estimated.

Beyond that, the minerals and other compounds on the ocean floor are much greater than those on the continental masses. As our technologies improve, who knows what the possibilities are for ocean floor mining of resources?

Coastal to Open-water Gradient

Coastal regions consist of several zones. **Shoreline** is the line marking the separation of water surface from land surface. It moves in and out with the tide. The coastline is the boundary between the shoreline and the coast. It is determined by an imaginary line at the highest point that is effectively reached by waves at high tide. The level halfway between mean high tide and mean low tide is known as **mean sea level**. That point is used as the standard for all other elevations on land in the world.

The **backshore** is that portion of the shore between normal high tide and the coastline. The foreshore is that portion of the shore between normal low tide and normal high tide. The nearshore is defined as the region of shore between where breakers form and the low-tide shoreline. Offshore is the region beyond the line of breakers (see Figure 17-3).

All of this is on the margins of the continents. It is generally in one hundred feet or less of water. Once

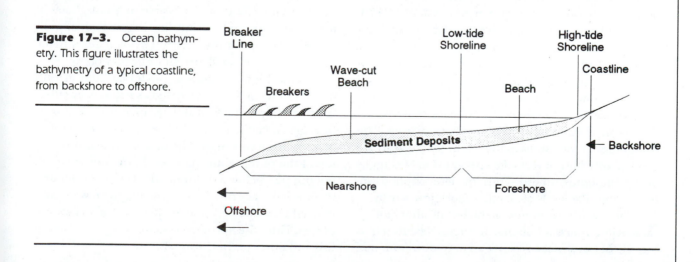

Figure 17–3. Ocean bathymetry. This figure illustrates the bathymetry of a typical coastline, from backshore to offshore.

beyond that depth, the coastal region continues outward along the continental shelf until the continental slope begins. The continental shelf ranges in width from less than three hundred feet to as much as eight miles. It typically begins to drop off at a depth of about 440 feet. It is generally at that point that the term open water or deep water is applied. The region above the continental shelf and below the foreshore (outside the littoral zone) is referred to as the **neritic** zone (named after a small coastal sea snail).

Latitude Gradient

The **latitude gradient** goes from the low latitudes (near the equator) to the high latitudes (near the poles). Solar energy is not distributed evenly over the globe.

At the equator, the rays of the sun reach an almost perpendicular surface. A square foot of level surface receives approximately a square foot of direct sunlight. The day length is nearly twelve hours year round. At the poles, a square foot of level surface would receive little direct sunlight except in midsummer. The day length varies from zero to twenty–four hours depending on the season.

This unequal distribution of solar energy has resulted in great differences in the amount of heat in the ocean water, even at great depths. Near the poles, ocean temperatures are at a near-constant 32.9° F from the bottom to within five hundred to six hundred feet of the surface. At the surface, temperatures range up to about 35.6° F. Near the equator, temperatures range from 35.6° to 41° F for deep water and average about 77° F at the surface.

The basis for all biological production in any part of the ocean is its level of photosynthetic activity. The beginnings of food webs in the ocean (as on land) are green plants that convert solar energy to food energy by means of photosynthesis. Thus, the level of photosynthesis limits the levels of all other biological activity.

Photosynthesis requires light, but all other biological activity requires additional nutrients. Near the equator, the ocean surface receives an even distribution of strong light and warm temperatures. As a result, the thermocline remains stable. Nutrients mix well in the photic layer and there is seldom either a nutrient-rich or a nutrient-poor environment. Photosynthesis occurs at an even but fairly low pace year-round there. As a result, the level of biological activity is moderate and stable throughout the year in tropical waters.

In the middle latitudes, seasonal variation affects the level of photosynthesis. In the winter, days are short and temperatures are cool. This means very little photosynthesis and in turn little **phytoplankton** production occurs in winter.

Plankton consists of all of the plants (phytoplankton) and animals (**zooplankton**) floating freely in the water and moved around by currents rather than being able to swim on their own. Phytoplankton are those plankton organisms consisting of green plants, mostly algae. Phytoplankton in heavy concentrations give water a green appearance and in nutrient-rich environments can even collect on the surface as a green scum.

The low levels of biological activity in the winter trap excess nutrients above the thermocline. In the following summer the presence of nutrient-rich water combined with long days produces a great initial surge of growth. That is followed by development of a seasonal thermocline which traps much of the nutrients below it. Thus, later in the summer, phytoplankton production slows because of a lack of available nutrients. Later in the fall, the seasonal thermocline begins to disappear. When that happens, the nutrients below it mix with the surface water and a fall growth surge may occur.

In the higher latitudes, phytoplankton growth would not seem to be supported as strongly by either light or temperature. Water temperatures are near freezing and light is never full and direct, yet the level of nutrients is very high. The thermocline is near the top of the photic zone. As a result, the nutrients are not trapped below a thermocline. Deeper water and surface water are free to mix. During the few months that light is available, a massive production of biomass occurs. Thus, the northern seas are very productive of fish and other marine animals.

Seas

In addition to the three great interconnected oceans, geographers generally recognize a number of marginal seas. These are smaller bodies of water, often located over the continental shelf, and always separated in some way from open water. There are two kinds of marginal seas: adjacent and enclosed.

Adjacent seas are generally open to the neighboring ocean along much of one or more sides. The Gulf of Mexico and the Sea of Japan are examples of adjacent seas. Oceanographers recognize 19 adjacent seas, including areas called straits and gulfs. The Arctic Ocean is usually considered an adjacent sea rather than a true ocean (in spite of its name). For the sake of the water volume computations in Chapter 16 and this chapter, adjacent seas are included in their larger adjacent oceans, except for the Arctic Ocean.

Enclosed seas are those with more restricted access to their neighboring oceans. In the case of the Mediterranean Sea, the connection to the Atlantic Ocean is through the Straits of Gibraltar and is indeed very narrow. Another example is the Persian Gulf with its narrow opening into the Indian Ocean at the Straits of Hormuz. But, that is not true in the case of the Gulf of St. Lawrence off the eastern coast of Canada or of the Bering Sea between Alaska and the former Soviet Union. For those two "enclosed seas" the interfaces are not so restricted and consist of strings of islands. Thus, the distinction between an enclosed sea and an adjacent sea can be quite arbitrary.

The marginal seas are much more impacted by human activity than are the open oceans. This is true for several reasons. The marginal seas are shallower. They are adjacent to coastlines. Some are almost totally enclosed by land, with only limited openings to open water. Water pollution from the land is more concentrated in seas than in oceans. And other pressures from human use, such as fishing and navigation, are more intense than in the open water.

Estuaries

At the point where a river empties into the sea, the water of the two becomes mixed. This **brackish** water is not as **saline** (salty) as sea water and is more saline than fresh water. It is a region of "in-between" (see Figure 17-4).

Estuaries make up a very important part of the ecosystem that is unique in many ways. According to the Chesapeake Bay Foundation, an estuary is a "semi-enclosed, tidal, coastal body of water open to the sea in which fresh and salt water mix." Fish and other organisms living in estuaries must be able to adjust to varying salt concentrations and temperatures. Many species live their entire lives in the estuaries. New

Figure 17–4. This river opening (mouth) is very close to sea level. At high tide, the sea flows in and the estuary becomes salty, almost like sea water. At low tide, the water flows out and the water becomes fresh, almost like fresh water. (Photo courtesy Bill Camp.)

England bay scallops (*Argopecten irradians*), American oysters (*Crassostrea virginica*), and some clams (*Mya arenaria* and *Mercenaria mercenaria*) are examples.

Other species spawn in fresh water and live much of their lives in salt water. Various species of salmon are examples. Fish that move from fresh to salt water or from salt to fresh water must pass through an estuary. The reduced salinity of the estuary allows the fish to adjust more slowly to that change.

Estuaries may be classified on the basis of their geomorphology. This refers to the estuary's geological origins—how it was originally formed. Using this classification, there are four major kinds of estuaries: drowned river valleys, bar-built estuaries, fjords, and deltas.

During the last ice age, the world's coastlines were at or near the edge of what is now the continental shelf. Ice age rivers flowed through broad valleys into the ocean. Runoff was relatively small as the ice cover took up much of the precipitation and held it. As the atmosphere warmed, ice began to melt. This rapid melting caused great valleys to be washed out (eroded) near the coastline. As the sea rose to current levels, those broad valleys became submerged. At their upper ends such river valleys became estuaries for the rivers that still exist today. Such formations are referred to as *drowned river valley estuaries.*

Bar-built estuaries form over broad, level, shallow submerged areas at river mouths. The rivers transport sand and silt from soil erosion. The sand forms submerged sandbars off-shore and partially enclose the broad areas within. Bar-built estuaries normally have slow fresh water flow and are protected from rapid ocean currents. The outer banks along the eastern seaboard of the Carolinas form many sandbar estuaries behind their protection (see Figure 17-5).

Fjords formed at the mouths of ice age rivers that flowed through narrow gorges or canyons. As the ice melted the gorges were deepened by torrential run-offs. Then the sea levels rose and narrow but deep estuaries developed. Some estuaries continue all the way to the open ocean with full depth. Others become shallow near their mouths from silt and sand deposits.

Deltas form at the mouths of some rivers. The rivers transport massive amounts of silt from upstream erosion. For those rivers that then flow into the ocean across a broad and shallow coastal plain, the sediment is deposited near the river mouth. The sediment builds up until the river mouth becomes very shallow. As the river fills it breaks up into many smaller channels. The system of channels can become very intricate and widespread. Over the millennia, the river mouth moves further and further out from the original coastline. The result is a very flat, wet, fertile delta such as is found where the Mississippi River meets the Gulf of Mexico and where the Nile meets the Mediterranean Sea.

Figure 17–5. The Bogue Inlet at the mouth of the White Oak River is separated from the Atlantic Ocean by the Outer Banks chain of islands. The result is this rich estuary along the coast of North Carolina. (Photo courtesy Bill Camp.)

Changing Sea Levels

News media stories would have us believe that rising sea levels result from environmental changes caused by human activity such as global warming. Such a belief is based on the assumption that sea levels have been stable in the past (Milliman, 1989). As we learned in Chapter 2, instability is a characteristic of everything in nature. Sea levels are always changing. If they were not rising, they would be falling. That is what instability means.

Sea level is simply the average point on a particular body of land where sea water reaches. Many factors affect sea level as it pertains to any body of land. We know that the Earth's crust is constantly moving. Land masses, even whole continents rise and fall as they move across the planet's crust. At a site that is uplifting, sea level will fall relative to that point of land—that means the land is rising. The land mass that contains the United States is moving slowly westward. Therefore, in general the western land areas in the coastal United States are rising while those in the eastern United States are falling.

As continental plates move, new canyons, trenches, and mountain ranges are formed at the bottom of the oceans. When that happens, the average depth of the ocean changes. That results in either a rising or falling sea level.

Shifts in the Earth's crust are geological changes. In addition to those, there are natural climate cycles that produce cooling and warming on the planet's surface (see Chapter 20). As global temperatures rise, polar and glacial ice melts and adds to the volume of water in the oceans (see Chapter 8). In periods of cooling, the opposite occurs. More ice is added to the ice pack and sea levels fall as a result.

According to Milliman, average sea levels at one representative United States port (Portland, Maine) have risen about four inches (10.2 cm) in the past sixty years. Over the past twenty-five thousand years, sea levels at that same location fell substantially (about 250 feet) as the peak of the last ice age approached. Beginning about fifteen thousand years ago, sea levels at that location began to fall. Milliman estimates that sea levels have risen since then by over 330 feet. The average worldwide sea rise could be as much as 440 feet. That means that what is now the continental break would have been a beach then. That would also explain why the continental shelf is so nearly level.

At times, the melting was particularly rapid. In some years, the rise in sea level could have been as much as an inch per year. When that happened in a very level region, the oncoming sea could have engulfed many miles of land in a single year.

The continental shelf underlies about eleven percent of the planet's surface. The average water depth at which the continental slope begins is about 440 feet. Thus, a rise in worldwide sea levels over the past fifteen thousand years would have flooded about ten to eleven percent of the Earth's surface.

Atmospheric models of global warming project worldwide temperature increases of from 4° to 9° F by the year 2100 (see Chapter 20). If that were to actually occur, ice could be expected to melt quite rapidly and sea levels could rise between one and seven feet. Such a rise would be catastrophic to the world's populations and economy.

Worldwide, human populations cluster around rivers and along coastlines. Areas where the land lies within a few feet of sea level could become completely covered. Just as an example, a rise of six feet in sea level would flood much of Virginia Beach, Savannah, Houston, and Biloxi in the United States. As an alternative, huge dyke systems such as those in Western Europe would be necessary to keep the coastal land dry.

IF THAT WERE TO HAPPEN . . . such flooding would move estuaries and salt wetlands inland from their present locations. It would eliminate the benefits gained from thousands of years of ecosystem development in a mere century. The new wetlands and estuaries created would not maintain the same level of biodiversity or productivity as the existing ones have. It would take thousands of years for the ecosystem to recover from such a human-driven, warming-induced sea level rise. The Louisiana bayous and marshlands could be eliminated by being covered with sea water by the year 2040 with a thirty–three-mile inland movement of the Gulf of Mexico. The salt marshes of the Chesapeake and Delaware Bays would be eliminated before new ones could develop.

IS IT LIKELY TO HAPPEN? *Nobody knows*!! The scientists who think global warming has already started believe it is. Most other environmental scientists do not think so. In fact, in the 1970s, climatologists were forecasting a new Ice Age.

THE AUTHORS' CONCLUSION: If you get a good deal on some beach property, go ahead and buy it. In ten thousand years it may be under water or it may be miles from the shore. But, do you really care about that? Regardless of whether the seas rise or fall, today's beaches will not be beaches forever.

Marine Catch

People from many nations and all of the continents rely heavily on the oceans as a source of food. The total marine catch in 1986 was estimated at 88,540,000 tons. The largest exploiter of marine products is Asia (about forty percent of the total), followed by South America and then Europe. Marine catches grew dramatically over the period 1970–1986 in all parts of the world, except Africa (see Table 17-2).

The largest percentage increase in marine catch was among the world's island nations, but the actual worldwide increase in tonnage was very small. Increases in marine harvests in Asia (over fourteen million tons) and in North America (3,736 tons) over that short period have been massive. The vast majority of the increase in world harvest from the oceans came in the Pacific (17,985 tons).

Although their total figures differ from those in Table 17-2, Barnes and Hughes provided estimates of the marine harvest by kind of organism. According to their estimates, marine fish comprised about 81.5 percent of the total worldwide marine harvest in 1986. They also provided estimates of the 1984 harvest of marine fish by species (see Table 17-3).

Table 17-2. Estimated Worldwide Annual Catch of Marine Organisms.

By Continent*	1970	1980	1986	16-year Change
	(thousand tons)			
Africa	3,450	3,003	3,183	-7.8%
North America	5,235	7,381	8,971	+71.4%
South America	16,121	8,197	15,096	-6.4%
Asia	21,437	28,936	35,700	+67.3%
Europe	13,020	13,410	13,332	+2.4%
Oceania	214	494	701	+227.8%
Former USSR	7,052	9,613	11,387	+61.5%
By Ocean				
Atlantic	25,930	28,046	27,330	+5.4%
Pacific	38,250	38,857	56,235	+47.0%
Indian	2,788	3,912	4,959	+77.9%
TOTAL**	67,698	71,024	88,540	+30.8%

*The former USSR is separated out because when these data were collected, it crossed major portions of two continents. Oceania includes noncontinental islands throughout the world.
**Totals exceed the sum of either the continental or the oceanic catches because of catches reported from other locations (e.g., Arctic Ocean, etc.)

SOURCE: Borgese, E.M., Ginsburg, N. & Morgan, J.R. (Eds) (1989). **Ocean yearbook 8**. Chicago: The University of Chicago Press.

The largest catch of fish is of herring, anchovy, and related species (about thirty percent) followed by cod, haddock, and related species (about 20 percent). Marine mammals make up only a small percentage (0.7 percent) of the world catch, but that still amounts to about 600 tons. By the same token, turtles make up only about 0.005 percent of the total marine harvest. But that still means about five tons of turtles, mostly endangered or vanishing species, were harvested in 1984.

Table 17-3. Estimated Marine Harvest by Type of Organism (1986 data) and by Species of Fish (1984 data), ****

Type of Organism	1986 Harvest (thousand tons)
Seaweeds	3,800
Benthic crustaceans	3,300
Pelagic crustaceans	100
Benthic mollusks	4,600
Pelagic mollusks	1,800
Other invertebrates	400
Diadromous fish*	1,100
Marine fish	69,100
Turtles	5
Mammals**	600
TOTAL	84,805

Species of Fish	1984 Harvest
Herring, Anchovy***	21,100
Cod, Haddock***	13,400
Jacks, Sauries***	9,500
Redfish, Sea Bass***	6,100
Mackerel***	4,600
Tuna***	3,400
Flatfish	1,300
Sharks***	800
Diadromous fish**	1,100
Miscellaneous species	9,000
TOTAL	70,300

*Fish that migrate between fresh water and the sea.
**Nondolphin whales, dolphins, and others.
***And related species.

****Source: Barnes, R.S.K., & Hughes, R.N. (1988). **An introduction to marine ecology**. Boston: Blackwell Scientific Publications.

An interesting point to be made is if the baleen whales of the South Atlantic Ocean were reduced in number by fishing, the krill they currently consume would be available for human consumption. Estimates are highly speculative, but perhaps as much as an additional 110 million tons per year of krill theoretically might be available if all of its predators were removed.

Worldwide fish harvests have generally grown less rapidly in the past few decades than overall marine harvest. The fish catch has been growing at over a million tons per year since the early 1970s. Estimates of the potential harvest of marine fish vary greatly among marine biologists. But, Barnes and Hughes estimate that a maximum sustainable fish harvest would be no more than 110 million tons. That would require more intensive fishing of waters not already overfished.

Most existing fisheries are already overfished. In some cases, entire fisheries have collapsed. In the past forty years, the Menhaden fishery off the coast of the United States was overfished to produce fish meal, as was a major fishery off Peru. The result in both cases was a collapse of the fishery resulting in an almost total loss of the predominant species.

Summary

Oceans contain all but a tiny fraction of the world's water. They range in depth down to 6.8 miles in the Marianas Trench. Beginning at its edge, the ocean floor consists of a broad, relatively level continental shelf. The shelf was almost certainly exposed as dry land during the last ice age. It receives most of the sediment that washes from the land as a result of soil erosion. Its width ranges from less than 330 feet to as much as eight hundred miles, averaging about forty–three miles. Its depth ranges from about one hundered feet to an average maximum depth of 550 feet.

At the edge of the continental shelf, the sea floor begins to angle downward. That point is called the continental shelf break. Beyond the break, water depth increases along the continental slope. As the ocean floor approaches, the slope begins to level off along the continental rise. The floor of the ocean is mostly a broad, relatively flat abyssal plain. The slope is broken by canyons formed by the cutting action of moving water. There are also trenches which occur around the margins of the abyssal plains and contain the deepest points in the entire ocean.

Three important gradients can be used to describe the ocean environment: depth, latitude, and shore to open water. Depth affects the marine environment in a number of ways. The first is in terms of light penetration. Light makes photosynthesis possible. The second major effect of depth is temperature. Surface water is affected directly by atmospheric temperature. But at greater depths (as much as 3,300 feet), a thermocline exists that separates warmer surface water from cooler deep water. The thermocline effectively prevents the mixing of surface water and deep water.

The primary effect of latitude on the oceans results from the angle at which light strikes the surface. The amount of solar energy reaching a surface varies directly with the area exposed. In addition, the further the light must pass through the atmosphere, the more is absorbed and reflected by the atmosphere. Finally, the steeper the angle of the sun to the surface, the more highly reflective is water.

All three conditions combine to produce a relatively cold polar climate compared to the lower latitudes. That results in the formation of extensive ice cover which is even more reflective of solar energy. The result is Arctic and Antarctic sea water very near freezing all the way from the sea floor to within a few feet of the surface. At the lower latitudes, the increased solar intensity and decreased light reflection produce a much warmer surface layer down to the thermocline. In the mid-latitudes, the changing seasons produce a second thermocline within a few hundred yards of the surface during summer.

Moving from the shore to open water, not only does depth increase, but several other changes occur. Breaker activity is associated with the nearshore and foreshore regions.

Waters over the foreshore are said to be the littoral zone. This zone contains transitional life-forms, because the zone is alternately covered and exposed by the tides. It is also the ocean zone receiving the most sunlight. Extending outward to the edge of the continental shelf is the so-called neritic province. Beyond the neritic zone lies open water—also called deep water.

In addition to the three great oceans, Pacific, Atlantic, and Indian, are many smaller marginal seas. Adjacent seas are separated from the open sea by strings of islands. They are relatively open and interchange of the sea and ocean waters are relatively free. Enclosed seas are more cleanly separated from the neighboring oceans. The interchange of waters may be more limited.

Estuaries occur at the points where rivers meet the sea. The waters of estuaries are more saline than river water upstream, but less saline than ocean water further out. This land of "in-betweens" is a critical part of the marine environment. It provides a buffer between land and sea. It allows a gradual merging of fresh water as it flows into the oceans. It provides habitat for many important organisms. Estuaries are usually one of four kinds: bar-built, drowned river valley, fjord, or delta estuaries.

Sea levels have never been stable. Whether sea levels rise or fall is dependent on many factors. Whenever geological processes cause a rise of a land mass, it produces a lowering sea level relative to that particular land mass. An increase in ocean water volume produces an overall rise. Melting ice from the polar regions and glaciers can cause such an increase in volume.

At the present time, general sea levels are continuing a fifteen thousand-year rising trend. Some climatologists are forecasting a significant global warming as a result of greenhouse gas emissions from human activity. If that occurs, an even faster rise in sea level could occur during the next century.

The sea provides humans with many things. One of the most direct is an annual catch of marine organisms estimated at more than sixty–seven million tons. The major harvester of sea life is Asia. The most important organism harvested is marine fish.

DISCUSSION QUESTIONS

1. What is the bathymetry of the ocean bottom? What are the depths associated with each part?

2. What effects does water depth have in the ocean?

3. How does latitude affect the waters of the oceans and their inhabitants?

4. How does the ocean change from the coastline to open water?

5. What are marginal seas? How do they differ from the open waters of the oceans?

6. How do enclosed seas and adjacent seas differ?

7. What is an estuary? Why are the estuaries so important in the environment?

8. List and describe the major kinds of estuaries. How was each formed?

9. What are the main marine organisms harvested for human use? Is the marine harvest increasing or decreasing? By how much?

ADDITIONAL ACTIVITIES

1. Construct a scale model of the ocean floor. The model could be either a hypothetical one or a model designed to show the actual topography of one of the major oceans. If it is a model of an actual ocean floor, this activity will require a marine atlas for specific details.

2. Using a sandbox, construct a simple "ocean" and line it with plastic. Be sure the sides are gently sloping. Estimate the area and average depth to a given level. Use those figures to estimate its volume. Measure that volume of water and pour it carefully into your "ocean" to see how close your estimates were. Raise the water level by placing ice above the "shoreline" and allowing it to melt and run into the "ocean." When the ice has melted, does the water rise or fall? Then estimate the effect on the land area surrounding your "ocean" in your model.

3. Select a specific major body of water for each class member. Prepare a detailed report on the body of water from resources in your library. Share the information with the class.

4. Organize a debate on the desirability of eliminating the baleen whales and other plankton eaters to release the krill harvest for human consumption.

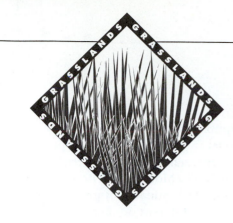

CHAPTER 18

Grasslands*

Courtesy Dr. Jay McKendrick, University of Alaska, Fairbanks.

Terms to Look for and Learn

Decreaser	Mixed-Grass Prairie
Forb	Overgrazing
Grassland	Shortgrass Prairie
Increaser	Tallgrass Prairie
Invader	Undergrazing

Learning Objectives

After reading this chapter and participating in the activities, you should be able to:

■ Define and discuss the characteristics of grasslands.

■ Differentiate shortgrass, mixed-grass, and tallgrass prairies.

■ Explain some of the natural and human-caused phenomena that produce grasslands.

■ Outline the history of the American prairie.

■ Describe the classes of vegetation based on grazing pressures.

* *Note to the reader.* Much of this chapter is taken directly from and reprinted with permission by Camp, W.G., & Daugherty, T.B. (1991). *Managing our natural resources*, 2nd Ed. Albany, NY: Delmar Publishers Inc.

Overview

Grasslands refers to the land areas of the world that tend to be naturally covered by grasses, grasslike plants, forbs, and shrubs as the primary vegetation instead of trees. Naturally, that excludes the true deserts and forested areas of the world.

Before the advent of agricultural settlement in the area that is the continental United States today, the treeline of the Eastern forests reached about as far westward as a line drawn roughly between Dallas, Texas and Chicago, Illinois. West of that lay open grassland stretching as much as fifteen hundred miles to the Rocky Mountains, where trees grew only along creeks, rivers, and wetlands. In your history books, the general area is frequently referred to as the Great Plains, a term which describes a geographical/physical region rather than a biome. Looking at the North American continent as a whole, the pre-European grasslands reached from the location of today's Edmonton, Alberta to Mexico City.

That huge expanse of land had many common characteristics. It provided habitat for many large ruminants. It is estimated that as many as forty-five million American bison lived there at one time. The animal, as tall as six feet at the top of its massive humped shoulders, weighed as much as a ton. The grasslands also swarmed with pronghorn antelope, perhaps in numbers as great as the larger bison. Mule deer, white-tailed deer, and elk thrived there. These and literally billions of smaller mammals provided food for carnivores of all sizes and habits from foxes to grizzly bears, and from rattlesnakes to eagles. Jackrabbits, cottontail rabbits, mice, voles, prairie dogs, and others abounded. One early naturalist estimated a single prairie dog town at twenty-five thousand square miles with a population of 400 million of the unique, barking animals.

Yet, its two most striking features were its topography and its dominant vegetation. The plains are known for their flatness. Although there are hilly regions, in general the topography is level with an almost imperceptible upward slope of about ten feet per mile (two meters per kilometer). That represents a slope of about 0.2 percent, from the Eastern treeline to the base of the Rocky Mountains. The vegetation was almost completely dominated by grasses. There were more than sixty different dominant species of grasses in various parts of the prairie. As many as two hundred different species of plants in total grew in the region.

In other parts of the world, there are different names for the great grassland regions. In the Asian continent and eastern Europe, the great Steppes stretch for thousands of miles. The African veld is a vast and rich grassland. We have all seen television documentaries about the huge herds of herbivores in Africa. What was usually not pointed out by these documentaries is the absolute reliance of those wild animals on the grasses of the veld for habitat and food.

In parts of South America, there are great regions of grasslands known as Pampas. In Brazil, the grasslands are called Campo. In other parts of the Americas, the term used to describe some grasslands is Savannah. In Spain, it is Protero—in the Philippines, Kogonales. The Arctic Tundra is another great grassland.

Grasslands in the United States

For the vast majority of Americans who live in the eastern or midwestern states and in large cities, the word "grasslands" is probably almost a meaningless term. But, the vastness of the original grasslands in this country and the significance of their contribution in the development of our nation may very well be surprising. Indeed, the size and importance of the grasslands on a worldwide basis today can hardly be overstated.

Over twenty-five percent of the U.S. land area was covered by grassland when the first European settlers landed on the east coast. In general, the grasslands in what is now the continental United States consisted of three broad types: tallgrass prairies, mixed-grass (also known as transition) prairies, and shortgrass prairies. In addition to these major kinds of vegetation there were bunch-grass prairies, mesquite prairies, "parkland" prairies (which consisted of a mixture of trees and grasses), chaparral, and sage (see Figure 18–1).

The most striking and most productive of the grasslands was the **tallgrass prairie,** which covered much of the area between the present states from Indiana to Nebraska and southward to Texas. The land that produced the tallgrass prairies is very rich. The topsoil is deep and abundant in available plant nutrients. The climate produces a fairly regular and adequate supply of rainfall with winters that vary from merely cold to extremely harsh.

The result of that combination of adequate rainfall and rich soil produces a lush growth of perennial prairie grasses that might reach six feet in length at the peak of a good growing season. It also makes for

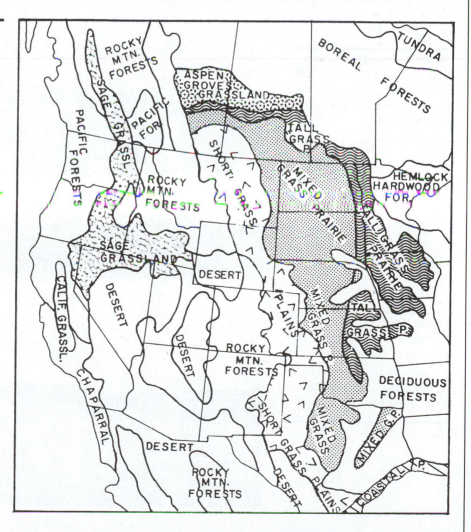

Figure 18–1. Major grasslands in what is now the Western continental United States and parts of Canada and Mexico, prior to European settlement. (**Source:** Bluemle, J.P. (1975). The prairie land and life. Washington, D.C.: United States Department of Agriculture. Forest Service and North Dakota Geological Survey.)

very productive agricultural land—in fact, some of the richest farm land in the world. When the settlers came, they recognized the richness of the land and cleared it for farm production to feed a growing and hungry nation. Thus the tallgrass prairies are almost gone today (see Figure 18-2).

The **shortgrass prairies** stretched from central Canada through parts of the Dakotas to central Texas. In general, the soils of that region were also very rich in plant nutrients with deep topsoils, as were the tallgrass prairies further east. However, this area has a much lower average annual rainfall and it is much less dependable for agriculture than that of the tallgrass prairie region further east. At the peak of a good growing season, these grasses might reach a height of up to three feet (see Figure 18-3).

Between the shortgrass and tallgrass prairies were the **mixed-grass prairies**. This region experiences more average rainfall but it tends to be less dependable than that of the tallgrass areas. The maximum heights of the grasses fell between three and six feet in good growing years, but good years are fewer and further between.

There still exist at least parts of all three original grassland types. However, at one time or another, the vast majority of the land in all three types has been

Figure 18–2. This Tallgrass Prairie is located in the Flint Hills area near Riley, Kansas. (Courtesy Dr. Jay McKendrick, University of Alaska Fairbanks.)

Figure 18–3. This is an example of a Shortgrass Prairie. It is located in the Custer State Park in South Dakota. (Courtesy Dr. Jay McKendrick, University of Alaska, Fairbanks.)

plowed under. There are still substantial areas of shortgrass prairie, primarily because it has less agricultural productive potential. Virtually all of the shortgrass region requires irrigation to make cultivated farming profitable. Almost all of the original tallgrass prairie land has been converted to farm production or other human development, primarily because of its high productive potential.

History of the Grasslands in the United States

Before the arrival of the European explorers and settlers, the center of this continent was a land of vast, uninterrupted grasslands. There has been a long-standing controversy among ecologists as to whether grasses are the true climax vegetation of the region. Many believe that much of the region could easily support forests. In fact, in much of the grasslands the rainfall is quite adequate for that. It may be that lightning-made fires caused the destruction of the forests that could have flourished there in prehistoric times. Later, the

combination of natural (mostly lightning-caused) fires and fires set by the Indians as a hunting and land management technique, caused the eventual elimination of the trees that would have been needed to provide seeds for natural reforestation.

Interestingly, spots of grasslands occurred all the way from the Pacific to the Atlantic, even before settlers began to clear the land. Wherever fire had destroyed the forest, rapidly growing grasses would quickly fill the burned-out areas until new trees returned to dominate the smaller grass plants. Often the fires were caused by lightning or other natural causes. But, just as often, they were set by Indians for hunting (as will be discussed later) or specifically to establish grasslands to attract bison and other grazing animals. In these cases, the grasslands were temporary and did not truly belong to the great grassland biome (see Figure 18-4).

In the final analysis, it does not make any difference whether the grasslands existed as climax vegetation or in response to the frequent prairie fires. The fact is that those vast areas were virtually empty of trees except along streams when the first European settlers arrived.

Figure 18-4. This range fire is in the Flint Hills area near Riley, Kansas. (Courtesy Dr. Jay McKendrick, University of Alaska, Fairbanks.)

At that time the only checks on the population of herbivores were predators, disease, and climate. Predators included the large cats, wild dogs, wolves, and native American Indians. The Indians took just enough animals for meat for food, hides for clothing and shelter, and bones for tools. Their meager needs would have hardly affected the population levels of wildlife on the prairie. On the other hand, their hunting methods could be quite destructive in one sense. To harvest enough wild game—much of it larger and faster than the Indians themselves—they often resorted to the use of prairie fires.

That may seem a remarkably destructive way to gather food, but there are a number of reasons that it was done that way. In the first place, the Indians were on foot. The Spanish were the first to introduce horses into this continent. A group of Indians on foot might not be very successful in harvesting deer, antelope, or bison in quantities large enough to provide a reliable food source for their people.

It is also important to remember that the grasses were not actually eliminated by the fire. The day after a prairie fire, the soil would be bare, but most of the nutrients in the plant matter remained after the fire as ash and were returned to the soil. Even more importantly, many of the seeds and roots remained alive in spite of the heat from the fire. Some seeds were released precisely because of the heat from the fires. When the growing season was favorable, the new crop of grass would begin to grow with the next rain, and the land would very soon become green and rich again.

Another side effect was that trees were virtually eliminated from most of the grasslands. Unlike grasses, trees take years to grow large enough to produce seeds. Prairie fires killed the trees that would otherwise have become dominant in the most productive of the grasslands.

Where prairie fires were fairly frequent and regular, the shrubs and **forbs** (broad-leaved, flowering, seed-bearing plants) were also eliminated for the same reason. A grassland without shrubs and trees to compete with grasses for sunlight, nutrients, and water produces lush grazing and has a much higher carrying capacity for herbivores.

Later, the English, French, Spanish, and finally the American settlers moved westward to make a living from those same great grasslands and herds of herbivores. In an earlier chapter, we discussed the effect of trappers and market hunters on game populations. They took some game animals for food, but by far the major use by people for the wildlife was furs, hides, and feathers.

Soon, however, other settlers moved westward from the American colonies along the east coast and northward from Mexico. At that point in time, agriculture became necessary for human survival. Even if that had not been the case, most of the new settlers came with the specific intent of earning a living from farming or ranching.

The grasslands offered one particularly important source of potential wealth for enterprising newcomers. Tall, abundant grass on the open prairies offered free grazing for potentially millions of cattle. How else could one make a profit more surely than by putting cattle into an area where they could eat, drink, and multiply freely at no cost to their owner? The cattle could then be slaughtered for meat and hides, with only a minimum of financial investment by the cattle rancher.

The only problem was how to get the cattle to the eastern and midwestern markets. The coming of the railroad offered a solution to that problem near the middle of the nineteenth century. As a result, a number of great fortunes were made by ranchers who were strong enough and ruthless enough to survive in the open-range cattle industry.

The Homestead Act of May 20, 1862 granted individuals 160 acres of land, mostly in the grasslands belonging to the federal government of the United States. Settlers were to live on the land or cultivate the soil for a period of five years. This act brought thousands of settlers to the grasslands.

A great part of American folklore comes from the time between the mid-1800s and the early 1880s. That was the time, particularly during the two decades just after the American Civil War, when the open range cattle industry became a very important source of wealth in this country. By this time, most of the tallgrass regions had been cleared for cultivation. The somewhat less productive, but still rich lands of the shortgrass prairies were still mostly in their original state.

It was also a time, coincidentally, when nature played a cruel trick on the people there. For over twenty years, there was a climatic cycle in which unusually dependable rainfall and mild winters prevailed. The cattle ranchers pushed their herds to sizes that overtaxed the long-term carrying capacity of the grasslands in the shortgrass prairies. Great ranches sprang up and the so-called "cattle barons" grew rich and powerful.

Farmers who wanted to clear the land for plowing were unwelcome and the cattle owners kept them away, often by force. Fencing material was largely unavailable in the region, and in any case, would not have been allowed by the cattle ranchers. After all, the grasslands were a vast region that belonged to everyone, but to no one in particular. The grass was a gift of nature to anyone who was able to take advantage of it. This was a huge commons area and the cattle industry grew up explicitly to exploit the wealth of the commons. In Chapter 3, the concept of the commons was explored in some detail.

Overgrazing occurs when animals eat too much of the desirable grasses. When that happens, less desirable plants (as you will see later in this chapter) replace the more desirable grasses.

By the mid-1880s, the effects of overgrazing were beginning to be obvious everywhere. The abuse of the grasslands occurred for a number of reasons. First, the cattlemen were not ecologists. They had little understanding of the impact of their activities on the ecosystem. Beyond that, the fact was that any single cattle owner could have little lasting effect on the whole grasslands biome. Finally, the motive of the cattlemen was profit, not conservation.

Thus, cattle populations were pushed beyond the ecological limits of the grasslands, with the cattlemen hoping each year for mild weather and adequate rain to make the grasses grow more rapidly. The land began to dry out more than usual because of the loss of soil cover of grasses. The organic matter in the topsoil was becoming depleted, thus further decreasing the soil's water holding capacity.

Then, when the vast herds of cattle were beginning to be less healthy because of degraded grazing, calamity struck. The winter of 1885–86 was a particularly cold and harsh one. Millions of cattle died or became sick as a result of the unusual cold combined with the lack of adequate grass for grazing. The following summer the grazing was still inadequate because of the lasting effects of the overgrazing of the previous decades. Then the winter of 1886–87 followed with even colder and more harsh conditions.

The open-range cattle industry was shaken to its foundations by this triple catastrophe. True, the ranchers had made such an event almost inevitable because of their management practices. At the same time, one cannot blame any single rancher for the disaster. One cannot even contend that any single rancher made an economically incorrect decision in overgrazing the grassland.

The whole thing occurred because the prairie was a rich commons area with value to many persons. It belonged to no one, but could be used by anyone who voiced a claim for it and was willing to stand behind that claim. Finally, it was an area in which no one was responsible for making and enforcing overall management decisions.

Barbed wire had been invented in the 1870s, but had been violently resisted by the cattlemen. They were the ones, they reasoned, who had homesteaded or taken the land from the Indians and who had labored to develop huge herds of cattle. They had "tamed the prairie." Settlers who wanted to clear the land for farming would have to build fences to keep cattle out of their fields. Thus fences were seen by the cattlemen as invasions of "their" land—most of which was not legally their land at all, of course.

The resistance to the settlement and fencing of the open ranges soon became so violent that laws against fence-cutting became extremely harsh. But, it was not until the disasters of 1885–87 that the final fate of the open ranges was settled. After that, fences came to be largely accepted as necessary to the future of the cattle industry.

Grassland Vegetation

Several major types of vegetation predominate in the grasslands: grasses, grasslike plants with fibrous root systems, forbs, and shrubs (many of which have tap-root systems). Each of the types of vegetation must compete for moisture, sunlight, and nutrients in the prairie regions.

Grasses are particularly valuable from two standpoints. First, the complex root systems are very adept at holding the soil in place and increasing infiltration of the rains, thus decreasing runoff and erosion. Second, grasses provide a high quality source of nutrition for grazing animals (see Figure 18-5).

The proportion of grasses varies substantially from year to year, and even during a single growing season. Grazing rates particularly affect the relative amounts of the four major types of vegetation present. Light grazing actually benefits the grasses in comparison to other types of plants. As long as over half of the length of a grass stalk is not eaten, grazing causes little damage to the grass plant. Unfortunately, overgrazing is very detrimental to the grasses. When most of the leaf is eaten, it is more difficult for the plant to recover—particularly if the plant is already stressed by dry, hot weather. In addition, **undergrazing** can be detrimental to grasses. Undergrazing occurs when too little grazing takes place to keep taller, but less desirable plants from growing in an area.

Varieties of grass that are easily damaged by even moderate grazing are called **decreasers**. These are often the ones that grazing animals find most desirable. Given the choice of which grasses to feed on, the grazers eat the palatable ones and leave the unpalatable ones. Thus, the more desirable grasses tend to be eaten until they are damaged. In other cases, the decreasers are simply more susceptible to damage from grazing by trampling because they are not very hardy plants.

Other grassland plants that tend to thrive under heavy grazing are called **increasers**. Many increasers are successful merely because the grazers find them unpalatable and so avoid eating them. In some cases,

Figure 18–5. This rangeland area near San Diego, California has been invaded by chaparral and other tap-rooted shrubbery. The brushy conditions developed because of extreme undergrazing. (Courtesy Dr. Jay McKendrick, University of Alaska Fairbanks.)

the plants are simply better able to get at the limited soil water and nutrients.

Plants that move into an area after it has been badly overgrazed are called **invaders** (see Figure 18-5). Oddly enough, badly undergrazed areas are also subject to being taken over by less desirable plants. Prickly cactus and some shrubs are particularly potent invaders.

Summary

Grasslands are land areas on which the climax vegetation is grasses, grasslike plants, forbs, and shrubs. Grasslands are some of the most productive land on Earth. They tend to grow in arid and semiarid climates. In North America, natural grasslands ranged from the midwest to the Rocky Mountains. Small pockets of grasslands occur in almost all parts of the planet. Actual grasslands are the result of limited supply of available soil moisture. Wherever disease or fire produces clearings in the forests, temporary grasslands spring up. In regions where fires are frequent—either natural or man-made—grasslands can become seemingly permanent fixtures of the landscape.

Grasslands are particularly productive of forage. Grasses and broad-leafed forbs tend to be more palatable and digestible to ruminants and other herbivores than trees and shrubs. The result is that animal life, particularly larger mammals, flourishes in grasslands much better than in forests. Climax forests are almost like deserts for most mammals, yet grasslands are ideal for such animals. From massive prairie dog towns to herds of bison reaching from horizon to horizon, the prairie could be called the crucible for mammalian life.

Yet, the dominant life-form on the prairie is grass—over 200 species in North America. Grasses prosper in harsh conditions. They become dormant in droughts. The plants may die to the ground during winter or as a result of wildfires. But, each spring or after each rain, life springs up again as the annual grass seeds germinate or the perennial grass roots produce new sprouts.

In the prairie regions with greatest precipitation, the grasses grow tall, sometimes averaging as much as six feet. In the most arid regions, only smaller grass plants can survive. Still, during the best growing seasons, even the shortgrasses may average as much as three feet in length.

Overgrazing has been a particular problem on the grasslands. This is true because of the limited agricultural uses for the land. Limited total annual rainfall or frequent droughts mean that without irrigation, row crops are not reliably profitable in the grasslands. At the same time, the rich grasscover was an obvious asset for cattle ranchers. Sadly, the pasture management practices developed in more humid regions did not work well in the arid grasslands. The result has been a tendency toward overgrazing which further lowers the grazing capacity. That in turn produces even more severe overgrazing.

In the tallgrass prairies, the soil was extremely rich and the annual rainfall was adequate to support cultivated crops. In addition, the land was not covered with forests that had to be cleared away before farming could be done. Virtually all of the former tallgrass prairies have been turned under for cultivated agriculture. Only a few limited reserves remain intact.

Grasses tend to be resistant to grazing and trampling by livestock. But not all grass species are equally tolerant to such use or abuse. Species that are readily damaged by grazing are termed decreasers. Increasers are species that tend to do well under range grazing. Invaders are species that tend to move into areas that have been overgrazed.

DISCUSSION QUESTIONS

1. Why do some parts of the world tend to be naturally covered by trees and other parts by shrubs, grasses, and other small plants?

2. What are the characteristics of grasslands? Consider rainfall, vegetation, soils, and animal life.

3. Discuss similarities and differences among tallgrass, mixed-grass, and shortgrass prairies.

4. Outline the history of the American prairie.

5. What happens to the vegetation in a grassland area if there is overgrazing for an extended period of time? Undergrazing?

6. Why are grasslands so productive in terms of mammalian species?

ADDITIONAL ACTIVITIES

1. Locate a grassed area that is not being grazed or mowed (i.e., in a natural state). Measure and mark off a typical 10' × 10' area. Estimate the percent of the 100 sq. ft. area that is covered by grasses and by shrubs. Estimate the numbers of plant species and animal species. Then, locate a grassed area that is being mowed or grazed moderately. Measure and mark off a typical 10' × 10' area. Determine the percent of the area that is covered by grasses and by shrubs. Estimate the numbers of plant species and animal species. How do the two areas compare? What do you think made the difference?

2. Look through the last several years of the *National Geographic* magazine. Find as many articles as you can about parts of the world with important grasslands. Bring the editions to class for an open discussion on the grassland ecosystems represented. What problems are they facing? How are people using them?

3. Find a natural grassed area. Dig up several grass plants and single examples of some of the other plants present there—particularly the woody ones. Try to identify the plants. Rinse off the root systems. As you compare the different plants, determine which would be more effective in preventing accelerated erosion and why? Which would be more useful to cattle and why?

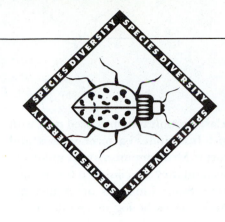

CHAPTER 19
Species Diversity

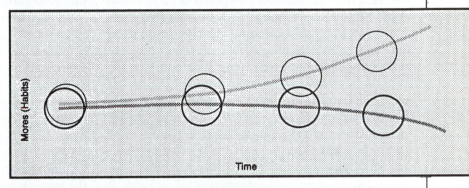

Terms to Look for and Learn

Binomial

Biodiversity

Character Displacement

Compartmentalization

Deforestation

Fuelwood

Green Revolution

Hybridization

Linnaeus

Monoculture

Mores

Niche Overlap

Rappoport's Rule

Speciation

Learning Objectives

After reading this chapter and participating in the activities, you should be able to:

■ Explain the concept of biodiversity.

■ Discuss why it is so difficult to estimate the number of species on Earth.

■ Discuss why many scientists believe that biodiversity contributes to ecological stability.

■ Outline the biological processes involved in species development and differentiation (speciation).

■ Explain why biodiversity is so much greater in the tropics than in cooler regions of the planet.

■ Explain and compare natural and human-caused species loss.

■ Explain the effects of tropical deforestation on biodiversity.

311

Overview

As you look around, you can see the great diversity of nature. There are animals and plants of all sizes and shapes. If you tried to name all of the different kinds of animals in your community, the task would seem impossible. Even limiting the list to mammals would still leave a gigantic task: dogs, cats, cattle, mice, humans, and many more. Just listing the different species of butterflies would be a big job. In fact, there are at least sixty-seven butterfly species in Britain alone. In the United States, there are more than eleven thousand species of butterflies and moths.

This chapter will outline the difficulties of estimating and understanding the diversity of life on Earth. It will outline the process of speciation (the development of new species). It will establish links between species diversity and the environment, latitude, and size of organisms. Finally, it will present information on the natural process of species loss and the effects of human activity in speeding up that process.

Counting Life-Forms

Imagine the difficulty of identifying and cataloging all of the species of living organisms on Earth! Scientists have been working on that job for thousands of years, yet they have identified only a tiny proportion of all the species on the planet. A natural historian in Sweden during the seventeenth century, **Linnaeus**, developed a **binomial** (having two parts) system of species classification using Latin.

Latin was important to use for two reasons. First, it is a "dead language," which means it is not in common use by any society. That means it is more stable than "living languages," which are constantly evolving through everyday use. Second, Latin had become accepted as the language of science. Scientists worldwide have studied and used Latin for many centuries.

Linnaeus' classification system is used only with higher organisms. It excludes bacteria, viruses, and other such life-forms. Microorganisms were unknown in Linnaeus' time and they are classified today using a different system. Since Linnaeus developed the system of classifying species, biologists have described and classified about 1.4 million higher organisms. But, even that number is not exact because many organisms have been classified under more than one name. That happens when more than one biologist discovers, describes, and names the same organism without being aware that others have already done the same thing.

In fact, various scientists have estimated that anywhere from one million to 1.8 million different species have been identified, described, and classified. Of that number, probably twenty-five percent are from the order Coleoptera (beetles). Only about two percent are vertebrates (having a backbone or spinal column), including all mammals, birds, reptiles, and so forth. About eighty percent are terrestrial (live on land) and the other twenty percent are aquatic (live in the water) (see Figure 19-1).

But, how may species are there? "Nobody knows," is the simple answer. Until recent years, most biologists who would venture a guess estimated the number at three to five million. But, one biologist recently estimated that there are at least thirty million species of insects in the tropics alone. To reinforce those figures, one can look at the number of new species being described each year.

There are only about three new birds and fifteen new mammals described each year. Yet, for each deep ocean sediment core and every tropical tree-top collection taken, at least ninety percent of the insects found are new to science. Probably a reasonable estimate would be that somewhere in the neighborhood of ten to fifteen million species of higher life-forms exist on the planet with the vast majority being insects.

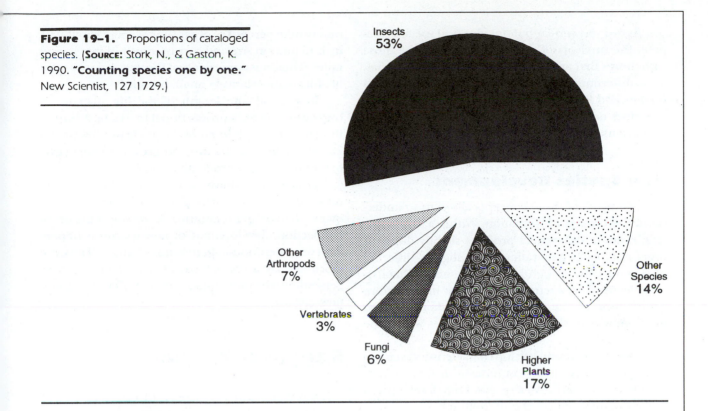

Figure 19–1. Proportions of cataloged species. (**Source:** Stork, N., & Gaston, K. 1990. **"Counting species one by one."** New Scientist, 127 1729.)

Insects 53%

Other Arthropods 7%

Vertebrates 3%

Fungi 6%

Higher Plants 17%

Other Species 14%

Speciation

Where do all of these species come from? The process by which new species develop and existing species differentiate is known as **speciation**. To understand speciation, refer back to Chapter 1 regarding the sections on competition and niche.

Review of Competition

Speciation is the development of new and unique species in the process of evolution. Closely related species can live close to each other. But the competitive exclusion principle says "one species—one niche."

Speciation occurs when similar species differentiate their niches within the same system. It may be based on the thickness of the branches each variety likes or the size of the fruit they prefer, or the height on a hillside or cliff they prefer for nesting. Something produces a unique niche. Closely related species develop muted competition by specializing some aspect of their **mores** (habits or preferences in behavior) within the system.

For its survival, every species must partition out some source for its essential resources for which it is the most efficient species. Within any ecosystem, each species will specialize on some unique aspect of the resources.

Niche overlap occurs when more than one species can use the same resource at the same time and in the same place. For instance, two species of birds may eat the same plant's seeds. But if that seed is only one of many foods for both kinds of birds, then there is only muted competition—there is a niche overlap.

Character displacement is the process by which two similar species with very similar, but not identical, niches change their living habits gradually over time. That allows for separation of niches.

As two very similar but not identical species compete, the survivors of each species tend to be those with mores that are different from the other species. That differentiation pushes the two species' niches further and further apart until the muted competition does not threaten the survival of either species (Colinvaux, 1986) (see Figure 19-2).

New Species Development

Combining all of those concepts allows us to understand the process of speciation. Within a single species, random differences produce slightly different organisms each generation. The slightly differing organisms exhibit slightly different mores (habits). That leads to character displacement in the future generations of the organism. That then leads to development of slightly different niches within the same species.

Later generations develop characteristics even further apart as their niches continue to change. Further evolution results in two new species, both closely related to the original but each slightly different.

Size and Diversity

Humans are among the largest 1/1000 of one percent of all organisms. In fact, songbirds and mice are among the largest one percent.   The remaining ninety-nine percent of organisms average about 0.01 inch (3 mm) in size and consist mostly of insects and mites. Thus, it is apparent that the vast majority of all life-forms are relatively small; but why?

In general, the length of the life span of an organism is directly proportional to its size. A butterfly may go through its entire life cycle in a few weeks, but an elephant must live and grow for many years before reaching reproductive maturity.

Because they have shorter life cycles, smaller organisms pass through generations much more quickly than larger organisms. As we saw in the previous section, development of new species is dependent on generation-to-generation changes. Thus, the more quickly a species passes from generation to generation, the more quickly it can differentiate into new species.

Rappoport's Rule

The number of different species in a given ecosystem is referred to as its biological diversity. Another name for that is **biodiversity**. It is well known that biodiversity is greater in the tropics than in the higher latitudes. The causes of this fact are less well known, however.

Stevens described what he referred to as **"Rappoport's Rule"** named after Eduardo Rappoport (1871–1950) who had mentioned the effect in earlier research. In scientific terms, Rappoport's Rule indi-

Figure 19–2.   Character displacement allows members of the same or very similar subspecies to become gradually more distinct through natural selection.

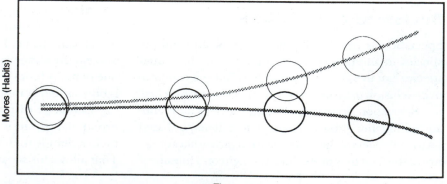

Time

cates that there is a relationship between an organism's range and the latitude in which it lives. An organism's range refers to the size of the area in which it lives. The latitude range of a species is the difference between its northern and southern limits.

In general, the latitude ranges of species becomes greater as the latitude becomes greater (i.e., the further from the equator, the greater the range). The result is that biodiversity is greater in the tropics than in cooler regions. But why?

In temperate and cold regions, plants and animals have relatively great ranges. That means, for instance, that a single species of maple can live in a wide variety of climates from the forests of Georgia where summer temperatures can often exceed 100° F to the forests of Central Canada where winter temperatures may fall to –70° F. They can survive such extremes, therefore they have a wide range.

This ability to survive in a widely changing environment became embedded in such species naturally. They evolved in regions where temperatures have always varied greatly from winter to summer. Tolerance to wide temperature ranges was an advantage to species evolving in the higher latitudes (see Figure 19-3).

At the same time, species that live near the equator evolved in environments where temperatures never vary widely. At Key West, the temperatures seldom drop below 50° F and seldom climb above 90° F. The ranges can become even more compressed closer to the equator. In species that evolved in such locations, the ability to tolerate wide temperature variations did not provide biological advantage. Thus, natural selection did not provide that characteristic. The result is that tropical species are not particularly tolerant of wide temperature changes. They have very limited ranges.

The Tropics and Diversity

What does all of that mean? Plants and animals far from the equator have greater ranges. They compete with each other over larger areas. Organisms near the equator have smaller ranges. They compete with each other over smaller areas. The result is that larger numbers of species are needed to populate the same amount of land or water in the tropics than in cooler regions. That means the tropics have more biodiversity than the rest of the planet.

Figure 19–3. In the higher latitudes, temperature extremes vary over a great range. Plants and animals evolving in such climates must be able to tolerate such environmental extremes. This photo shows the same location in July and January. (Photo courtesy Bill Camp.)

Rappoport's Rule holds true for trees, mollusks, amphibians, and generally all forms of life except migratory birds, which move great distances as a part of their life cycles. When we combine the difference in biodiversity discussed above and the difference in solar energy reaching the surface for the production of food energy (see Chapter 1), the result should be clear.

Tropical regions produce both a greater diversity of life-forms and a greater quantity of living organisms than do the cooler regions of the planet. The result of Rappoport's Rule is that a much greater diversity of life-forms can live in a tropical rainforest ecosystem than in other ecosystems of the same size.

In addition, water is probably the single most important limiting factor on biomass development in an ecosystem (see the section on Population Limits in Chapter 2). So you can see that tropical rainforests are a special place. They combine both the greatest biodiversity and the most lush production of biomass on Earth. It is little wonder that biologists generally consider tropical rainforests to be the greatest storehouse of genetic diversity on the planet.

Diversity and Stability

Historically, biologists have believed that the more complex an ecosystem, the more stable it should be. Think back to the first two chapters. In an ecosystem with only two species, a producer and a consumer with a predator-prey relationship, the outcome would be inevitable. First the consumer (predator) species would increase in numbers until it destroyed the producer (prey) species. At that point the predator species would also disappear for lack of a food energy source.

In real-world ecosystems, such a simple situation would be unlikely to ever occur. The introduction of more members of a food web should produce buffering effects that would prevent wild swings in numbers for any single species. The result, biologists have argued, is that diversity is inherently good because of

its stabilizing effect. McNaughton quoted a biologist from the 1960s as saying that simple ecosystems are "more subject to destructive oscillations in population… and more invasions" (McNaughton, 1988, 204).

In recent years, scientists have used computer modeling to show that simple diversity may not produce greater stability after all. McNaughton pointed out that complex ecosystems tend to become effectively blocked, or segregated, into smaller subsystems, each of which is actually a more simple ecosystem. The term used to describe this segregation of simple ecosystems as parts of a more complex, larger ecosystem is **compartmentalization**. Basically what that means is that even though they may interact, food webs are important in ecosystems. In effect, each food web becomes almost an ecosystem in itself. In a complex ecosystem, there may be many different food webs (see Chapter 1). Often the food webs overlap (are connected). One food web can fluctuate wildly. In so doing, it affects connected food webs in unforeseeable ways.

Effects of Species Loss

The effect of the loss of a single species may be insignificant to human life—or it may be very important. For instance, a seemingly unimportant species of fungus in a tropical rainforest may someday prove to produce the only effective preventive medication for some form of human cancer—*probably not, but who knows?* Bread mold produced penicillin, after all. The antibiotic, aureomycin, came from an organism in a soil with low fertility at the University of Missouri at Columbia. One of the most promising treatments for some forms of human cancer involves toxol, a substance produced only by the Pacific yew. What if something had happened in the past to one or all three of those species? What if one of the species we lose today had even greater potential for use to cure AIDS, but we had not discovered it yet?

The effect of species loss on food webs is more direct and clear, however. We have seen (in Chapter 1) how each organism is part of one or more food webs. Every member of the food web is at least partially related to every other member of the web. For instance, a grizzly bear would hardly seem dependent on a species of alga for food. Yet, if the alga were gone, the herbivorous fish might also be gone. In that case, the carnivorous fish would become less available. That, in turn could adversely affect the grizzly bear population. As we have seen, food webs are always interconnected with other food webs. Thus, the impact of the loss of even a single species could have unforeseeable consequences to the ecosystem.

Diversity and Human Life

Why is this argument so important? Scientists who believe that diversity leads to stability argue that species loss is dangerous to human survival. They contend that the more biodiversity in our ecosystem, the more stable it will be. Those who argue that diversity and stability are not linked find less reason to preserve species from extinction.

The issue is not settled. Until it is, we should probably take the more cautious approach. The danger of being wrong about biodiversity and ecological stability is too great to take a chance. Until (and unless) science ultimately confirms that diversity does *not* lead to stability, we should work to maintain as much biodiversity as possible on the planet. We should do that to make more likely the long-term survival of the human species.

Even if we eventually confirm that biodiversity is unnecessary for stability, we should work to maintain as much biodiversity as possible on the planet. We should do that because of human ethics and because we appreciate the richness of life—remember Chapter 1. That is an environmentalist's position, not an environmental scientist's position. The question is not whether we should work to prevent extinctions caused by human activity. The only question is what the primary motivation should be—human survival or human values.

Species Loss as a Natural Process

New species are constantly being added to the planet through mutations and natural selection (speciation). This is a natural process that has been going on since the beginnings of life on Earth.

By the same process, species are constantly becoming extinct. There are a number of biological causes of species loss. Biological causes of extinctions can be thought of as changes in relationships—predator-prey relationships or relationships among competitors. Some examples of relationship changes include the following:

- Population fluctuations of predatory species lead to extinctions of prey species.

- New diseases resulting from genetic mutations of existing pathogens produce species extinctions.

- Existing competitors become more effective.

- New competitors emerge from existing species which were not competitors in the past. This can result from evolutionary changes in other organisms. Or, it can result from adaptations forced on the new competitor when its traditional prey becomes unavailable.

- Existing predators become more effective or lose alternative prey.

- New species are introduced into the ecosystem.

In general, changes in the environment are by far the most important cause of extinctions. As we will see in Chapter 20, climatic changes are a normal part of life on Earth. Worldwide temperature changes occur because of fluctuations of our orbit around the sun, wobbles in the axis of the planet's tilt toward the sun, and many other reasons. Such temperature changes occur with some degree of regularity. But unpredictable changes also occur as a result of volcanic eruptions, changes in atmospheric make-up, and other causes including human activity.

The only thing that is certain is that the world's climate is constantly changing. More importantly, local climate changes are even more frequent and extreme. Short-term droughts can give way to flooding in the same locations. Temperature highs and lows can fluctuate greatly from year to year.

Now, think back to the definition of niche, and the effect of these natural changes should become clear. Each species must have its own niche in which it has competitive advantage over every other species. When the environment changes, even for a short time, the nature of the competitive advantage changes and that niche may no longer exist. For an extreme example, imagine a drought in which a large lake dries up completely. Only those species that can survive in a nonaquatic environment survive. Other species in the lake, like most fish, simply die.

In fact, when the environment of a given ecosystem changes, the organisms in it have only three choices:

- They can adapt to the change and make adjustments in their mores.

- They can move to a location where the environment is more suitable.

- Or they can become extinct. (Bluntly put, they can adapt, move, or die.)

During the early part of the Pleistocene period (from 12,000 to 600,000 years ago), worldwide climate changes produced a marked cooling in the surface layers of the oceans. As a result, as many as twenty percent of the mollusk species along the Eastern seaboard from Florida to Maryland became extinct. During the same period, the loss of mollusk species in the regions of the Netherlands was as great as fifty percent (Vermeij, 1986).

The effects of environmental changes can be most pronounced in the tropics. As we have seen, tropical regions have the greatest biodiversity on the planet. Tropical plants and animals also tend to have very limited ranges. Many tropical organisms exist in only a single region, forest, lake, or similar small ecosystem. Droughts, fires, or other natural occurrences need not be very large to cause the complete extinction of such species.

Human Activity and Species Loss

Species losses and development of new species are natural phenomena, as we have seen. Both processes have been going on as long as there has been life on Earth. They will continue as long as the planet continues to support life of any kind. That is not a result of human activity, but a potential problem emerged when humans developed technology.

As we have seen, it is environmental change that causes most species loss. Only a few events in the past million years have affected the environment of the planet more than human technology. An earthquake may reshape part of the Earth's surface. It may even change the direction of a river. But, the drainage and pumping of groundwater from the Everglades has completely changed the ecosystem of south-central Florida. Human-caused emissions of CO_2, largely from internal combustion engines, has measurably increased the concentration of that major greenhouse gas in the past century. The burning of oil from the wells in Kuwait during the year after the Gulf War in 1991 slightly increased atmospheric CO_2 levels.

Environmental changes are not the only human-induced cause of species loss. A classic example of change in relationships came about with the improvement of human weaponry. When the earliest humans had only their hands and teeth as weapons, they were hardly a threat to most large animals. But, when they discovered that stones and clubs could be used as weapons, their relationships with many animals changed. Humans became a new predator for those new prey animals.

With the development of spear throwers and later bows and arrows, even more animals became prey for human hunters. With the domestication of horses, humans could travel farther and faster and could outrun much faster animals. The invention of guns changed even more predator-prey relationships, with people becoming much more effective as predators and as competitors.

The result of those technology changes has been the extinction of a number of species of large birds and animals. Improvements in hunting technology have contributed to the elimination of at least forty-five genera of land mammals in South America and thirty-two genera in North America alone in the past fifteen thousand years (Vermeij, 1986).

Many scientists believe that the extinction of the great wooly mammoth was the result of hunting pressure from prehistoric humans. The most notable modern example of human-caused extinctions from hunting is the passenger pigeon. At one time, the population of passenger pigeons in North America was thought to be in the billions. The bird flew in enormous flocks. John James Audubon once estimated a single flock at over one billion. Such huge flocks would move into an area, strip the vegetation, and leave the land bare. Americans "declared war" on the birds, killing all they could. They were killed for food, for sport, and just to get rid of them. See Figure 3-2 for a photograph of the last passenger pigeon.

Soon the flocks began to disappear. Several state legislatures passed laws protecting the remaining birds, but their action came too late. A few birds were captured, but failed to breed. As a result, the last known passenger pigeon died in 1914 in the Cincinnati zoo (see Chapter 3).

Other North American birds that have been hunted to extinction are the Carolina parakeet, heath hen, and Labrador duck. The Carolina parakeet lived in a wide range of the eastern United States. It also traveled in huge flocks. With a screaming cry, it liked to eat the seeds of cockleburs. Sheep herders liked the birds because they helped control the cocklebur populations, thus reducing the problem of cockleburs getting into the sheep's wool. Unfortunately, the birds had beautiful feathers which were prized for

ladies' hats in this country and Europe. The feathers were worth more to hunters than the birds, so the Carolina parakeet was hunted to extinction (see Figure 19-4), also in 1914.

The heath hen was prized for food and it was easy to find and kill. By the time people realized the bird was getting scarce, it was almost too late. A heath hen sanctuary was established and the flock began to grow, but a wildfire destroyed the sanctuary and killed most of the remaining birds. The last known heath hen died in 1932. The Labrador duck was hunted for its meat and to provide feathers for pillows and feather beds. The duck was actually extinct before biologists even noticed it was becoming scarce.

Human-caused extinctions of larger mammals, birds, and fish from hunting and fishing are dramatic. And they certainly produce noticeable impacts on worldwide ecosystems. But, the impacts of other human activities are even greater.

The intentional introduction of animals such as rabbits and goats into an ecosystem can dramatically alter the food system. Goats multiplying freely and without effective predators have stripped vegetation from mountainsides producing seriously degraded carrying capacities and erosion in many parts of the world. Rabbits were introduced and have become a major environmental hazard in parts of Australia because they have few natural diseases and predators there.

Accidental introductions of new species can present major problems, too. Mosquitos (*Culex quinquefasciatus*) were accidentally introduced into Hawaii in 1826 when they arrived on sailing ships. The mosquito carried several diseases of birds to which the native bird populations were particularly susceptible. In addition, roof rats escaped from the same ships and preyed on bird eggs and young birds. As a result, fully eighty-five percent of native Hawaiian bird species have become extinct from the Islands.

A totally different kind of problem can arise from changes in the ecosystem resulting from human activities. Humans have all but eliminated wolves and large cats as predators in most of the United States because they preyed on domestic livestock. As a re-

Figure 19–4.   Carolina parakeet, orginal watercolor by John James Audubon. (Courtesy of The New York Historical Society, N.Y.C.)

sult, by early 1992, deer populations in parts of this country threatened to outgrow the ecosystem's carrying capacity. A major die-off from starvation or disease, or both, was threatened in the deer herd. Not only that, but accidents involving automobile-deer collisions were becoming a major problem on the nation's highways. Some states have extended their hunting seasons and allow hunters to take both bucks and does to reduce deer populations.

But, these impacts are still minor compared to the effects of agriculture, industry, urbanization, and highway construction on the environment. When the land is cleared for any of those uses, the entire ecosystem undergoes fundamental change. Clearly, agriculture, industry, cities, and highways are essential for human survival at the population levels we see today.

Civilization and organized society could never have developed without agriculture. Yet, we must realize that by farming the land, we alter in many ways the planetary ecosystem. Cities are essential for the numbers of humans on the planet, yet cities have major negative impacts on the ecosystem. Industries are essential, as are highways, yet the effects on the ecosystem can be very great.

Thus, we see that humans can cause extinctions in a number of ways. The greatest impact of humans on the environment historically has been a combination of agriculture, urbanization, industry, and transportation—all things that are essential parts of civilization as we know it. Clearing of forests and draining of swamps have reshaped entire ecosystems. Introduction of animals for food production or for other reasons has often produced unforeseen results. Accidental introductions of species have also had major impacts. Hunting for food or feathers or simply to eliminate "pests" or dangerous animals has had important side effects.

Destruction of Tropical Forests

One environmental issue that began to draw a great deal of attention in the mid-1980s was the destruction of tropical forests, and particularly tropical rainforests. The process of conversion of a forest to another use is referred to as **deforestation**. Deforestation is sometimes necessary and often produces major improvements in the human welfare. Other times, it can be very destructive in an ecological sense.

As we have seen, tropical rainforests are the most biologically productive and diverse regions on the planet. Not only that, but many of the species that live there are limited to very small ranges and are frequently very rare.

At the same time, human populations in many tropical regions are large and growing rapidly. Sadly, most tropical nations tend also to be underdeveloped and poor. In order to provide food and firewood for their growing numbers, these peoples frequently must clear their forests for farming or to sell the lumber. At the same time, tropical soils tend to be relatively poor because the heavy rainfalls leach water-soluble minerals away from the surface. They also tend to be low in organic matter because the warm moist climate promotes rapid decomposition of plant and animal tissue. As a result, the land that is cleared is not very productive. Moreover, the climate is conducive to growth of plant and animal pathogens, so crop and livestock production tends to be low.

Thus, more and more forests must be cleared to provide only marginal agricultural production. Lacking adequate transportation systems (highways and railroads), much of the trees are not even used for lumber. Rather, they are simply burned. That results in additional CO_2 emissions into the atmosphere. The lack of adequate machinery and capital for implementing soil conservation measures means that the land can quickly become damaged by erosion. The end result is that still more tropical rainforest must be cleared to provide for still more people.

The leaders of the developing tropical nations look at the model of exploitation that made America and Europe wealthy. They want the same for their people. Their arguments are forceful and their emotions are strong. On the most fundamental level, much of the deforestation is a direct result of poverty. The poor of the developing world must have farm production to survive. They have little resources except for forest land with which to secure that farm production—so they exploit that. It is very difficult for people in developed, wealthy nations to effectively argue against the process.

About seventy percent of the people in developing nations rely on wood for heating their homes and cooking their food. Their low incomes restrict their ability to purchase other sources of energy. At least eighty percent of all wood harvested in developing countries is used for fuel. Such countries rely on the forests for over half of their total energy use. That can be compared to less than twenty percent of wood harvested in developed countries, which rely primarily on fossil fuels, hydroelectric, or nuclear energy (Burley & Hazlewood, 1988).

In addition, the forests provide poles, fencing materials, construction materials, furniture, and other wood products such as paper pulp. The forests provide grazing for livestock, fruits and nuts for human and animal consumption, gums, oils, resins, and other nonwood products. In short, in many developing nations, the poor depend almost completely on the forests for their subsistence and shelter.

In addition, tropical forests play a role in increasing agricultural production in a number of ways. Trees help maintain soil fertility by holding the soil in place and controlling erosion. Nitrogen-fixing tree species actually increase soil nitrogen. They further increase soil organic content by growing roots in the soil and depositing leaves on top of the soil. In arid and semiarid regions, they help maintain soil moisture, decrease surface winds, and provide a direct cooling effect.

Over twenty-seven million acres of tropical forests are cleared annually for agricultural and other uses. In most developing nations, the rate of forest clearing is rising. Only in the developed nations are forests being replaced at a net gain in the early 1990s.

According to Burley and Hazlewood, about two-thirds of the developing nations are faced with chronic shortages of **fuelwood** (wood burned to provide heat for cooking and for warmth during cold weather). Most of the countries with severe fuelwood shortages are in sub-Saharan Africa. The most severe fuelwood shortages are in semiarid and mountainous regions. It is in such ecosystems that trees grow most slowly and wood productivity is lowest. Some of the countries most severely affected are listed in Table 19-1.

It is also in such regions that overexploitation of the forests produces the most devastating effect on the ecosystems. Remember that many life-forms live in the tropics, particularly the tropical rainforests. Also, remember that many tropical species inhabit small ranges and as a result are sometimes quite rare. Thus, clearing tropical forests, particularly rainforests for fuelwood or for farming affects many species of plants and animals—often rare species with limited ranges.

Rhoades and Johnson estimate that by the middle of the 21st century, as much as one-quarter of the planet's plant species may become extinct. It is important to point out that their estimate is based more on guesses and assumptions than on scientific fact. But, it is clear, that with the anthropogenic changes that are occurring in the ecosystem, many species will be unable to adapt quickly enough, or move. Thus, many species will face extinction (Rhoades & Johnson, 1991). How many? Nobody really knows.

This would result partially from deforestation, but also because of a shift to agricultural monoculture. In addition, overgrazing, urbanization, highway construction, and drainage projects will contribute to the process of species loss.

Table 19-1. Developing Nations with Severe Fuelwood Shortages*

Africa	Asia	Latin America
Botswana	Afghanistan	El Salvador
Burkina Faso	China	Haiti
Burundi	India	Bolivia
Cape Verde	Nepal	Peru
Chad	Pakistan	
Comoros	Turkey	
Djibouti		
Ethiopia		
Kenya		
Lesotho		
Malawi		
Mali		
Mauritania		
Mauritius		
Namibia		
Niger		
Reunion		
Rwanda		
Senegal		
Sudan		

*Source: Burley, F.W., & Hazlewood, P.T. (1988). Tropical forests: A resource in jeopardy. In T. Kristensen & J.P. Paludan (Eds). **The Earth's fragile systems: Perspectives on global change.** Boulder, CO: Westview Press.

Genetic Diversity and the Human Food Supply

In early civilizations, almost everyone was a farmer, hunter, or fisherman. The level of productivity of the farmers was relatively low. There was little use of technology—human power and later animal power were the only sources of energy to till the soil and husband the crops and livestock.

Seeds and animal species were found in nature. Early farmers collected wild grain for planting. They selected seeds from the most successful plants each year to use for succeeding crops. It was through this process of selection that the genetic potential of crop plants and livestock improved over the millennia.

With the dawn of scientific agriculture and an understanding of genetics, came systematic plant and animal breeding and selection. Plant and animal breeders developed pure species for genetic uniformity. They could then cross the pure lines (a process called hybridization) to produce extremely uniform and high-yielding hybrids. Probably no other single development in agriculture has had a greater impact on human life than the development of scientific methods of plant and animal breeding and hybridization.

The result of this process has been the replacement of traditional varieties of such crops as corn (maize), rice, and wheat with very productive and uniform hybrids of the same crop. The uniformity of genetically pure strains and hybrids produces plants of all the same size that mature at one time. It is that uniformity that makes fertilization and mechanical harvesting so effective. It is the great productivity of highly selected varieties and hybrids that make it possible to produce so much food on so little land. The agricultural revolution formed the basis for human civilization, and the development of mechanized and scientific agriculture formed the basis for the modern world.

During the 1960s, plant breeders produced a few genetically superior varieties of wheat and rice. The "miracle" wheat and rice have meant higher yields and improved living conditions. The result of these developments has been called the "**green revolution**." The green revolution effectively meant that the carrying capacity of the planet for humans increased dramatically. It enabled several nations which had been food importers to become food exporters. It prevented literally millions of humans from starving.

Genetic uniformity in our crops and livestock make possible great productivity. Yet, in such uniformity there is also great danger. The use of mechanization encourages production of single crops. A farmer can produce more goods and earn more income by specializing in only a few crops or animals. Production of large fields of a single crop or large numbers of a single species of livestock is referred to as **monoculture**. Monoculture is much more efficient and profitable than production of many different crops and animals. This is particularly true with highly productive pure varieties and hybrids.

At the same time, monoculture places large numbers of the same variety of plant or animal in the same location. More importantly, it can mean that a large proportion of a nation's food supply is produced from only a few varieties of farm plants and animals. As long as the ecosystem cooperates, that is good. But, we have seen that the most constant thing about the environment is change. The only thing about a given ecosystem of which we can be certain is that it will not remain constant.

An example of the dangers of monoculture can be found in Sri Lanka. In 1959, local farmers produced about two thousand different varieties of rice. Today, practically all rice production in Sri Lanka comes from just five varieties. But, the introduction of a single virulent disease or insect could spell disaster for the people of that tiny island country. On a much more vast scale, India once grew as many as thirty thousand varieties of rice. Today a mere ten varieties account for 75% of India's rice production (Rhoades & Johnson, 1991).

An actual event that occurred in the United States a few decades ago affected the production of corn in this country and was of critical importance to our economic system. Corn production was highly monocultured in the United States and only a few varieties dominated the farm landscape. Farm production was then, as it is today, in a potentially dangerous situation.

United States farmers learned that the hard way in 1970, when an unexpected epidemic of corn leaf blight wounded the pride of the world's most agriculturally advanced nation. A virulent new strain of fungus appeared in south Florida that winter and raced north like a killer flu. Since each ear of corn was a (virtual genetic) copy of every other, there was no margin for safety. The fungus destroyed half the crop from Florida to Texas. Nationwide losses amounted to fifteen percent, at a cost of perhaps one billion dollars. (Rhoades and Johnson, 1991, 84).

The same authors contended that a similar calamity may have destroyed the Mayan civilization about AD 900. A virus apparently destroyed the corn (maize) crop that year, dooming the civilization to mass starvation. The Irish potato famine in 1845–1848 resulted from a fungus (*Photophthora infestans*) accidentally introduced from Mexico. However, there are twenty wild, inedible potato varieties resistant to this fungus. Scientists have used this biological diversity to cross the domestic potato with the blight-resistant wild potato. The result was a blight-resistant, edible potato.

To guard against the danger of losing the genetic pool from those thousands of wild plant varieties, scientists have developed "gene banks." Specimens of as many varieties of plant seeds as possible are collected and preserved in laboratories around the world. The seeds are saved as "insurance" against the dangers that crops produced in monoculture potentially face. For the interested reader, Rhoades and Johnson published an article entitled "The world's food supply at risk" in Volume 179, No. 4, of the *National Geographic Magazine*, 1991.

Summary

There may be as many as thirty million separate and distinct species of plants and animals on Earth. Of that number, somewhere between 1 million and 1.8 million species have been found, described, and classified by biologists. The classification system in use today was developed by the Swedish scientist, Linnaeus, during the seventeenth century. The vast majority of known species are very small. In fact, humans are among the largest 0.00001 (1/1,000 of one percent) of all species.

The largest single group of life-forms known is insects. The Coleoptera (beetles) make up about twenty-five percent of all known species, both plant and animal, on Earth.

New species are constantly being added to the ecosystem. Old species are constantly being lost. The process by which new species develop is known as speciation. The process of species loss is extinction. Extinctions result from natural changes in the ecosystem.

The most common cause of a species extinction is some change in the ecosystem resulting from an environmental change. Changes in temperature or water availability result in changes in relationships within the various food webs in a biotic subsystem. When that happens, the affected organisms must either adapt, move, or become extinct.

The regions of greatest biodiversity are near the tropics. Particularly in regions of high rainfall, there are not only a high density of species, but also a high level of biological productivity. Thus, tropical rainforests represent the greatest concentration of biological diversity on the planet.

Biodiversity may well be critical to stability in the biotic subsystem. For many years, biologists have believed that the more complex an ecosystem is, the less subject it is to extreme swings in population levels. In other words, the more stable it is likely to be. Recent computer models suggest that such an assumption may not be valid. But, until more definite research is available, we must assume that biodiversity helps stabilize the ecosystem.

When a species becomes extinct, its genetic pool is lost forever. Science has only begun to discover the wonderful compounds that are contained within the plants and animals of the planet. Unnecessary extinctions caused by human activities waste those potentially valuable things. With the changes humans have caused in the tropical rainforests and on all other parts of the planet, such extinctions appear to be increasing rapidly.

At the same time, we cannot accept total responsibility for maintaining all species on the planet. The biological realities are clear—the ecosystem is changing. That is natural. Some of the changes are increasingly anthropogenic (caused by humans). We cannot anticipate or correct most natural changes. But, we need to be careful as we cause anthropogenic changes, that species important to human survival and well-being are not lost.

We tend to think of larger and more colorful animals when we discuss human-caused extinctions—the passenger pigeon, for instance; but the many thousands of varieties of wild rice and maize that are in danger of being lost as a result of scientific plant breeding and monoculture are undoubtedly more important for human survival.

DISCUSSION QUESTIONS

1. What is biodiversity and why is it important to human survival?
2. How do new species develop?
3. How do existing species become extinct in nature?
4. How do human activities contribute to species loss?

5. Why is it so difficult to estimate the number of species on Earth?
6. Why is biodiversity so much greater in the tropics than in cooler regions of the planet?
7. What impact does deforestation have on biodiversity?

ADDITIONAL ACTIVITIES

1. Organize a field trip to a natural area near your school. Find and identify as many different species of plants and animals as you can. Select at least a few of each to identify by scientific name. Your biology teacher may be able to assist you in this process.
2. Make a detailed study of one extinct and one endangered species. Report on your results to the class.
3. Organize a debate in your class about whether it would be right to build a dam for hydroelectric power for a developing country, even if a very rare fish or insect would become extinct as a result.
4. Organize a debate in your class about protecting the spotted owl as opposed to allowing logging on old-growth forests in Oregon.

CHAPTER 20
Global Temperature

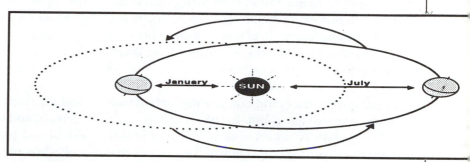

Terms to Look for and Learn

Axis of Spin
Computer Simulation
Ecliptic Plane
El Niño
Equator
General Circulation Model

Greenhouse Effect
Greenhouse Gases
Ice Core Records
La Niña
Milankovich Model
Sediment Core Records

Learning Objectives

After reading this chapter and participating in the activities, you should be able to:

- Explain how global temperature changes over time can be inferred from the sediment core and ice core records.

- Discuss the reasons for seasonal variations in the climates of various parts of the world.

- Discuss the three Milankovich motions and explain their effects on the planetary climate.

- List the major greenhouse gases and explain the mechanisms by which they influence atmospheric temperatures.

- Describe the greenhouse effect and explain its relationship to global warming.

- Explain the effects of major tidal changes on global temperatures.

- Discuss the effects of human activities on global warming in the past and potential effects in the future.

Overview

Probably no single environmental issue in recent years has sparked more discussion than global warming. Yet, there is widespread disagreement on whether global warming, as a result of human activity, presents any particular hazard to the environment, or that it even exists to any appreciable extent. In this chapter we will attempt to put the issue of global temperatures into a rational context and examine the current state of human knowledge about this phenomenon.

Petroleum was formed many years ago by decomposition of organic plant residues. This means that where oil is now being pumped, at one time plant life was very luxurious. Oil wells at Prudoe Bay, Alaska, at 71° N latitude are evidence that where there is now permafrost, there was at one time a tropical or semitropical climate. Such strong evidence is clear proof that the planet's climate pattern has changed drastically over the ages.

The issue of global warming is not nearly as simple as many would argue. Global temperatures have never been stable. There are many factors that influence the level of global temperatures. One important factor is the amount of solar energy reaching the planet's atmosphere—the so-called solar constant, which is not exactly constant.

The changing albedo of the planet is another factor. Orbital variations have been shown to affect not only the global mean temperature, but just as critically, the distribution of temperatures around the globe. Finally, the greenhouse effect, without which life on Earth would be difficult if not impossible, produces warming in a very measurable way.

Indeed, the decade of the 1980s was the warmest on record. There was a brief period approximately six thousand years ago when the global temperature appears to have been slightly higher, but that is based on estimates rather than actual records. The global temperature today appears to be near its warmest in at least 160,000 years. Nevertheless, there have been warmer periods in Earth's history. Geological evidence indicates that we should be in a cooling period, but historical evidence is clear that global temperatures are on the rise.

The remainder of this chapter will examine the geological and historical global temperature trends and the impacts on global temperatures of seasonal temperature variations, variations resulting from orbital fluctuations of the planet, the effects of ocean currents, and the greenhouse effect. Finally, the historical and potential impacts of human activity on global warming through greenhouse gas emissions will be explored.

Global Temperatures
Factors Affecting Global Temperatures

There are many factors that have affected the global temperature of the Earth over the past few billion years. After the formation of an atmosphere and the forming and cooling of the surface of the planet, the global temperature tended toward equilibrium. However, the forces acting on the planet have continued to fluctuate. Thankfully, the global temperature has varied within a fairly narrow range. That situation has allowed for the development and continuation of life on the planet. The effects that human activity have had, and are predicted to have, on those fluctuations are at the heart of the global warming debate in the scientific and political communities today.

H.H. Lamb identified nine major classes of effects on global mean temperatures and the distribution of warmth in the form of regional climates around the planet. His nine classes of effects result largely from the following six factors.

Solar Fluctuations

Chapter 1 discussed the solar constant; however, recent observations by scientists, in particular measurements by the satellite Solar Max, confirm that the luminosity (brightness) of the sun varies over time. Relatively minor fluctuations in the solar constant in the short run can be more profound over time.

Sunspots are generally considered to be an indicator of increased solar activity. During the period AD 1645–1715, there was an absence of sunspots reported in historical records. Interestingly, that corresponded fairly closely with the so-called "Little Ice Age" that took place in Europe at about the same timeframe.

Geographic Changes

Geologists believe that the continents are not stable and that they have shifted in size and position over the ages. They base this on the theory of plate tectonics. Movement of major land masses would greatly affect the climates of the continents in numerous ways, not the least of which is drifting toward or away from the equator. Mountain ranges would have formed and disappeared. The albedo of the planet would have varied as a result.

Atmospheric Changes

Circulation patterns evolve over the ages. This could affect the mixing or segregating of cold polar air and warm equatorial air. Changes in the composition of the atmosphere affect the absorption or reflection of solar energy. Greenhouse gases and cloud cover are important here. The effects of greenhouse gases on global warming will be discussed in detail later in this chapter.

Volcanic Activity

Volcano eruptions release massive amounts of heat from the Earth's core. In addition, they form new land from magma (molten rock) and ash. Finally, they inject huge quantities of gases and dust directly into the atmosphere, increasing the albedo of the Earth. The higher albedo results in lower global temperature. The eruption of the Indonesian volcano of Krakatoa in 1883 was followed by about a 1° F (0.5° C) drop in the global mean temperature the following year. A massive eruption of Tambora, also in Indonesia, seventy years earlier resulted in the

famous "Year Without a Summer" in New England and Northern Europe in 1816. There were frosts and light snows there throughout what should have been summer (Kerr, 1989).

The Pinatubo volcano near Manila, Philippines erupted on June 12 and 13, 1991. As a result, the U.S. Clarke Air Force Base and the U.S. Subic Bay Naval Base were abandoned. More than twenty thousand Filipinos had to be evacuated and 435 were killed.

Fluctuations in Ocean Currents

Periodic shifts in the patterns of ocean currents can have drastic effects on the climates of affected regions and even on the total global temperature. The famous El Niño and la niña currents will be discussed later in this chapter.

Orbital Fluctuations

The Earth's orbit about the sun is not fixed. Instead, it fluctuates in observable and predictable ways. It may well fluctuate in ways of which we are not yet aware. Orbital fluctuations, as explained by the Milankovich Model, will be discussed later in this chapter.

Estimating Global Temperatures

Today

Estimating global temperature today is a relatively complex matter. At any point in time, half of the Earth is in winter while the other half is in summer. For half, it is day while for the other half it is night. Almost seventy-one percent of the Earth's surface is covered by water while just under twenty-nine percent is covered by land. The part that is covered by land ranges from almost barren desert to lush rain forest to tundra to ice-covered mountain ranges.

But, with the existence of modern communications, it is possible for scientists and other weather observers worldwide to simply measure the temperature in different locations and transmit their readings to collection points for aggregation. Based on this procedure, climatologists estimate that the average, current global temperature is about 59° F (about 15° C). Even with extreme local and regional temperature fluctuations, the worldwide temperature mean is remarkably steady at that level.

For the recent past there are historical records. Fortunately, for several centuries naval and commercial ships of the major maritime nations have practiced methodical recordkeeping regarding their journeys. One important part of any journey was the weather that the ship encountered. Ships' logs are a valuable source of temperature records from all over the world.

By taking the temperatures recorded in ships' logs along with the position records (latitude and longitude), it is possible to estimate worldwide temperatures fairly accurately. Based on those records, climatologists estimate that global temperatures have risen by about 1° F (about 0.5° C) in the past century.

Drilling into the Past

Estimating global temperatures for periods of time before the advent of thermometers and carefully controlled records is more subjective, however. Basically, there are two major sources of data for such estimates and both involve drilling for geological evidence: ocean sediments, and glacial ice.

Ocean Sediment Core Records

The bottoms of the oceans are constantly receiving deposits of sediments of all kinds. Simple dust that settles from the atmosphere onto the ocean surface, and that does not dissolve, eventually makes its way to the sea bed. Plankton (microscopic plants and animals) and larger sea creatures die and their remains fall to the bottom. Soil particles washed into the ocean from eroding soil surfaces settle there. Volcanic ash thrown into the atmosphere makes up parts of the sediment.

The newest deposits are near the surface. As the deposits become older they are buried progressively deeper under the surface. As the sediments form, they lock away many kinds of materials, some of which carry magnetic charges from the Earth's magnetic field.

The magnetic field of the Earth changes over time; sometimes in intensity and sometimes completely reversing directions. Fairly accurate estimates are available for when those changes have taken place. When a core is removed from an ancient sea bed by drilling, a cross-section of this geologic record is removed and it can be examined in detail. By reading the magnetic "fingerprints" of the core at a given depth, the age of the materials at that depth can be estimated with some degree of accuracy. These are called **sediment core records** (see Figure 20-1).

One organism whose shells make up part of the sea bed sediment is the protozoan, *Forminifera*. The biology of this protozoan is particularly useful in climate study over geological time, because the proportion of its shell [comprised of a heavier oxygen isotope (^{18}O)] tends to decrease relative to the lighter oxygen isotope (^{16}O) during periods of colder temperatures. Conversely, the proportion of its shell comprised of the heavier isotope compared to the lighter isotope tends to increase in warmer periods.

By matching the proportions of ^{18}O and ^{16}O in the *Forminifera* shells to the estimated age of the sediment, it is possible to determine when climate changes took place and even to estimate the approximate mean temperature of the location where the sediments were deposited at a given point in time.

Ice Core Records

As snow falls, it lands gently on the ground and includes many air pockets. As the snow is buried under subsequent snows, pressure turns it into ice. Eventually, if the climate is cold enough and precipitation is adequate, the ice thickens and ice-packs form, as in the Arctic, Antarctic, and on the tallest mountains. Over the ages, the ice cover of the coldest regions of the Earth have thus captured pockets of air from the atmosphere and held them in place.

As in taking sediment cores and deciphering them, **ice core records** can be collected and studied (see Figure 20-1). By examining the air pockets captured in the snow at a given period, scientists can estimate the amounts of various gases in the atmosphere at that time. Researchers using this technique have determined approximate concentrations of the major greenhouse gases CO_2 (carbon dioxide), N_2O (nitrous oxide), and CH_4 (methane) at various times

Figure 20-1. Sediment cores and ice cores are removed by drilling into the seabed and into glacial ice pack deposits. Such cores are then examined for many kinds of geological information, including evidence of changes in global temperatures. (Photo courtesy Bill Camp.)

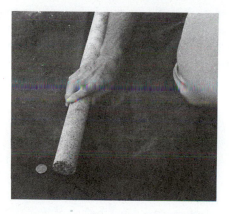

Global Temperatures Over Time

Nothing in nature has ever been found that is forever constant. The temperature of this planet is no exception. As was pointed out in Chapter 1, the amount of radiant energy reaching the Earth from the sun has been relatively constant in terms of a human lifetime, but in terms of geologically important time spans, it varies somewhat.

But, even if the solar constant were actually "constant," there are other factors that would cause fluctuations in Earth's mean temperature. So, global mean temperatures have varied measurably over the ages. In addition, the distribution of temperature ranges around the planet is also important to consider.

As should become abundantly clear from reading the next few pages, and from examining Figure 20-2, global temperatures have varied greatly in the past million years. All but a small fraction of that time was before modern human activities began to affect the atmosphere, yet the changes took place anyway. Only a totally uninformed person could contend that human activity has resulted in more than a very tiny fluctuation in that long-term pattern of variability.

Figure 20-2. Global temperature variations over the past 150,000 years. **Source of Data**: Lamb, H.H. (1977), **Climate: Past, present, and future,** Vol 2. New York: Barnes & Noble Books.

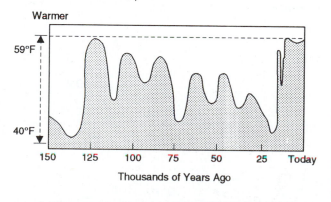

throughout the past 800,000 years. Using those data and temperature estimates derived from the sediment record, climatologists have linked low concentrations of those gases with cold periods in the Earth's history and high concentrations with warm periods. As you will see later, that is exactly what would be predicted by the theory of greenhouse warming.

Global temperatures today are at their highest in over one hundred years, yet they are only about at the same level as in the 1940s. Indeed, global temperatures are as high as they have been since just before the Little Ice Age that occurred in the seventeenth century. On the other hand, the Earth's climate was slightly warmer in the days of ancient Greece and Imperial Rome. The trend over the past 150,000 years has been generally upward and the present situation simply serves as a continuation of that trend.

Seasonal Variations

The Earth is a roughly spherical globe. The angle at which the surface faces the sun differs the further one moves North or South from the equator. Solar energy is radiated outward from the sun in a relatively uniform way. Thus, for every square foot of Earth's surface, the amount of solar energy reaching it varies inversely with the angle at which that surface is tilted from perpendicular to the sun. That is the basic reason that the polar regions are much colder than the equatorial

regions and that the middle latitudes (as in the United States) fall in between those extremes (see Figure 20-3).

The **Ecliptic plane** is an imaginary plane that passes from the center of the sun and through the center of the Earth. It describes the annual orbit of the Earth around the sun (see Figure 20-4). As the Earth spins through its daily rotation, the axis of the spin is tilted at about 23½° from the Ecliptic plane. The **axis-of-spin**, like an axle in a wheel, is an imaginary line through a planet, around which the planet rotates.

The so-called "true" poles are imaginary points on Earth's surface corresponding to the centers of the axis of spin of the planet. They differ rather substantially from the "magnetic" poles which result from the Earth's magnetic field. As we have seen, the magnetic fields of the planet are somewhat unstable over long periods of time. Geological evidence indicates that the polarity of the Earth's magnetic fields actually reverses about every million years (Lamb, 1977). Yet, the magnetic poles are stable enough to be useful in navigating because of the effect they have on free-floating or swinging magnets in such devices as compasses.

The North Pole is arbitrarily defined as that planetary pole which points in the general direction

Figure 20–3. The angle of the Earth's surface to the sun determines the amount of solar energy it receives.

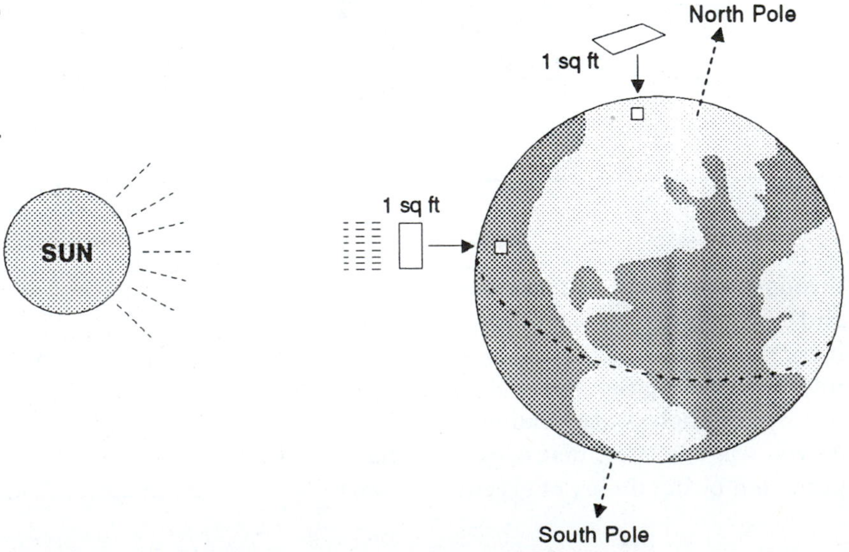

(within about 1°) of the North Star (Polaris). The Earth's **equator** is an imaginary plane passing around the planet at a 90° angle to the axis of spin and therefore also at the halfway point between the true North Pole and true South Pole. The northern hemisphere is all of the Earth's surface lying between the equator and the North Pole. Conversely, the southern hemisphere lies between the equator and the South Pole.

As the Earth moves around the sun, its spin produces a gyroscopic effect in that the axis continues to point in the same basic direction throughout the year (see Figure 20-4). The result is that during one half of the year, the portion of the Earth north of the equator receives more direct sunlight. During the other half of the year, the southern hemisphere receives more direct sunlight. As we have seen, the more direct the sunlight, the more solar energy is received and the warmer the climate.

Thus, the changes in seasons result from the tilt of the axis combined with the rotation around the sun. When it is winter in North America, it is summer in South America, and vice versa. In addition, more extreme seasons occur as one moves further from the equator toward the poles as a result of the increased angle from the Ecliptic plane.

Orbital Variations

If the Earth's axis of spin were completely stable, and the solar constant were absolutely steady, and the Earth's orbit around the sun were completely fixed, then one would expect a stable climate throughout the world. Such a stable situation should eventually lead to an equilibrium in the ecosystem. As we saw in Chapter 2, change is the only true constant in nature. As a result, regional climates have changed over the centuries. In fact, during the 1800s, scientists developed a theory that at various periods of time, great sheets of ice had covered parts of what were then Europe, Asia, and North America. Those periods of time came to be known as the Ice Ages.

In 1864, a Scottish astronomer published a paper proposing that the Ice Ages were results of changes in the Earth's orbit. Just prior to World War II, a Yugoslav astronomer named Milutin Milankovich worked out the details of the orbital changes and proposed a

Figure 20–4. Earth's axis of spin tilts from the ecliptic plane approximately 23.5°. As the planet rotates around the sun, that tilt produces seasonal variations.

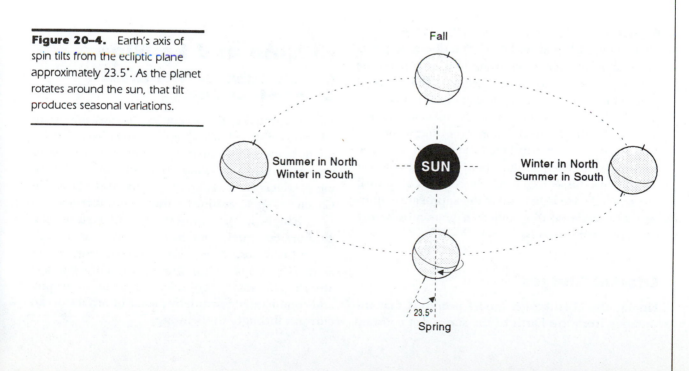

theory that has come to be known as the **Milankovich Model**. The Milankovich theory proposed three irregularities in the Earth's orbit to explain the cyclic nature of the Ice Ages: the circle of precession, "wobbles" in the tilt of the axis of spin, and "bulges" in the plane of the Ecliptic (Gribbin, 1989).

"Wobbles" in Tilt

The tilt of the axis of spin is not constant. Milankovich proposed that the angle of tilt of the axis varies between 21.8° and 24.4° from the plane of the Ecliptic. According to this part of the theory, the Earth's current tilt of about 23.5° is changing and over the next twenty thousand years or so, will continue to decrease. This change in angle toward the sun would result in a variation in seasonal extremes apart from the changes caused by the previous phenomenon. As the tilt becomes more extreme, the seasonal variations would also become more extreme. Milankovich believed that this cycle lasts approximately 41,000 years, see Figure 20-5A.

Circle of Precession

Milankovich proposed that the axis of spin of the planet was not constant in its direction relative to the Ecliptic plane. Instead, he believed that it describes an approximate circle of precession around an imaginary line drawn perpendicular to the Ecliptic. The circle of precession represented a cycle that would periodically reverse the seasonal effects of the tilt and rotation described in the previous section. According to this part of his theory, the North Pole of about ten thousand years ago would be pointing about 46.8° from the star Polaris, and the seasons would be reversed from their present cycle. He believed that the approximate duration of this cycle was about nineteen thousand to twenty-three thousand years (see Figure 20-5B).

Orbital "Bulges"

Finally, the Milankovich model proposed that the distance from the Earth to the sun is not constant throughout the year but varies from a near-circle to elliptical in shape. In fact, the present orbit varies by about 3.1 million miles during the year. Today the Earth is about 91.5 million miles from the sun in January and about 94.6 million miles from the sun in July. But that bulge changes over the years and its orientation reverses. Someday, the greatest distance from the sun will be in January rather than July. Again, the effect of this change would be to produce greater seasonal variations in climatic extremes. The approximate duration of this third Milankovich movement was estimated at 100,000 years, see Figure 20-5C.

Effects of Milankovich Movements

Recent geological evidence indicates that for at least 2.5 million years, there has been a series of ice ages on the Earth. It also appears that the ice ages have come and gone or have strengthened and weakened, roughly on a series of cycles conforming to the Milankovich model. According to the Milankovich Model, at the present time the Earth should be in a cooling period on the way to the next ice age.

El Niño and La Niña
Earth's Oceans—The Great Reservoir of Energy

The deep waters of the oceans are the most important reservoirs of stored solar energy on the planet. Permanently hidden from the sun, the depths of the ocean remain at a remarkably steady temperature. Little mixing of waters from different depths takes place. See Chapter 17 for discussions of the oceanic thermocline.

Changes in the atmospheric and surface temperatures are moderated by the ability of the oceans to gradually absorb or release latent energy in the form of heat. Just as importantly, the surface waters absorb great amounts of solar energy in the tropics and conduct that warmth by means of surface ocean currents throughout the world.

Figure 20–5. Orbital fluctuations according to the Milankovich Model. (A) On a cycle of about forty-one thousand years, the axis of tilt wobbles in a minor circle. The axis forms an angle from a perpendicular line drawn to the ecliptic plane that ranges from about 21.8° to about 24.4°. The effect is to change the angle of the sun's rays that strike the planet's surface at various seasons slightly. (B) On a cycle of about nineteen thousand to twenty-three thousand years, the axis of tilt (wobbles and all) rotates in a great circle that substantially changes the planet's orientation. In ten thousand years, what is now the North Star will be pointing 46.8° away from what is now the North Star (Polaris). (C) Earth's orbit is elliptical, not circular. At its furthest point (during July) the Earth is about 94.6 million miles from the sun. At its closest point (in January) it is about 91.5 million miles away. On a cycle of about ninety-six thousand years the furthest point will change from July to August to January and so on back to July.

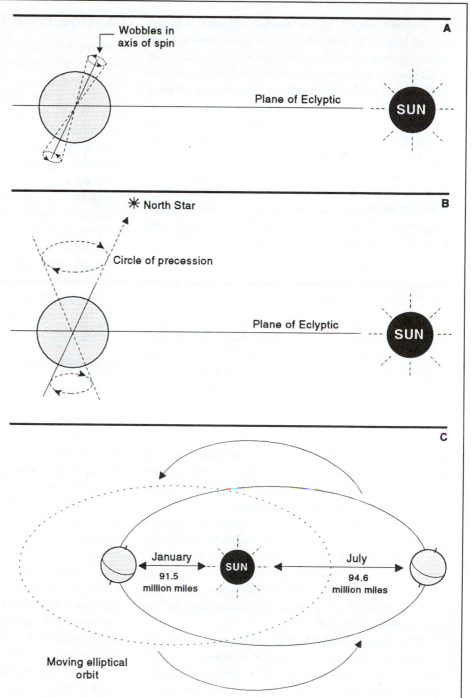

Indeed, much of the world's climate is determined largely by the effects of ocean currents. The warming and other weather effects of the water are well known. One example is the Olympic rainforest of northwestern Washington State. The region receives over 140 inches of annual rainfall and has reliably moderate temperatures as a result of the nearby Pacific Ocean. At the other end of the United States, Key West's temperature has never been recorded below the mid-40s (° F) or above the mid-90s (° F). That stability results directly from the warm waters of the Gulf of Mexico which wash past its shores. The monsoons of India are born in the warm currents moving across the tropical oceans.

El Niño (The Boy)

Normal tropical Pacific currents and their accompanying surface winds move at a leisurely pace from west to east along the equator. The climates of much of the world depend on that flow for warming breezes and moisture to feed regular rain systems as far away as Northern Africa.

About every 4 years the direction of that flow reverses for some unknown reason for as long as several months. The effects of that reversal have been known for centuries as **El Niño**. The result is a severe disruption of rainfall patterns that can affect major parts of the planet (Pearse, 1989).

Droughts in parts of Africa have been linked directly to the effects of El Niño. In Southern California, the results of this El Niño shift can be seen as extremely hot and windy weather accompanied by extreme drought. The global effect, after a delay of about 6 months, is a measurable warming of the global mean temperature.

The Spanish name El Niño was coined because of the phenomenon's usual habit of appearing near Christmas. The connection is with Christ's traditional birthday, and refers to The Boy (Christ). Because it refers to the Christ child, El Niño is customarily capitalized.

La Niña (The Girl)

The opposite of the boy is the girl. Because la niña does not refer to a specific girl, it is not customarily capitalized. La niña is also the opposite of its fiery counterpart. **La niña** results in an increase in the speed of the ocean currents and the surface winds. The effect is to bring cooler water from below the surface to the top. There it produces a general cooling of the regions affected. The result is also a general lowering of the entire global temperature.

In fact, it almost exactly offsets the effects of El Niño and traditionally occurs at about the same frequency. Interestingly, during the period 1975–1988, there was no la niña, yet El Niño continued as usual. Many people believe that the warming effects of the latter without the offsetting cooling effects of the former explain much of the global warming that occurred during that timeframe.

The Greenhouse Effect

The **greenhouse effect** is the absorption of solar energy by gases and particulates in the atmosphere. When the energy is absorbed, much of it is expressed as heat. Thus, the greenhouse effect produces a warming of the atmosphere by absorption of solar energy.

As we have seen, not all of the energy reaching the planet is absorbed. Based on the varying albedos of the atmosphere, clouds, atmospheric dust, and differing planetary surfaces, the Earth reflects thirty-five to thirty-seven percent of the solar energy reaching the upper atmosphere. Much of the energy that is not reflected is subsequently radiated into space as long-wave radiant emissions. Much of the rest is captured in chemical reactions in various forms of organic and inorganic compounds. Part of the energy retained by the planet appears to us as latent and perceptible heat.

The energy that is not returned to space as reflected or emitted energy, is absorbed by various surfaces. That part of the energy that is absorbed by the atmosphere produces the greenhouse effect (see Figure 20-6).

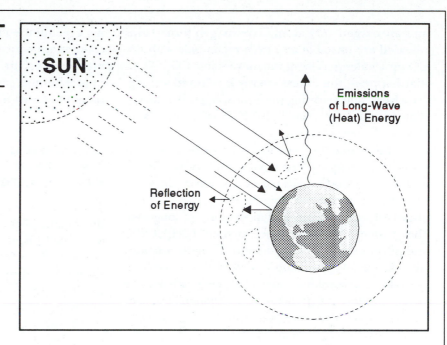

Figure 20–6. The greenhouse effect.
Net Planetary Absorption = total solar
energy – reflection – emission

Given the amount of solar energy reaching the Earth, climatologists estimate that the planetary mean temperature should be about –0.4° F (about –18° C). In the absence of the greenhouse effect, that is what it most likely would be. But, as we have seen, the current mean global temperature is actually closer to +59° F (+15° C). The difference of about 59.4° F or 33° C has made life on Earth possible (Lorius, Jougel, Raynaud, Hansen, & LeTreut, 1990).

In an Earth without the greenhouse effect, there would be little free water. Instead of oceans, the world would be covered by massive bodies of ice, with the possible exception of the tropical regions receiving most direct sunlight where small amounts of liquid water might accumulate during the warmest summer days, only to freeze again at night.

The Greenhouse Gases

Fortunately for us, the gases that make up the atmosphere act much like the panes of glass on a greenhouse, allowing solar energy to enter but preventing escape for much of the energy. If you sit in the sunlight passing through a window on a sunny day, you will understand what the effect feels like. The most important greenhouse gases are H_2O (water vapor), CO_2 (carbon dioxide), CH_4 (methane), O_3 (ozone), a range of CFCs (chlorofluorocarbon compounds), and N_2O (nitrous oxide).

The primary effect of water vapor is in its requirement for energy to remain in a gaseous state. Water vaporizes only with the capture of radiant energy as heat. It then releases that heat into the atmosphere as it liquifies into rain and releases even more heat as it solidifies into ice.

Although clouds reflect much solar energy, they also absorb part of it. Some of that absorbed energy is used to maintain water vapor in its gaseous state. Part of it is emitted as warmth into the atmosphere. And part of it is carried from the clouds, stored within the falling precipitation.

The other greenhouse gases absorb energy in slightly different ways than atmospheric water. When carbon dioxide is struck by solar light, some of the energy is absorbed. It then forces the CO_2 molecule to break down into a carbon monoxide (CO) molecule and an oxygen (O) atom. The oxygen atoms thus released are unstable and then recombine with the CO molecules to reform the more stable CO_2. When that happens, the excess energy is released as long-wave heat, thus warming the atmosphere. We can see that the radiant energy (light) is transformed first to chemical energy, then to heat.

When ozone (O_3) is struck by ultraviolet (UV) light waves, it absorbs the UV energy and breaks apart into O_2 and O. Again, the O ion is unstable and recombines with O_2 to reform ozone, releasing the energy that had been absorbed as ultraviolet light in the form of long-wave (heat) energy. In a like manner, methane (CH_4), CFCs, and nitrous oxide (N_2O) undergo chemical reactions that absorb solar radiation to produce unstable compounds that recombine and release energy as heat. The basic difference is that these gases, although in smaller concentrations than water vapor and carbon dioxide, are much more effective in absorbing solar energy than is either of the more plentiful gases.

Molecule for molecule, methane absorbs 20–30 times as much energy as CO_2. CFCs absorb as much as 16,000 times as much energy as CO_2.

To further complicate matters, CFCs react specifically with ozone to produce oxygen gas. As you will recall, ozone absorbs ultraviolet light and converts it to heat energy. But, as you read in the chapter on air pollution and noise (Chapter 16), UV light reaching the Earth's surface can be very harmful to living organisms. Thus, CFCs not only produce excess warming, they also reduce the UV screening effect of atmospheric ozone.

Thus, the **greenhouse gases** are those gases that react in the presence of solar energy (light waves) by breaking down into less stable compounds that subsequently recombine into the more stable original compounds with the release of the transformed light energy as heat.

The result of all these chemical reactions is a warming of the atmosphere—the greenhouse effect. The clear implication is that the concentrations of the various greenhouse gases determine the strength of the greenhouse effect. As the greenhouse gases become more concentrated, more warming of the atmosphere occurs. And atmospheric concentrations of greenhouse gases in general are at their highest point in at least the past 160,000 years.

Combining Milankovich Movements and the Greenhouse Effect

The climatic changes that would have resulted from the Milankovich fluctuations of the planet's orbit would never have produced the Ice Ages. The cooling and warming would not have been great enough to produce the dramatic climate changes necessary for ice sheets (glaciers) to form as far south as the Ohio Valley and northeastern Kansas.

By the same token, naturally occurring changes in atmospheric greenhouse gases would not have produced such effects on global climates. But, taken together, slight warming resulting from orbital fluctuations produce increased organic activity. That in turn produces more greenhouse gases which then causes more warming which then promotes still more organic activity, and so on. The cumulative effect of the two factors is thus seen to be potentially much greater than the effect of either, alone.

Human Generation of Greenhouse Gases

Air captured in the ice core from various periods in the Earth's history allows scientists to study how concentrations of greenhouse gases have changed.

Methane

The ice record indicates that methane levels have normally ranged from 350 ppb (parts per billion) during the coldest parts of the ice ages to 650 ppb during the warmest parts of the interglacial periods. Calculations indicate that the effect of the change in methane concentration alone would be expected to produce about $+0.14°$ F $(0.08°$ C) change during the warming period.

Increased methane concentrations have come from a number of sources. Wetlands have been drained for farming, urbanization, and highway construction. But, wetlands often contain great quantities of organic matter (think of peat bogs). As the water levels lower, that organic matter decays producing much methane. Many industrial activities produce large quantities of methane. Methane is a by-product of internal combustion engines such as automobiles.

In addition, agricultural development over the past few centuries has been responsible for a substantial increase in the generation of methane. Methane is generated when the organic residue of crops is allowed to decay under anaerobic conditions. In particular, rice grown in paddies encourages the development of bacteria in the water and a by-product of the bacteria is methane. Finally, as much as fourteen cubic feet of methane per day is produced in the digestive tract of a grazing cow. Thus, the concentration of methane in the atmosphere has risen from its historic normal range of 350–650 ppb to about 700 ppb by AD 1700 and to approximately 1700 ppb today.

Carbon Dioxide

Carbon dioxide concentrations, like those of methane, have tended to vary greatly with global temperatures. Ice core records indicate that the concentration of CO_2 was 190 to 200 ppm at the height of the last Ice Age about twenty thousand years ago. Air trapped in ice pack about a century ago had about 280 ppm. Air measured at Mauna Loa Observatory in Hawaii regularly over the past thirty-two years has shown a gradual rise from 315 ppm in 1958 to a predicted 355 ppm by the mid-1990s.

Carbon dioxide is produced continuously in nature by respiration in humans and animals and at night by respiration in plants. It results from natural fires and is a by-product of organic decay. At the same time, CO_2 is essential for plant growth and is absorbed by plants during photosynthesis in daytime. It also dissolves in water where it is used in the formation of many organic compounds in the soil and in the oceans.

Humans generate CO_2 emissions into the atmosphere largely as a result of burning. We burn coal, oil, and natural gas for heat energy. We burn gasoline and other petroleum distillates in our automobiles, trains, airplanes, and buses. We burn forests to clear the land in tropical regions of the world. We burn grasslands in much of the world to clear the land for planting and even to improve hunting. Beyond that, when we burn grasslands or rainforests, the agricultural crops planted are somewhat less effective at converting atmospheric CO_2 to sugar through photosynthesis. That is, their biomass is somewhat less per acre than that of a forest.

Largely as a result of our energy consumption and to some degree as a result of agricultural activities, humans are generating levels of carbon dioxide greater than at any time since the geological record began. At current rates, we can expect CO_2 concentrations in the atmosphere of 550–600 ppm by the middle of the next century with an actual doubling of the current level before AD 2100.

Chlorofluorocarbons

CFCs in particular are not naturally occurring compounds. They have been introduced into the atmosphere by industrial processes. CFCs are used in refrigeration and in aerosol cans as a propellant. When they are released into the atmosphere, they are extremely stable until they reach the stratosphere where they react with ultraviolet light. The entire effect of CFCs both on global warming and on the destruction of upper-level ozone are a direct and sole result of human activity—we are all guilty!

Predictions of Global Warming

Orbital fluctuations should be producing a mild global cooling according to Milankovich's theory. Greenhouse gas concentrations are at their highest in the life of modern humankind and should be producing substantial global warming. The current warming trend has lasted only since the early 1980s. It was preceded by a cooling period from about 1940 to 1970, which led to predictions by scientists of a coming Ice Age. The effects of volcanic activity, such as the eruption of Mount Pinatubo in 1991, may counteract the warming trend with some cooling. Moreover, a resumption of cooling la niña currents, beginning in 1989 may produce some cooling.

But, the more recent predictions of greenhouse warming come on top of an actual return to a warming trend that had held sway before 1940. Thus, the current thinking in most (but not all) scientific circles is that the next century will see significant global warming. It is that assumption that we will next pursue.

A doubling of CO_2 concentrations in the next century (a reasonable estimate based on current trends) could produce a rise in global temperatures of at least 2° F (1.2° C). A rise of as much as 9° F (5° C) has even been predicted. Naturally, those predictions assume that all other factors affecting global temperature are held constant, which of course we know cannot happen in nature. Even a 5° F (3° C) rise would put worldwide temperatures at the highest point in at least 100,000 years.

Climatologists, using **General Circulation Models** (GCMs), have developed a number of predictions for what the climate of the Earth may look like after another century of increases in greenhouse gas emissions. GCMs are **computer simulation** programs that use known and suspected cause-and-effect relationships to provide mathematical models of atmospheric conditions. Using those models, it is possible to explore "what-if" situations, such as "what if CO_2 concentrations were doubled?" (Matthews & Sugar, 1989).

Using GCM results, Schieder has developed a series of predictions for climate changes that might be expected over the next century, again assuming that the only effects on the global climate during that time period are those of the greenhouse gases. The distribution of the changes would probably not be uniform. Temperatures in the southern hemisphere are moderated because of the lower levels of industrialization and because of the larger proportion of the surface that is covered by ocean. Regional temperatures could actually drop in some places, because of climate changes, by as much as –5° F (–3° C). In other places, they could increase by as much as 18° F (10° C). Substantial changes in rainfall patterns can be expected as can changes in cloud cover which will in turn affect the amount of direct sunlight reaching the ground (see Table 20-1).

Table 20-1. Predicted Changes in Global Climate Resulting from Greenhouse Warming *

Predicted Change	Global Average	Regional Average
Temperature	+4° to +9° F	-5 to +18° F **
Sea Level	+4 to +40 in	
Precipitation	+7% to +15%	-20 to +20%
Direct Solar Radiation	-10 to +10%	-30 to +30%
Evaporation/Transpiration	+5 to +10%	-10 to +10%
Soil Moisture	?	-50 to +50%
Runoff	increase	-50 to +50%

*SOURCE: Schieder, S. (1990). Prudent planning for a warmer planet. **New Scientist**, 128(1743).

**To interpret, if the greenhouse effect produces the results that some scientists predict, these kinds of changes could occur. The average global temperature could increase between 4° and 9° F. Regional changes could range from a drop of 5° to an increase of 18° F in different parts of the world. Other categories of possible changes are given.

Agriculture and Global Warming

Agriculture is more weather-dependent than perhaps any other major world industry. The supply of water, temperature ranges, and length of growing period determine the range of possibilities for agricultural production. A major decline of rainfall in the grain-producing regions of China, the former Soviet Union, or the United States could mean disaster for the world's human population. On the other hand, a large-scale thawing in the Steppes or in the Tundra regions of Alaska or Canada could open up whole new possibilities for agricultural production.

The GCM predictions are so speculative in nature and so unspecific in their detail that no real meaningful prediction for the future of agriculture is possible. It is certain, however, that the climates of the world will change over time—*with or without the greenhouse effect*. As those changes take place, agricultural producers and scientists can be expected to react aggressively to take advantage of opportunities and to minimize the effects of problems.

The burning of tropical rainforests for agricultural use adds massive quantities of carbon dioxide to the atmosphere. But the subsequent growth of green grasses, production of agricultural crops, and regeneration of the rainforests removes much of that additional CO_2 in photosynthesis. One also must consider that in a mature rainforest, the production of new growth (added organic matter) is almost exactly balanced with the death and decay of old growth, so that the net removal of CO_2 from the atmosphere by photosynthesis in a mature forest is approximately zero.

The annual burning of the grasslands in Africa probably puts three times as much CO_2 into the atmosphere as does the burning of the rainforests (de Groot, 1990). But, it makes possible the subsequent food production that keeps much of Africa's population alive. Beyond that, the new biomass produced in the next growing year almost exactly equals that

burned from the previous year. Thus, the net emission of CO_2 is again minimal, over the long term. In truth, the real long-term culprit in the generation of CO_2 emissions is the burning of fossil fuels rather than agricultural practices.

CO_2 and Agricultural Production

There is one very major silver lining in the cloud of greenhouse gases. Plants require CO_2 in the process of photosynthesis thus, in one very real sense the increase in CO_2 concentrations in the atmosphere is like a very inexpensive fertilizer. In fact, as much as "5 to 10% of the actual rate of increase in agricultural productivity worldwide can be ascribed to the fertilizing effect of rising atmospheric CO_2" (Goudriaan & Unsworth, 1990, 111).

A comparison of the rising worldwide average grain yield between 1959 and 1986 and the rising CO_2 levels in the atmosphere during the same period shows an almost perfect correlation. Although a cause-effect relationship can be inferred it cannot be proven.

Regional changes produce uneven effects. In the tropics, even higher temperatures would be harmful, but in the temperate and cooler regions, a warmer climate would boost agricultural productivity. Crop production changes in various parts of the United States' important corn belt could range from -40% to +80% as a result of the various GCM-predicted climate changes.

Summary

The most talked-about environmental issue of the latter half of the 1980s and so far in the 1990s is global warming. Climatologists differ over whether the warming of the late 1980s and early 1990s is an artifact of the greenhouse effect and their arguments are very powerful. Some environmental extremists argue that greenhouse warming will lead to the imminent collapse of the global ecosystem as it currently exists, and there are strong reasons for concern. Other scientists are rightfully skeptical.

The truth is that the Earth's climate is at its warmest point since temperature records have been kept. Solid historical records imply that it is warmer now than it has been in a millennium. Yet, in mankind's long history, global temperatures have apparently been higher still. The direction of the long-term trend of global temperatures depends on the starting point. In the past five to seven thousand years, global temperatures are either level or perhaps slightly down. Over the past twenty thousand years they are substantially up.

There are many factors that affect average global temperatures: solar fluctuations, geographic shifts in the Earth's crust, atmospheric changes, volcanic activity, changes in the ocean's currents, and fluctuations in the orbit of the Earth around the sun. Nobody knows exactly how each of the factors will affect the future climate of the Earth because in some cases they are unpredictable. In other cases, nobody knows how the factors will interact with each other. It is possible to predict how a doubling of CO_2 levels in the atmosphere would affect the global climate, if we can assume that no other factors will change.

Unfortunately, or perhaps fortunately, we cannot know whether or when another great El Niño or la niña or another massive volcanic eruption will occur. We can predict when orbital fluctuations will occur, but we cannot know when the next major period of solar sunspot activity will occur.

The point of concerned environmental scientists looking at global warming resulting from the greenhouse effect is that concern is warranted. Although we do not know that the greenhouse effect is contributing to global warming, we can be sure that it produces a tendency toward warming. All other things being equal, the greenhouse effect presents a potential problem that we humans are bringing on ourselves. Moreover, it is a problem that we can solve by aggressive and concerted action. Just as importantly, the solutions to the problem of greenhouse gas emissions could actually contribute to the improvement of the ecosystem in other ways than just global warming.

The next chapter will present a number of suggestions that will address this encouraging thought.

Are we really experiencing global warming as a result of greenhouse gases? Certainly we are, that is one of the main reasons that we can exist.

Is excess global warming resulting from anthropogenic greenhouse emissions? Probably not.

Is the planetary ecosystem on a collision course with global warming that will produce a devastating effect of life on Earth? It is very unlikely, in spite of the hysterical cries of some people who earn their livings by creating excitement to get greater attention, more research grants, or increased contributions from the public.

Will the next century see substantially warmer temperatures? Perhaps. In fact, it is very likely. The planet has been on a warming trend for over 100,000 years. The increasing concentrations of greenhouse gases MAY even contribute to that warming trend.

Would a warmer world be good or bad? It probably depends on where you live and how you earn your living. After all, good and bad are philosophical questions—not questions for science to answer.

Finally, should people make a concerted effort to curb greenhouse gas emissions? Absolutely, and for many reasons, one of which is the unproven potential of greenhouse gases to alter the global climate in unforeseeable ways.

DISCUSSION QUESTIONS

1. How do drilling cores of ice and ocean sediments provide evidence of Earth's changing climates?

2. What causes seasonal variations in climate?

3. What are the Milankovich Movements and what impacts do they have on the Earth's climate pattern?

4. What is the greenhouse effect?

5. What are the major greenhouse gases and how do they produce a warming effect in the atmosphere?

6. What is the normal pattern of ocean tides in the tropical Pacific? Why is its stability important?

7. What are the El Niño and la niña? What effect do they have on regional and global climates?

8. What effects do human activities have on the global climate? How?

ADDITIONAL ACTIVITIES

1. Place two similar pieces of wood that have been painted black in direct sunlight. Select a place that is sheltered from the wind. Tilt one at 90° to the sun and the other at 45° to the sun. After a few minutes, check the temperatures of the two surfaces. Relate the difference to the effect of latitude on regional climates.

2. Place a series of different colored pieces of wood in direct sunlight and at 90° to the sun in a place that is sheltered from the wind. Be sure to include white, black, and at least a few other colors. After a few minutes, check the temperatures. Repeat the experiment with containers of clear water, slightly tinted water, and darkly tinted water. The containers should all be clear or very light in color. After a few hours, check the temperature of the water. Relate your findings to the concept of albedo.

3. Place a globe of the Earth near one side of a darkened room. Place a source of light in the center of the room. Tilt the globe at 23½° from a line perpendicular to the floor with the North Pole of the globe leaning toward the North. You may need a compass, or you may simply designate an arbitrary direction as North. Move the globe around the room while holding its axis constantly pointing North. Repeat the demonstration with the axis held perpendicular to the floor. Relate your observations to the effects of orbital fluctuations.

4. Get two identical cardboard boxes and remove the tops. Place a thermometer in the bottom of each. Cover one with clear plastic or glass and tape the edges to minimize heat loss. Place both boxes with the tops, one open and the other sealed, exposed to direct sunlight. Place a third thermometer beside the boxes on a piece of cardboard the same color as the bottoms of the boxes. Beginning as soon as they are placed in the sunlight, make a record of the temperature indicated on each thermometer every 5 minutes until there is no change in either thermometer. What happened? Why? How does this relate to the greenhouse effect?

CHAPTER 21

Wetlands and Deepwater Habitats

Photo courtesy Bill Camp.

Terms to Look for and Learn

Draining
Emergent Hydrophyte
Estuarine System
Filling
Hydric Soil
Hydrophyte
Lacustrine System
Marine System
Mesophyte

Obligate Hydrophyte
Palustrine System
Reclamation
Riverine System
Section 404, Clean Water
    Act of 1972
Swamp Acts
Swamp-Buster Act
Water Regime

Learning Objectives

After reading this chapter and participating in the activities, you should be able to:

■ Define and identify wetlands.

■ Explain the importance of wetlands within the global ecosystem.

■ Describe a hierarchical structure for the classification of wetlands and deepwater habitats.

■ Explain the common water regimes of tidal and nontidal wetlands.

Overview

Almost everyone has seen wetlands in one form or another and most people have developed a general concept of what a wetland is like. Many of us have walked in an open field, or in a forest, or alongside a road, or somewhere else where the soil was soggy wet or had standing water on it long after the most recent rain. The intuitive definition of a wetland probably would include standing water or saturated soil, but that is a gross oversimplification (see Figure 21-1). Wetlands actually range from land permanently covered by water to uplands that are never covered by standing water. As we shall see later in this chapter, defining wetlands is not a simple task and identifying and delineating them is even more difficult.

Early American Perceptions

At the time of European settlement, the area that is now the continental United States had an estimated 215 million acres of wetlands. Today only about ninety-four million acres remain, and the conversion of wetlands to other uses continues at the rate of several hundred thousand acres each year. Almost three-fourths (seventy-four percent) of all U.S. wetlands are on private property, with the remaining fourth on federal and state property.

With a people struggling for survival, and a nation still in the process of being built, there was little room for environmental concern in early America. To the early European settlers, wetlands, like forests, were simply something to be overcome in taming the land. An early president of the American Public Health Association stated in 1876, "The state cannot afford to be indifferent to [the presence of swamps] because they check production, limit population, and reduce the standard of health and vigor." (Steinhart, 1990, 18).

Wetlands served as breeding places for mosquitos, snakes, and other vermin. Crops could not be grown on them. Travel across them was difficult or impossible. Wetlands were simply problems to be solved and the best solution was to drain them. Indeed, since wetlands were considered useless, the act of draining or filling them to convert them to other use was thought of as **reclamation**.

Toward that end, Congress enacted a series of so-called **Swamp Acts** between 1849 and 1860. By this legislation, the federal government gave some 65 million acres of wetlands to the states in which they were located. The primary charge to the recipient states was that they drain the wetland to reclaim it for productive and healthful use. Even as recently as the twenty years from 1940 to 1960, the federal government subsidized the drainage of as much as 60 million acres of wetlands for agricultural use.

To a preservationist, that would sound like massive destruction of the ecosystem. We must remember that the wetlands were drained so humans could live better. Almost all of southern Florida had to be drained before cities could be built. Of the 121 million acres that were

Figure 21–1.   This permanently water covered area is a swamp. This is only one example of a wetland. (Photo courtesy Bill Camp.)

drained, most have become farms, highways, or urban areas. The draining of the wetlands was one of those opportunity cost "trade-offs" that were discussed in Chapter 3. To a developer, the wetlands were improved for human use by draining them.

The main American wetland preservation activity during those years came from an interesting source. Waterfowl hunters realized that wetlands were necessary for a healthy population of ducks and geese. It was through their efforts that wetlands were preserved as breeding, migrating, and wintering areas for migratory waterfowl. Indeed, major American wetland preservation policies prior to 1972 are attributable to this group of hunters, which traditionally has included many wealthy and influential people in this country.

The first major national policy for the protection of wetlands, not motivated largely by game management concerns, was the **Clean Water Act of 1972. Section 404** of that act charged the Army Corps of Engineers with responsibility for issuing permits for filling of wetlands on both public and private lands in the United States.

Wetlands can be reclaimed either by filling or by draining. **Filling** involves adding solid material to displace the water, thereby raising the soil level. Fill can consist of soil, but it can also consist of rocks, old tires, solid wastes, or anything else that is solid. **Draining** involves removing standing water and lowering the water table. Draining can consist of creating ditches to conduct water away from the wet site. It can also consist of constructing underground tile or corrugated pipe drainage systems (see Figure 21-2).

The Clean Water Act of 1972 regulated only one aspect of wetland reclamation—filling. Draining was not prohibited. The vast majority of wetland conversion has been by drainage and of that, between eighty

A

B

Figure 21–2. (A) Filling a wetland involves adding solid materials such as soil to displace standing water or to raise the soil level relative to the water table. (B) Standing surface water can be drained away by ditches. (C) Excessive water can be removed by means of underground drainage systems such as this corrugated pipe.

C

and eighty-five percent of conversions have been for agricultural use over the past two decades.

Provisions of the 1985 Food Security Act prohibited crop support payments and crop subsidy payments to farmers for crops produced on newly converted wetlands. It also provided for payments to farmers for maintaining wetlands on their land. This act came to be known as the **"Swamp-Buster Act"** because, in effect, it was aimed at discouraging the conversion of wetlands to farmland. Subsequent farm legislation, including most notably the 1987 Agricultural Credit Act and the 1990 Farm Bill, have continued to place emphasis on minimizing the conversion of wetlands for farming.

In 1989, Congress passed the North American Wetland Conservation Act. This law is reminiscent of the earlier legislation aimed at preserving habitat for game birds such as ducks and geese. It authorizes federal funds for migratory waterfowl management. It also provides for expenditure of U.S. tax dollars for wetland preservation in Mexico and Canada, on a matching basis.

Definition of Wetlands

As we have already discussed, swamps, bogs, and marshes have been well known for many centuries. From the famous moors of England to the Florida Everglades of the United States, we all understand these concepts. But, the more encompassing concept of wetlands is more recent. There is no single definition of wetland on which every expert agrees. One of the most quoted current definitions is that given in the United States Fish and Wildlife Service publication *Classification of wetlands and deepwater habitats of the United States.*

> Wetlands are lands transitional between terrestrial and aquatic systems where the water table is usually at or near the surface or land is covered by shallow water. For purposes of this classification wetlands must have one or more of the following three attributes: (1) at least periodically the land supports primarily hydro-

phytes; (2) the substrate is predominantly undrained hydric soil; and (3) the substrate is nonsoil and is saturated with water or covered by shallow water at some time during the growing season of each year. (Cowardin, Carter, Golet, & LaRoe, 1979, 3)

For the sake of this definition, two terms must be explained. First, a **hydric soil** is one that is very poorly drained for at least part of the year. In that condition, the soil is saturated with water to the extent that anaerobic conditions are present. Recall from Chapter 5 that a healthy soil always contains mineral matter, organic matter, water, and air. Anaerobic means absence of air. For a soil to be hydric, that anaerobic condition must last long enough to adversely affect normal growth for most agricultural crop plants except rice and cranberries.

Second, **hydrophytes** are plants typically found in wet habitats. They grow in soil covered by standing water or in hydric soil. Most terrestrial plants require relatively constant aerobic conditions such as those found in well-drained soils. Hydrophytes either tolerate extended periods of water-logged soils or require constant submersion to grow. **Emergent hydrophytes** are typically herbaceous plants. They may be found away from permanent standing water growing in soil that is sometimes water logged. **Obligate hydrophytes** may be woody plants such as alder. But they are almost never found outside wetlands (see Figure 21-3).

Directly opposite to hydrophytes are **mesophytes**. Mesophytic plants cannot tolerate water-saturated soils for extended periods of time. Such plants are not normally found growing in wetland areas. Most agricultural crops, such as corn and soybeans are mesophytic.

Wetlands can fall into any one of five broad categories:

- Areas with both hydrophytes and hydric soils, such as bogs, marshes, and swamps
- Areas with hydric soils but generally lacking in hydrophytic plants, such as flats, which are affected by tide action or high salt concentrations to the extent that hydrophytic plant life is prevented

Figure 21-3. Hydrophytic plants either tolerate or require water saturated soils or standing water in order to survive. (Photo courtesy Bill Camp.)

- Areas where hydric soils have not yet developed but in which hydrophytes have already appeared, such as newly created marshes, impoundments, or excavations such as road ditches.

- Areas without true soils but in which hydrophytes are growing, such as the rocky shores or rocky bottoms of shallow water with seaweeds

- Areas with neither hydric soil nor hydrophytes, such as gravel beaches or rocky shores without any vegetation (Cowardin, et al., 1979, 3).

Hierarchical Classification System

The United States Fish and Wildlife Service of the Department of the Interior published a hierarchical classification system for wetlands and deepwater habitats. The hierarchy is a three-tiered classification scheme that includes systems, subsystems, and classes.

The largest classification category is the system. It is used to mean a major grouping of wetlands and deepwater habitats that share similar hydrologic, soil, chemical, and biological characteristics. There are five major systems: marine, estuarine, riverine, lacustrine, and palustrine. In all five cases, the systems refer to the land at the surface of the Earth or the solid crust underlying water.

Marine System

The **marine system** (in this classification scheme) consists of those land areas underlying the continental shelf and including its high-energy coastline. That includes all of those coastal areas affected by wave action and tidal movements. The landward limit is either: (1) the upper reaches of the extreme high water mark of maximum spring tide, including the splash zone from wave action; (2) the seaward limit of wetland vegetation; or (3) the seaward limit of the estuarine system. The seaward limit of the marine system is the outer edge of the continental shelf. Deepwater habitats beyond the continental shelf are not considered in this classification scheme.

The marine system has two subsystems: subtidal and intertidal. The **subtidal subsystem** is constantly submerged. The **intertidal subsystem** is alternately submerged and exposed by changing tides and includes the coastline affected by the splashes of waves at the maximum high tide (see Figure 21-4).

Estuarine System

The **estuarine system** includes both deepwater and the associated tidal wetlands. As we learned in Chapter 17, estuaries are at least partially enclosed by land. In addition, estuaries are periodically or constantly diluted by fresh water from surface runoff or groundwater seepage from the land. That produces water salinity conditions of waters that are typically between those of fresh water and ocean water, and known as brackish water.

Figure 21–4. The marine system includes land underlying open water and its associated shoreline including the splash zone at extreme high tide. (Photo courtesy Bill Camp.)

In some estuaries the salinity can become greater than that of the open ocean. That condition occurs when the water body is completely enclosed and less fresh water is introduced than evaporates.

Stream-fed estuarine systems extend up the channel landward to the point that the average concentration of ocean-derived salts is below 0.5 percent during the average low flow period. In general, the estuarine system extends landward to the point that the growth of hydrophytes associated with the estuary ends. In the opposite direction, the estuary extends seaward to

an imaginary line across the mouth of the river or other partial enclosure. The seaward extent of the estuary can also be defined as the limit of continuously diluted sea water (Cowardin, et al., 1979).

As with the marine systems, estuarine systems also include both subtidal and intertidal subsystems. Subtidal lands are continuously submerged. Intertidal lands are those that are alternately submerged and exposed by the rising and falling tides. Because estuarine systems normally have low-energy shorelines, the splash zone is not an important consideration (see Figure 21-5).

Riverine System

Streams and rivers move through fairly well-defined channels. A channel can be defined as an "open conduit either naturally or artificially created which periodically or continuously contains moving water, or which forms a connecting link between two bodies of standing water" (Langbein & Iseri, 1960, 5, cited in Cowardin, et al., 1979, 9).

The **riverine system** includes all wetlands enclosed by the channel banks, either natural or man-made, such as levees. It also includes all of the associated land on which hydrophytes are the dominant vegetation. The upper limit of the riverine system is the top of the channel bank or the end boundary of the riverine-associated wetland vegetation. Some streams or rivers

Figure 21–5. This estuarine system along the coast of North Carolina is created by the mixing of fresh water from the White Oak River mixing with ocean water from the Atlantic in an area partially enclosed by an offshore island. (Photo courtesy Bill Camp.)

have multiple branches (such as in the Mississippi Delta region). Those branched (or braided) streams often have relatively continuous wetlands between the branches. In such cases, the entire area enclosed within the braids is part of the riverine system.

The lower limit of the riverine system ends when the stream enters a standing body of water. In the case of entry into the ocean, the riverine system ends where the average salinity concentration is greater than 0.5 percent at the average low water flow period. In the case of the stream that terminates in a lake, the riverine system ends at the mouth of the channel.

Riverine subsystems include tidal, lower perennial, upper perennial, and intermittent wetlands. The **tidal subsystem** consists of that part of the channel where the water level is influenced by the ocean tides. The **lower perennial subsystem** is flooded year round, but the water flow is slow because the channel is nearly level. The **upper perennial subsystem** is characterized by a steeper gradient along with year-round water flow. The **intermittent subsystem** contains flowing water only part of the year. During the dry season, the water flow stops and any remaining water is contained in isolated pools, or there is no surface water remaining at all.

Lacustrine System

If the riverine system is characterized by moving water, the **lacustrine system** is characterized by standing water. It consists of wetlands and deepwater habitats with relatively large areas. Lacustrine lands may be tidal or nontidal, but waters are always below 0.5 percent ocean-derived salinity. Lacustrine system includes lands:

- In topographic depressions or in dammed river channels
- A total area of more than twenty acres
- If less than twenty acres, minimum water depth at low water of at least 6.6 feet
- Less than thirty percent coverage by trees, shrubs, emergent hydrophytes, or emergent mosses and lichens (Cowardin, et al., 1979).

Emergent plants live in wetlands which may be periodically covered by water but do not tolerate extended periods of submersion. An example is cattails, which grow in wetlands in all areas of the world (pandemic). Some emergent plants can be found in permanent shallow water. Others can be found in soil that is merely waterlogged much of the time.

There are two lacustrine subsystems: limnetic and littoral. The **limnetic subsystem** includes all deepwater habitats within the system. **Littoral subsystem** includes all lacustrine wetlands, extending out to 6.6 feet of water depth at average low water in lakes and empoundments (see Figure 21-6).

The limit of a lacustrine system is at the point where the land becomes dominated by upland trees, shrubs, or other nonaquatic plants. If the lacustrine system borders a riverine, estuarine, or marine system, the boundary of the two is at the point where the two systems meet.

Palustrine System

There are no subsystems in the palustrine system. Palustrine system wetlands may be permanently or

---

Figure 21–6. The lacustrine (lakes) system includes nonocean deepwater habitats and the associated wetlands. Lakes may be fresh, brackish, or salty. (Photo courtesy Bill Camp.)

Figure 21-7. Wetlands of the palustrine system includes mostly upland areas. They are not associated directly with deepwater but they may be flooded temporarily or permanently. (Photo courtesy Bill Camp.)

temporarily flooded by shallow fresh or brackish water, or they may be merely saturated to the surface at times (see Figure 21-7).

Palustrine wetlands include all:

- nontidal wetlands
  - □ not directly associated with standing water of 6.6 feet (two meters) or more depth at low water, and
  - □ dominated by trees, shrubs, and emergent hydrophytes;
- tidal wetlands
  - □ with ocean-derived salinity always below 0.5 percent, and
  - □ dominated by trees, shrubs, and emergent hydrophytes.

Classes within Each System

The classification system has four levels: system, subsystem, class, and subclass. Within each subsystem,

there is a series of classes. As Table 21-1 shows, there are ten classes, each with characteristics distinct from the others. The classes and a description of each are given below (Cowardin, et al., 1979):

- Rock Bottom—at least seventy-five percent covered by stones, boulders, or bedrock; submerged at least on a semipermanent basis.

- Unconsolidated Bottom—no more than twenty-five percent covered by rocks, boulders, or bedrock; submerged at least on a semipermanent basis.

- Aquatic Bed—dominated by plants that grow either under watercover or floating in water for all or most of the year; either permanently covered by water, or at least seasonally flooded.

- Reef—rocky structures formed by the growth of sedentary invertebrates; may be permanently underwater (subtidal) or alternately covered and exposed by water (intertidal).

- Streambed—water channels that are completely drained of water during dry periods or low tides.

- Rocky Shore—at least seventy-five percent covered by rocks, boulders, or bedrock; not permanently flooded.

- Unconsolidated Shore—less than seventy-five percent covered by rocks, boulders, or bedrock; less than thirty percent covered by vegetation other than pioneering plants; not permanently flooded.

- Moss-Lichen Wetland—areas, other than rocks, covered by mosses and lichens; soil saturated but not flooded on a regular basis.

- Emergent Wetland—dominant vegetation emergent hydrophytes, not mosses or lichens; any water conditions except subtidal possible.

- Scrub-Shrub Wetland—dominated by woody shrubs under twenty feet tall; any water conditions except subtidal possible.

Table 21-1. United States Fish and Wildlife Service Classification System for Wetlands and Deepwater Habitats

System	Subsystem	Class	System	Subsystem	Class
Marine	Subtidal	Rock Bottom			Rocky Shore
		Unconsolidated Bottom			Unconsolidated Shore
		Aquatic Bed			Emergent Wetland
		Reef		Upper Perennial	Rock Bottom
	Intertidal	Aquatic Bed			Unconsolidated Bottom
		Reef			Aquatic Bed
		Rocky Shore			Rocky Shore
		Unconsolidated Shore			Unconsolidated Shore
Estuarine	Subtidal	Rock Bottom		Intermittent	Streambed
		Unconsolidated Bottom	Lacustrine	Limnetic	Rock Bottom
		Aquatic Bed			Unconsolidated Bottom
		Reef			Aquatic Bed
	Intertidal	Aquatic Bed		Littoral	Rock Bottom
		Reef			Unconsolidated Bottom
		Streambed			Aquatic Bed
		Rocky Shore			Rocky Shore
		Unconsolidated Shore			Unconsolidated Shore
		Emergent Wetland			Emergent Wetland
		Shrub-Shrub Wetland	Palustrine		Rock Bottom
		Forested Wetland			Unconsolidated Bottom
Riverine	Tidal	Rock Bottom			Aquatic Bed
		Unconsolidated Bottom			Rocky Shore
		Aquatic Bed			Unconsolidated Shore
		Rocky Shore			Moss-Lichen Wetland
		Unconsolidated Shore			Emergent Wetland
		Emergent Wetland			Shrub-Shrub Wetland
	Lower Perennial	Rock Bottom			Forested Wetland
		Unconsolidated Bottom			
		Aquatic Bed			

Source: Cowardin, L.M., Carter, V., Golet, F.C., & LaRoe, E.T. (1979). **Classification of wetlands and deepwater habitats of the United States.** Washington, D.C.: United States Fish and Wildlife Service.

The Fish and Wildlife Service classification system also includes subclasses within each class. Because of space limitations, this chapter will not describe the subclass schemes. But the reader should be aware that subclasses do exist.

Water Regimes

In classifying wetlands and deepwater habitats, water regimes must be described. Water regime for a particular land area refers to the status of water in its liquid form in, on, or covering the land. The commonly noted water regimes for those regions influenced by ocean tides are (Cowardin, et al., 1979):

- Subtidal—ocean, sea, estuary; permanently underwater.

- Intertidal—ocean, sea, estuary; alternately exposed and covered by rising and falling tides.

- Irregularly Exposed—exposed by low tides less often than daily.

- Regularly Flooded—routinely covered and exposed by typical tides every day.

- Irregularly Flooded—covered by high tides less often than daily.

Nontidal wetlands are affected by direct precipitation, surface runoff from rainfall, surface runoff from snow melt, or subsurface discharge. They are usually fresh water, although many have high concentrations of nonocean salinity. In the United States, nontidal wetlands represent the vast majority of all wetlands. Common names for nontidal wetlands include shrub swamps, wooded swamps, bottomland forests, bogs, marshes, seeps, springs, streams, rivers, ponds, and lakes. For wetlands that are not influenced directly by ocean-driven tides, the commonly cited water regimes are as follow (see Figure 21-8):

- Flooded—covered by standing or moving water, other than moving runoff, at some predictable frequency.

  □ Permanently Flooded—covered by water constantly.

  □ Semipermanently Flooded—flooded throughout the growing season in most years.

Figure 21–8. Schematic representatation of various water regimes. (**Source:** Tiner, R.W. (1988). **Field guide to nontidal wetland identification**. Annapolis: Maryland Department of Natural Resources, and Washington, D.C.: United States Fish and Wildlife Service, Department of the Interior.)

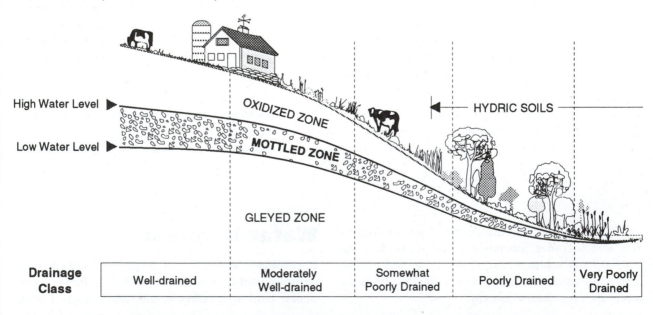

- Seasonally Flooded—flooded for at least two weeks sometime during the growing season, but exposed during the remainder of the growing season.

- Temporarily Flooded—flooded periodically during the growing season, but usually for less than two weeks at a time.

■ Saturated—soil in which the water table is at or very near the surface.

- Permanently Saturated—constantly saturated at or near the surface.

- Seasonal—high water table during wet periods only.

- Seeps—frequent water discharges from the soil surface.

- Wetland Floats—organic soil masses floating on the surface of water bodies.

■ Poorly Drained—saturated soil associated with slow water movement rather than high water table.

- Very Poorly Drained—saturated most of the time; usually, but not always level or in a depression; nonhydrophytic (i.e., mesophytic) crops cannot be grown except with artificial drainage.

- Poorly Drained—saturated to the surface during much of the growing season; mesophytic crops can be grown effectively only with artificial drainage.

- Somewhat Poorly Drained—wet periodically during growing season; mesophytic crops frequently damaged.

Hydric Soils

The United States Soil Conservation Service defines hydric soil as "soil that is saturated, flooded, or ponded long enough to develop anaerobic conditions in the upper part" (Tiner, 1988, 233). As was discussed earlier, anaerobic conditions refers to an absence of air—more specifically, an absence of oxygen. Soil contains both air and water in fairly large quantities. When the soil becomes saturated with water, the air, with its oxygen gas, is forced out (see Figure 21-9).

In soil containing a normal amount of oxygen in gaseous form, plant roots obtain oxygen by absorption. Most plant roots must have ready access to free oxygen for normal respiration, growth, and reproduction. In a saturated condition, that growth and

Figure 21-9. A "normal" soil is about twenty-five percent water most of the time. In a hydric soil, the air is replaced by water all or much of the time. This hydric soil shows extensive mottling (gray discoloration). (**Source:** United States Department of Agriculture.)

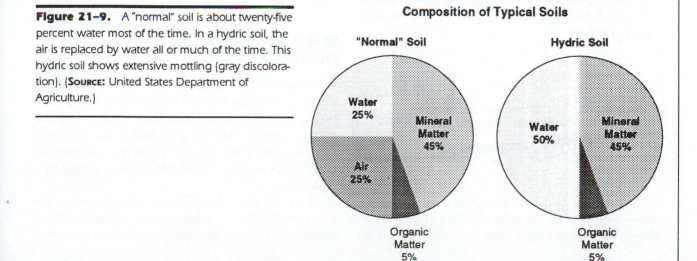

normal respiration is prevented for most plants, except hydrophytes. The result is that such plants cannot grow in saturated soil, and in some cases, they cannot even survive.

To be classified as hydric, a soil must be saturated or flooded long enough to damage most plants and to favor hydrophytes. In addition, the flooding or saturation normally must occur during a season when the plants are not dormant. During winter, many plants become dormant, and flooding or saturation may cause little or no harm.

In general, there are two major kinds of soils: organic soils and mineral soils. Organic soils form from the accumulation of plant and animal (organic) materials on the soil surface. In saturated or flooded conditions, the aerobic decomposition of organic matter is almost stopped. Thus, anaerobic conditions promote the formation of thick layers of organic matter in the form of peat or muck. Peat and muck soils are often "mined" for other uses. Peat moss and organic matter purchased at garden centers are commercial examples of the agricultural or landscaping use of organic soils.

Organic soils are defined as peat when at least two-thirds of the organic matter has not been completely decomposed and can still be identified. Organic matter that is more than two-thirds decomposed beyond recognition is called muck. Peaty muck and mucky peat fall between the two extremes. The degrees of decomposition of plants could be illustrated as follow:

Decomposed --------------------------- Recognizable
Muck   Peaty Muck   Mucky Peat   Peat

Where organic matter does not accumulate to appreciable depths, the soil is said to be a mineral soil. Mineral soils may have a layer of organic matter on the surface, but the composition of the soil itself is inorganic. Mineral soils that are well-drained tend to be brown or black near the surface. The color comes from decayed organic matter. Further down they tend to have a yellowish or reddish color. That color comes from the oxidation of iron (ferric oxide) contained in the mineral soils. Rusting iron has a reddish color. In the extended absence of free oxygen, iron returns to its reduced (ferrous) state.

In frequently saturated hydric mineral soils, the aerobic oxidation of iron is prevented. The result is often a mottled or gleyed soil. Mottling of a mineral soil is indicated by the presence of gray spots or splotches in the soil structure. Gleying is indicated by a predominance of gray, greenish, or bluish-gray color in the soil, sometimes all the way to the surface. Mottling results when the soil is periodically saturated. Gleying occurs when the soil is permanently or semipermanently saturated.

Identifying and Delineating Wetlands

Before wetlands can be wisely managed, they must be identified. In many cases that is a fairly simple process. In other cases, determining whether an area is a wetland or an upland is problematic.

National Wetlands Inventory Maps

The United States Fish and Wildlife Service of the Department of the Interior is involved in a long-term project to locate, catalog, and map all of the wetlands in this country. The project, called the National Wetlands Inventory, involves developing large-scale (1:24,000) maps showing the location, size, and type of wetlands.

By 1988, over half of the wetland maps for the lower forty-eight states had been completed. About fourteen percent of Alaska had been catalogued and mapped. The mapping for Hawaii was completed. The map shows outlines of the wetland boundaries. For each wetland area drawn, the system, subsystem, class, subclass (not described in this chapter), and water regime are given.

Soil Surveys

For many years, the United States Soil Conservation Service has been involved in developing soil surveys. Soil surveys are documents that detail the soils within given geographic areas, most often counties. The surveys include maps indicating soil series classifications. See Chapter 5 for more details.

The soil surveys are not aimed at identifying wetlands; however, they do provide information on hydric soils that can indicate obvious wetland areas. They also include written descriptions of climate, topography, and other information useful in locating wetlands.

Field Research

Probable wetlands can be identified by examination of aerial photographs, particularly infrared. They can be located from soil surveys, or from National Wetlands Inventory maps. Once a probable wetland area has been located, its actual existence as a wetland must still be confirmed. That involves field research.

A walk-through of the suspected wetland will usually provide access to the needed information. Presence of hydrophytes, saturated soil, debris lines at the base of plants, hydric soils, or standing water are primary indicators.

It is important to remember that wetlands occur along a continuum of wetness. Some wetlands are obvious. An area with standing water or saturated soil weeks after the latest rain during mid-growing season is almost certain to qualify. An area with many hydrophytes but few or no mesophytic plants is also clearly a wetland. But, for less clear cases, it is more difficult to determine. For a much more detailed description, see Tiner's *Field guide to nontidal wetland identification.*

Note: A draft manual (entitled *Revised Federal Manual for Identifying and Delineating Vegetated Wetlands*, one hundred pages) for the identification of wetlands was made available for review in April, 1991. At the time of this writing, that manual had not been approved. Once approved and disseminated, it can be expected to replace the Tiner manual.

Once an area has been confirmed as a wetland, its exact boundaries still need to be determined. That is also done by field research. Abrupt changes of topography, limits of hydrophyte domination, or limits of hydric soils are good indicators.

Summary

Until recent years in this country wetlands were regarded as liabilities rather than assets. As such, public policy encouraged draining and filling of wetlands. It was not until recent decades that the ecological value of wetlands was recognized. Today we know that wetlands contribute to both the amount and quality of groundwater; provide breeding grounds and habitats for countless birds, mammals, and reptiles; and serve many other valuable purposes.

Wetlands occur along a continuum of wetness. They range from land permanently covered by water to soils only periodically saturated with groundwater. They are characterized by some degree of hydric soils, hydrophytic plants, and a water regime from somewhat poorly drained to permanently flooded or subtidal.

Emergent hydrophytic plants are those that tolerate water-saturated conditions. With rare exceptions, obligate hydrophytic plants grow only on wetlands. Mesophytic plants do not tolerate saturated soils well. Hydric soils are soils that experience saturation or flooding frequently enough to cause problems for mesophyte production.

The Fish and Wildlife Service of the United States Department of the Interior has developed a classification system for wetlands and deepwater habitats. It consists of five major systems: marine, estuarine, riverine, lacustrine, and palustrine. In general, those systems are in order as they occur beginning below sea level and continuing inland.

Within each system except the palustrine, there are subsystems. And within each subsystem there are classes. The subsystems are determined largely by the surface water regime. The classes are determined by the nature of the soil or land surface.

Wetlands are being identified and mapped by the Fish and Wildlife Service in the National Wetlands Inventory project. Much of the mapping is completed, but as of the date of this writing, much mapping remains to be done, particularly in Alaska. Even after a wetland is mapped by this project, much field work is still needed. Effective wetland management cannot be planned until the exact nature, location, and extent of the wetlands are determined by field research.

DISCUSSION QUESTIONS

1. What is a wetland?
2. Outline the major systems included in the U.S. Fish and Wildlife Service Wetland classification system.
3. What does water regime refer to, in discussing wetlands?
4. What are the most commonly noted water regimes for tidal regions?
5. What are the most commonly noted water regimes for nontidal regions?

ADDITIONAL ACTIVITIES

1. Contact your local United States Soil Conservation Service Conservationist and request to see a copy of your local soil survey. As an alternative, the soil survey should be available through the college of agriculture at your state land grant university. Examine the soil survey to determine likely wetland areas in or near your community.
2. Locate a likely wetland in your local area. Visit the area and use the key provided in this chapter to determine whether it is actually a wetland. If you determine it to be a wetland, classify it using the Fish and Wildlife classification system. Determine its system, subsystem, class, and water regime.
3. Visit a local wetland area to identify and catalog as many of its species of plants as possible.
4. Dig a hole several feet deep in the wetland soil and observe:
   - mottling (gray and brown flecks or spots), and
   - a water table (as the hole fills with water).

CHAPTER 22

A Sustainable and Livable World Order

Photo courtesy Vernon E. Doyle, Missouri Highway & Transportation Department.

Terms to Look for and Learn

Anthropocentric	Responsible Environmentalism
Livable	Sustainable
Naturalist	TINSTAAFL

Learning Objectives

After reading this chapter and participating in the activities, you should be able to:

■ Compare and contrast the perspectives of naturalists and anthropocentrists.

■ Explain the importance of responsible environmentalism.

■ Explain the basic conflicts between what humans want and what is possible in the long run.

■ Discuss the concept of a sustainable world order.

■ Discuss things that responsible environmentalists can do as individuals to promote a sustainable and livable world order.

359

Overview

We live in a finite world. In many ways, Earth could be thought of as a single, planetary life-support system—a global ecosystem. In one way or another, all of its parts are interconnected. Yet, within that massive and infinitely complex system, there are countless smaller subsystems—each itself an ecosystem.

We have learned that nature is not, and has never been, stable or kind. Indeed, we can never expect it to be stable. Earth's ecosystems, from a drop of water to the entire planetary ecosystem, are constantly changing. Humanity is not responsible for everything that happens within the global ecosystem. But at the same time, human activities affect local and even the planetary ecosystems in both direct and indirect ways.

In the ancient past, human activity could have little impact on the planetary ecosystem. But early humans invented tools, and then weapons, and then fire. Eventually humans could live more comfortably and safely. Their numbers began to grow exponentially, as we learned in Chapter 15. The ecological relationships among humans and the other members of the biotic subsystems changed drastically. Over the millennia, the population of humans has grown and our technology has advanced. Today, human activity very directly and profoundly affects not only local ecosystems but the global ecosystem as well.

It All Depends on Your Perspective

Much literature on the environment would have us believe that all human-caused environmental changes are automatically bad—that any new highway, shopping center, housing development, or oil well should be opposed. But, not all environmental changes are bad. In fact, from a purely philosophical perspective, changes are neither good nor bad; they simply are changes. It is human thinking and value judgments that make a thing good or bad. So what is good or bad in terms of ecological changes must be defined from some perspective—from how we look at events as humans.

There are several different perspectives that could be used in addressing the question of what is good or bad in the environment. From one perspective, all creatures on Earth are of equal worth and the effects of environmental changes on all of them must be considered equally. In this view of life, nature is central and humans are merely co-equal inhabitants of the ecosystem. These people believe that nature is good in and of itself, and is worth preserving simply because they think it is inherently good. This is the basic philosophy of many people who think of themselves as **naturalists**—this could be characterized as a "back to nature" dream.

The viewpoint from which most humans operate is the **anthropocentric** perspective. From that outlook, humans are central and nature is there for human use. In that system of thinking, changes our activities cause in the ecosystem are good or bad only to the extent that they result in positive or negative results for humans. That perspective is labeled anthropocentric (human-centered). The basic philosophy is that the environment should be used for human wants and needs. Only in recent years have we come to understand how hard it is to exploit the environment without degrading it. Interestingly, the teachings of both Judaism and Christianity, among many other religions, are very much anthropocentric.

Three potential anthropocentric problems arise when human activities cause changes in the planetary ecosystem.

■ The first problem is when human activity produces a change in the ecosystem that degrades its ability to sustain long-term human survival. In an ecological sense, that translates into a diminished carrying capacity for humans, either in the near term, or more importantly in the long term.

■ The second problem arises when human population growth exceeds our ability to feed, clothe, and shelter ourselves without damaging the ecosystem. That translates into a different, but equally serious ecological problem. When the population of a species exceeds the ecosystem's carrying capacity, only one result has ever occurred in nature. The carrying capacity is damaged by overfeeding. Then the species experiences a "die-off" (or a population crash) that brings its numbers in line with the reduced carrying capacity of the system.

■ A third problem, closely allied to the second one, occurs when human activities degrade the environment in a way that diminishes the quality of life for the humans who populate the planet. It is in this area that anthropocentric and naturalist thinking come together—and often they clash.

Responsible Environmentalism

It is from the anthropocentric perspective that this chapter will address the most fundamental question of our time. The question is how can humans survive as a species within the finite, yet infinitely complex, planetary ecosystem? We will look at that question not just in the near term, but into the foreseeable future.

The question of human survival is not as simple as people once thought, as environmental science has taught us in the past few decades. If the human species is to continue at its present numbers, or as is almost definite, in larger numbers, something must be done, and immediately.

It is clear that simple preservation is not the answer. There are precious few (if any) places on Earth that are still in their pristine condition. Even in the few places where humans have never gone, our by-products in the atmosphere and in the waters have gone. By the same token, simple exploitation is clearly not the answer. Americans, Europeans, and other world inhabitants have used economic exploitation

for many thousands of years, and we are just now getting really "good" at it. Another two or three centuries on our present human course, and the planet could be literally uninhabitable, using our present technology and mode of living.

The Free Lunch Syndrome

Perhaps you have heard the old saying, "There is no such thing as a free lunch." This is the so-called **TINSTAAFL** paradox. The term "opportunity cost" was used in Chapter 3 to mean basically the same thing.

If you have read the rest of this book, by now you should understand much more about the environment than before. You should understand that every action by one species affects its environment—its ecosystem. Human activities are no exception. Indeed, with modern technology, human activities drastically impact the ecosystem, not just locally, but globally.

But, that technology is used to get things we want. We want a comfortable home. We want a reliable means of transportation. We want to be warm in winter and cool in summer. We want a reliable supply of clean, nutritious food. We want, we want, we want…. There really is no end to what we want. That is the nature of the human animal. *We have limitless wants.* The reality is that we cannot have everything we want, with today's technology. How many of our "needs" are really just "wants"?

As humans have sought to exploit the ecosystem for their own use, they have sometimes degraded the ecosystem's carrying capacity for humans. Emitting lead into the air so our gasoline will burn more smoothly is an example of lowering the planet's carrying capacity. In many other ways, its carrying capacity has been increased. Clearing prime agricultural land for farming to raise food, feed, and fiber is an example of increasing Earth's carrying capacity to supply our wants and needs.

TINSTAAFL applies to human activities and the ecosystem, too. Every time we exploit some aspect of the ecosystem, we give up some other aspect. *That is because we have limited resources.*

For instance, a quiet river cannot remain unaltered while humans build a dam on it to generate electricity. One of the biggest controversies in central Europe today is over whether to complete building the Gabcikovo Dam complex on the Danube River. Some environmentalists claim it will result in a terrible ecological disaster. Other environmentalists claim it will be a wonderful addition to that part of the world by generating electricity and controlling floods. Each side believes they are right and that the other side is being stubborn. The environmentalists who are naturalists bemoan the loss of the Danube as they know it. The environmentalists who are anthropocentrists look forward to more electricity and greater safety from floods for humans (*The Danube: Nature's resilience put to the test*, 1991).

Ironically, both sides are right because of the TINSTAAFL paradox. The people who live along that part of the Danube cannot have both the flood protection and electricity that will be generated by the dam AND the Danube River as it exists now. They must decide. They must give up one or the other. There is no such thing as a free lunch.

Sustainability

Humans cannot exist in the numbers alive today and in the near future without fundamentally affecting the ecosystem. All too often, our activities have been environmentally destructive. In the long run, our only hope is to learn to exploit the ecosystem without degrading its carrying capacity for humans. If we can get to that point, we will have established a **sustainable** world order.

A sustainable world order does not mean maintaining some fictional "natural order of things" or "balance of nature." It does not necessarily mean retaining large areas of pristine forest or grasslands. It does not necessarily mean preserving endangered species. It does not necessarily mean beautiful sunsets or flowers. Most certainly it does not mean converting the world into a museum. A sustainable world order simply means that human activity would not degrade the planet's carrying capacity forever for other humans.

Beautiful sunsets, clear skies, rare birds, and pristine forests are attributes of QUALITY of life, not sustainability. Such things may be desirable—they may even be essential to true humanity. But, they are not an absolute requirement for a sustainable world order.

That is the difference between environmentalism and environmental science. Environmental science can lead us to a sustainable world order, but environmentalism can help make that sustainability **livable**. Livability refers to the quality of life. It refers to those things that make life worthwhile. Sustainability refers only to those things that make life possible on a continuing basis—supplying most human needs but not all of their wants.

In the process of making environmental decisions, BOTH sustainability and livability should be considered. We must consider environmental science, but we should consider environmentalism. Only by becoming active, responsible environmentalists can humans contribute to a livable future for ourselves and those who will come after us. At the same time, the scientific realities of our actions to assure a sustainable society should be taken into account.

Responsible environmentalism means being an environmentalist, but doing so within the framework of environmental science. It means making economic and political decisions that take into account all sides of issues. Finally, it means taking into account the concept of opportunity cost—the TINSTAAFL paradox.

There are many things that only governments can do. Mandating a change from fossil fuel-driven automobiles to alternative fuels is a governmental decision. Changing the emphasis between highways and railroads is a national policy decision. At the same time, there are many things that you can do as an individual responsible environmentalist.

Be a Responsible Environmentalist

It is said that the news media are interested mostly in the negative, the catastrophic, and trivial events, and that they characterize anyone who is *opposed to develop-*

ment as an environmentalist. That is an unfortunate generalization. Being an environmentalist does not automatically mean you are against all development. Development can be very environmentally beneficial if it is done in a responsible manner (see Figure 22-1).

The authors have developed several categories of positive things a true environmentalist *can do*—things that are realistic, that are scientifically valid, and that are environmentally responsible. These are presented in a way that anyone can determine if he or she can be considered an environmentalist.

You can be called an environmentalist if you actively participate as a volunteer in at least one-third of the following 11 categories of environmental actions:

- Five of the fifteen community goals

- Three of the nine steps to avoid waste

- Four of the nine steps on ways to alleviate disposal problems

- Three of the nine ways to recycle waste

Figure 22–1. Photo of adopt-a-highway sign shows that responsible environmentalists do more than just complain about environmental problems. One way to do something about the environment is the "Adopt-a-Highway" program. One of the authors of this book organized a group to keep 2.93 miles of highway clean by picking up trash at least four times a year. (Photo courtesy Vernon E. Doyle, Missouri Highway and Transportation Department.)

- Two of the four better ways to use household products

- Three of the seven better ways to garden

- Four of the nine ways to reduce air pollution

- Four of the seven ways to conserve water

- Two of the five ways to conserve electricity

- Two of the six ways to reduce excess noise

- Six of the eleven ways to get more involved in environmentalism

Community Goals

1. Balanced community-wide planning and zoning
2. Effective conservation agencies
3. Effective pollution control measures for air and water
4. Modern methods of solid waste disposal; no more open burning
5. Effective sign and billboard control
6. Junkyard screening and control
7. Adequate open spaces (green belts)
8. Playgrounds, parks, and other recreation areas
9. Attractive, convenient, downtown malls
10. Protection of watercourses and natural areas, especially in or near places of heavy use by people
11. Protection of wildlife
12. Trail systems for walking, bicycling, horseback riding, snowmobiling, and jogging
13. Promoting conservation and ecological and environmental education in schools
14. Underground utility lines
15. Proper maintenance of parks and city streets

Steps to Avoid Waste

1. Use disposable paper and plastic goods only as necessary.
2. Buy products in returnable bottles or cans.
3. Wash and reuse plastic bags, glass, and plastic containers.

4. Try to buy items that are not excessively packaged.
5. Buy one large-size product rather than several smaller sizes.
6. When you shop, take a reusable mesh shopping bag with you and refuse the "paper or plastic bag syndrome."
7. Pack children's lunches in lunch boxes or reusable bags.
8. Try to buy good-quality, long-lasting toys. Broken toys add to the disposal problem.
9. At the gas station, do not let the attendant "top off" your gas tank; this means waste and polluting spillage.

Ways to Alleviate Disposal Problems

1. Do not litter. This should be the easiest pollution to stop—but it is not.
2. Make garbage compact. Flatten cans, boxes, and cartons or stack them inside one another to conserve space.
3. Do not use colored tissues, colored paper towels, or colored toilet paper. The paper is disposable in water, but the dye lingers on.
4. Use decomposable (biodegradable) containers such as pasteboard, cardboard, or paper. Soft plastic bottles made of polyvinyl chloride (PVC) give off lethal hydrochloric acid when incinerated and almost never decay in a landfill.
5. Never flush away what you can put in the garbage, especially organic cloggers like cooking fat (give it to the birds) or coffee grounds and tea leaves (put them on the garden).
6. Filter tips on cigarettes are plastic and practically indestructible. Do not flush them down the toilet because they will ruin your plumbing and clog up pumps at the sewage treatment plant or septic tank drain fields.
7. Disposable diapers clog plumbing and septic tanks.

8. Drain oil from automobiles, power lawnmowers, or snowplows into a container and bury or recycle it; do not hose it into the sewer system. Oil also makes a biodegradable herbicide; it kills all plants and lasts for about two years.
9. Observe parking regulations so that sanitarians can clean the streets.

Recycling

1. If there is a recycling station for them, save newspapers, magazines, cans, and glass containers.
2. Encourage manufacturers to establish recycling centers.
3. Encourage money-raising programs for young persons or organizations through collecting papers, cans, and glass for recycling.
4. Reuse your scrap paper.
5. Use live Christmas trees and replant them. If you use cut trees, use them for mulch later or as a shelter for birds.
6. Donate magazines and paperback books to libraries, hospitals, or similar organizations.
7. Bring old usable clothing to thrift shops for resale, or donate it to charitable organizations.
8. Be willing to purchase and reuse recycled products.

Use Household Products Carefully

1. Use detergents with care. Detergents high in phosphates can upset the ecological balance of aquatic life. Substitutes are being developed.
2. If you live in a soft-water area, use soap and not a detergent for washing all but heavily soiled items.
3. Do not use products that contain solvents (petroleum distillates) if alternates are available, such as water-based paint.
4. All products labeled "poison" and *all* medicines (including aspirin) should be kept out of reach of children.

Garden Carefully

1. Use organic fertilizers, chemical fertilizers, and animal manures in amounts according to recommendations of the County Extension Agent based upon a soil test.
2. Build a compost heap. Include vegetation such as lawn clippings and leaves. Eventually, you can spread the compost as fertilizer; this is nature's way of recycling refuse.
3. If possible, try to work fertilizer deep into the soil so it will not pollute surface waters.
4. Avoid the use of persistent and toxic pesticides. These include the organochlorines (DDT, aldrin, dieldrin, endrin, heptachlor, chlordane, and lindane) plus compounds containing arsenic, lead, or mercury.
5. Helpful insects (ladybugs, praying mantises, aphis lions, and parasitic wasps) may be used to control many destructive garden pests.
6. Strong-smelling herbs such as mint, sage, chrysanthemum, and basil, may repel some insects and help keep down your use of chemical pesticides.
7. When you spray, use the right insecticide in the recommended amounts and at the right time (see Chapter 11).

Reduce Air Pollution

1. Motor vehicles contribute half of this country's air pollution. When possible, walk, ride a bicycle, use mass transit systems, or form carpools to help reduce the number of cars on the road.
2. Keep your car in good operating condition.
3. Do not idle the car unnecessarily.
4. Check to confirm that the car is equipped with one or more emission-control units and that they operate properly.
5. Buy a lightweight, low-horsepower car.
6. Buy lead-free gasoline.

7. Do not burn leaves or garbage; put them in a compost pile.
8. When you use your fireplace, burn wood and not high-sulfur coal, leaves, or garbage. Charcoal should never be used indoors because it releases poisonous carbon monoxide gas.
9. Stop smoking cigarettes, and we will all breathe a lot easier and have less lung cancer.

Conserve Water

1. Do not leave water faucets open; if one leaks, fix it.
2. Keep a jar of water in the refrigerator for drinking to avoid the practice of running excess water from the tap until it is cool.
3. Repair leaking hoses.
4. Take reasonably short showers instead of baths. The average bath uses more water than a seven-minute shower.
5. Run the dishwasher and washing machine with one full load rather than several small ones.
6. Ask your plumber to adjust the toilet for the minimum amount of water. Some people put a concrete brick or a plastic bottle filled with water inside the toilet tank to conserve water. A clay brick may cause stains in the plumbing.
7. At each watering of the lawn, wet the soil only to the depth of rooting of the grasses, usually three to six inches. Do not water again until the driest part of the lawn shows drought symptoms, as indicated by partial wilting of some grass blades during the afternoon.

Conserve Electricity

1. Use as little electrical power as possible. Power production pollutes by creating smog and heat.
2. Turn off lights in rooms not in use; turn off appliances such as the radio and television when not in use.

3. Turn off air conditioners if not really needed.
4. In winter, turn house thermostat down at night to 50° F to 60° F, and no higher than 68° F during the day.
5. Try to avoid using major appliances during periods of peak power demands.

Avoid Excess Noise

1. Keep radio, television, or phonograph tuned to a reasonable level.
2. Do not buy excessively noisy toys for children.
3. Keep household appliances in good condition.
4. Do not use you car horn unless necessary.
5. Run the power lawnmower and power chainsaw at as slow a speed as possible.
6. Repair or replace noisy mufflers and tail pipes on cars.

Get Involved

1. Know and obey your community's laws concerning pollution control, zoning, building regulations, and beautification standards.
2. Support improved municipal sewage treatment and solid waste disposal facilities.
3. Write to pertinent officials about your concern and ask them what they are doing about environmental problems that interest you.
4. Work with existing groups for pollution control. Help organize groups to fight such problems in your neighborhood.
5. Encourage neighborhood clean-up campaigns.
6. If you work with children, alert them to the dangers of polluting the environment.
7. Try to get an environmental reading shelf established in city and school libraries.
8. Complain about unsightly billboards and signs to advertisers and local officials.
9. Report instances of abandoned cars or offensive industrial wastes to local officials, and note the time, location, and nature of the offense.

10. If you see something wrong and do not know who to contact, write to newspaper editors, television stations, and radio stations.
11. Write to one or both of the following agencies for factual information on environmental enhancement:
    a. A university in your state (see Table 23-2).
    b. Public Enquiries Branch, Environmental Protection Agency, 401 M Street SW, Washington, DC 20460.

Summary

As humans we cannot do anything about the eruptions of volcanos, sunspots, orbital fluctuations, storms, or tidal shifts like El Niño. Much of the environment is beyond our control, but there are many things that have been done that have drastically affected our environment. More importantly, there are many things that we, as individuals, can do to affect the environment in the future.

How the environment is used depends on how we look at things. If the perspective is taken that human wants and needs are the primary concern (anthropocentric) then decisions are made based on what is best for people. If the perspective of the naturalist is taken, the environment is of greatest importance. According to that view, our activities must be planned to minimize human impact on "nature." In this view, human activities are almost always bad because they "upset the natural order of things."

If the anthropocentric perspective is taken, people are of the greatest importance. By that view, our activities must be planned to produce a sustainable and livable world order. There are basically three possible problems that arise with this view. Human activities can change the environment in a way that lowers its carrying capacity, such as excessive soil erosion. Human population exceeds the planetary carrying capacity, causing a "crash." Human activities lower the livability of the ecosystem, making those who are alive less comfortable.

Responsible environmentalism is anthropocentric. It involves using the environment in such a way that neither its carrying capacity nor its livability is degraded. In all of this, it is important to remember the TINSTAAFL (*There Is No Such Thing As A Free Lunch*) paradox. We simply cannot have things as they are now and change them too.

A sustainable world order requires some modifications of how business is done. For instance, solid waste disposal is becoming a major problem. Because that is the case, the volume of our solid wastes must be reduced. That means recycling. Human wants may be limitless but our resources are limited (at least with current technology).

Livability is a totally different concept than sustainability. Many people who would like to be called "environmentalists" confuse the two. Both are important, but we should not confuse the two. It is possible to be a responsible environmentalist. To do so, one must consider both sustainability and livability. Without a sustainable world order, humankind cannot continue forever as a species. Without a livable world, humankind would have little reason to do so.

DISCUSSION QUESTIONS

1. What are the basic differences in naturalist and anthropocentric perspectives?

2. What are the basic environmental problems that arise from human activity.

3. What does it mean be a "responsible" environmentalist?

4. What is the TINSTAAFL paradox? What are its implications for human activity in the ecosystem?

5. What would a sustainable world order mean? What does it not necessarily mean?

6. How do sustainablity and livability differ?

ADDITIONAL ACTIVITIES

1. Develop a personal plan of action for your own use in improving your environment. For each activity, be able to justify why it is the right thing to do, from the perspectives of environmentalism AND environmental science.

2. Develop a class plan of action for improving your school or community environment. For each activity, be able to justify why it is the right thing to do, from the perspectives of environmentalism AND environmental science.

3. Invite a member of your local planning district, or of your city or county planning commission to speak to your class about the land use planning process and the zoning process.

4. Organize groups of concerned persons to pick up and properly dispose of solid waste around the school yard or in other public places.

5. Organize a group to "Adopt a Highway."

6. Work with the news media to educate people to refrain from littering public places.

CHAPTER 23

Education for Jobs and Careers in the Environment

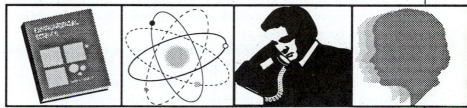

Environmentalism

It is a social good to pick up soft drink and beer cans along a highway, but it is an environmental good to never throw them away — they should be recycled for the aluminum.

The Authors

Terms to Look for and Learn

Art	Paraprofessional
Career	Professional
Employability Qualities	Science
Job	Scientist

Learning Objectives

After reading this chapter and participating in the activities, you should be able to:

- Compare and contrast jobs and careers.

- Find sources of information about education and training for paraprofessional and professional careers in the environment.

- List career opportunities in the environment.

- List important characteristics of a scientist.

369

Overview

Environment means all of the conditions, circumstances, and influences surrounding living things: plants, animals, and humans. A **job** is anything one has to do by agreement for pay, such as taking orders in a fast food place.

A **career** is one's progress through life or in a particular vocation; a profession or occupation that one trains for and pursues for a lifetime. Examples are a nurse, a biology teacher, a forester, or an environmental control technician.

It is difficult and mostly impossible to name any job or career that is not related to the environment. A part-time or temporary job in a fast food establishment affects the human environment by serving healthful or contaminated foods, or by aesthetics—the extra care needed to serve customers gracefully.

A career in health care, such as nursing, affects the environment of everyone cared for. Examples are mental attitude transmitted from nurse to patient, physical health due to correct medication, and attention to human body temperature.

Since the first Earth Day in 1972, more and more persons in various careers claim to be environmentalists. Ecologists were among the first professionals to want to be called environmentalists. When the U.S. Environmental Protection Agency was established in late 1969, many more persons in various other careers said "me too" because that was where the money for new research was and therefore the professional environmental prestige. Engineers, agriculturists, botanists, zoologists, economists, sociologists, and meteorologists all claimed to be environmentalists—and rightly so—because environment means us and our surroundings.

Goals in Life and How to Attain Them

If your primary goal in life is to have a good time, probably a job is all you want, and no special schooling is needed. Most people, however, have higher aims. A person should aspire to be his or her best. In general, the more years a person studies in high school, vocational-technical school, and college, the higher the salary and the greater the quality of life.

Your **employability qualities** are your attributes and characteristics that affect your ability to get and to be successful in a job. Be honest with yourself and respond to the following 10 attributes on your "Employability Qualities:" (see Figures 23-1 through 23-10.)

- Ability
- Talent
- Interest
- Physical Make-up
- Educational Aspirations
- Attitudes and Values
- Personality
- Self-concept
- Flexibility
- Previous Experience

When you are a junior or senior in high school you are mature enough to establish goals in life and to test these choices by asking questions and possibly working in the field of the preliminary career choice (see Figure 23-11). For example:

- If you think you want a career in biology, go to a scout camp and start passing tests in botany, wildlife, and nature (see Figure 23-12).
- If you think you want a career in business, get a summer job in some business and "test the waters."
- If you think you want to be a computer operator, enroll in a course in computers and when you have sufficient skill, get a job using a computer.

Figure 23–1. **Ability** must be adequate to achieve your goals in life. (**Source:** Texas A&M Instructional Services.)

EMPLOYABILITY QUALITIES

❖ **Ability**

Are you knowledgeable in the following subject matter areas?

Mathematics
Language Arts
Science
Social Studies
Agriculture
Environmental Science

Figure 23–3. **Personal interest** must be compatible with your life goals. (**Source:** Texas A&M Instructional Services.)

EMPLOYABILITY QUALITIES

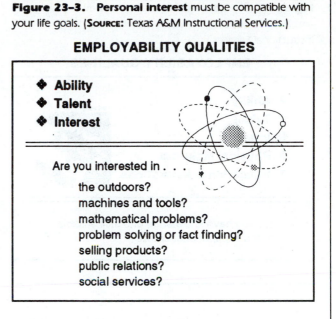

❖ **Ability**
❖ **Talent**
❖ **Interest**

Are you interested in . . .

the outdoors?
machines and tools?
mathematical problems?
problem solving or fact finding?
selling products?
public relations?
social services?

Figure 23–2. **Talent** must be properly developed to achieve your goals in life. (**Source:** Texas A&M Instructional Services.)

EMPLOYABILITY QUALITIES

❖ **Ability**
❖ **Talent**

Do you read and write well?
Do you type proficiently?
Are you a good speaker?
Can you sketch drawings?
Do you enjoy working outdoors?
Do you enjoy working indoors?
Are you mechanically minded?
Can you perform physical feats?
Do you enjoy working with others?
Are you scientifically minded?
Do you like working with plants?
Do you like working with animals?

Figure 23–4. Your **physical make-up** must be adequate to achieve your life-time goals. (**Source:** Texas A&M Instructional Services.)

EMPLOYABILITY QUALITIES

❖ **Ability**
❖ **Talent**
❖ **Interest**
❖ **Physical Make-up**

Are you healthy?

If you have physical limitations, can you adapt to the work environment?

Figure 23–5. **Educational aspirations** must equal your ambitions to achieve your goals (**Source:** Texas A&M Instructional Services.)

EMPLOYABILITY QUALITIES

❖ **Ability**
❖ **Talent**
❖ **Interest**
❖ **Physical Make-up**
❖ **Educational Aspirations**

Do you have knowledge and skill suitable to the occupational choice?

Figure 23–6. **Attitudes and values** on the job must agree with life-time goals for job satisfaction (**Source:** Texas A&M Instructional Services.)

EMPLOYABILITY QUALITIES

❖ **Ability**
❖ **Talent**
❖ **Interest**
❖ **Physical Make-up**
❖ **Educational Aspirations**
❖ **Attitudes and Values**

Does your choice of a job agree with your attitudes and values?

Can you express your attitudes and values on the job?

What attitudes and values are expressed by others holding the same job status?

Figure 23–7. **Personality** and career goals in life must mesh to achieve job satisfaction. (**Source:** Texas A&M Instructional Services.)

EMPLOYABILITY QUALITIES

❖ **Ability**
❖ **Talent**
❖ **Interest**
❖ **Physical Make-up**
❖ **Educational Aspirations**
❖ **Attitudes and Values**
❖ **Personality**

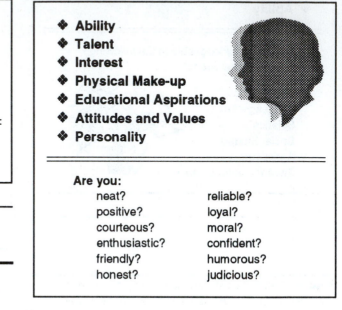

Are you:

neat?	reliable?
positive?	loyal?
courteous?	moral?
enthusiastic?	confident?
friendly?	humorous?
honest?	judicious?

■ If you know you need a college education to achieve your career goals in environmental science, write to the state of your choice for information (see Tables 23-1 and 23-2).

Sources of Information

There are two primary sources of information on environmental and other education and training:

■ **Paraprofessional** (also known as post-secondary) schools in each state offer courses in environment and related subjects. Paraprofessionals perform many of the same tasks as professionals, but paraprofessionals generally work under the supervision of a professional. The training required for

Figure 23–8. **Self-concept** to achieve life's goals means to see yourself as others see you. (**Source:** Texas A&M Instructional Services.)

EMPLOYABILITY QUALITIES

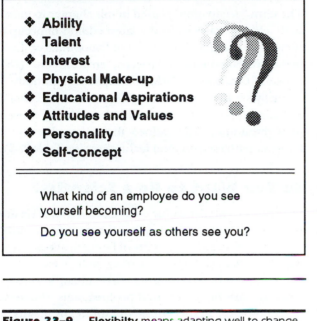

- ❖ Ability
- ❖ Talent
- ❖ Interest
- ❖ Physical Make-up
- ❖ Educational Aspirations
- ❖ Attitudes and Values
- ❖ Personality
- ❖ Self-concept

What kind of an employee do you see yourself becoming?

Do you see yourself as others see you?

Figure 23–9. **Flexibilty** means adapting well to change in your job. (**Source:** Texas A&M Instructional Services.)

EMPLOYABILITY QUALITIES

- ❖ Ability
- ❖ Talent
- ❖ Interest
- ❖ Physical Make-up
- ❖ Educational Aspirations
- ❖ Attitudes and Values
- ❖ Personality
- ❖ Self-concept
- ❖ Flexibility

Can you take constructive criticism?

Can you adapt to changes around you?

Figure 23–10. **Previous experience** in the kind of work you aspire to do in life will give you confidence in your selection of an occupation. (**Source:** Texas A&M Instructional Services.)

EMPLOYABILITY QUALITIES

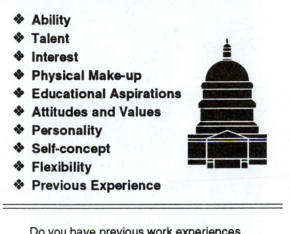

- ❖ Ability
- ❖ Talent
- ❖ Interest
- ❖ Physical Make-up
- ❖ Educational Aspirations
- ❖ Attitudes and Values
- ❖ Personality
- ❖ Self-concept
- ❖ Flexibility
- ❖ Previous Experience

Do you have previous work experiences related to your occupational goal?

What work experiences do you need to find employment in your occupational field?

becoming a paraprofessional is usually less than a four-year college degree, but more than high school graduation.

- ■ **Professional** jobs generally require high levels of education, both general and job-specific. Professionals work with little direct supervision and make many of their own decisions. Professional jobs and careers in the United States usually require at least four-year degrees.

The 130 colleges in the United States listed in Table 23-2 offer special short courses and degrees in environment and related subjects. Write to obtain specific information. It is usually cheaper to attend a college in your state of residence.

Figure 23–11. A conservation aid is learning to use a transit in land surveying to lay out field boundaries. He may decide to become a District Conservationist. (Courtesy New York State College of Agriculture at Cornell University.)

Titles of Jobs, Occupations, and Careers in Environment

The term "occupation" is used in this chapter to mean both jobs and careers. In the latest edition of occupations from the U.S. Department of Labor three of these titles specifically mention the word, "environment" (*Dictionary of Occupational Titles and Supplement*, published annually by the United States Department of Labor). Following this list of environmental titles will be thirty-seven titles the authors have added that relate directly to environmental science and technology (see Table 23-3).

So You Want to Be a Scientist?

Wise persons in history have concluded that **art** is the pursuit of beauty and **science** is the pursuit of truth. Truth is defined as the quality of being in accordance with experience, facts, or reality; actual existence; agreement with a standard; correctness; accuracy. **Scientists** are highly trained professionals who seek new knowledge in specialized disciplines by means of careful research—they are seekers of truth.

Figure 23–12. This 4-H Club member is interested in small engines as a part of his project program. From this initial interest and an opportunity to develop it, a new career in Engine Emission/Pollution Control may result. (Courtesy National 4-H Youth Clubs, United States Department of Agriculture.)

Table 23-1. Information on Careers and Jobs in the Environment and Related Subjects and Paraprofessional (Post-secondary) Training in the Environment and Allied Subjects

For information, write to the following state of your choice:

Alabama: State Supervisor
Agribusiness Education
Department of Education
State Office Building
Montgomery, AL 36104

Alaska: Superintendent,
Trades, Industries, and Fisheries
State Department of Education
Alaska Office Building
Juneau, AK 99801

Arizona: State Supervisor,
Agricultural Education
1333 W. Camelback Road
Suite No. 111
Phoenix, AZ 85013

Arkansas: Director,
Agricultural Education
Division of Vocational, Technical
 and Adult Education
Arch Ford Education Building
Little Rock, AR 72201

California: Division of
 Occupational Education
California Community Colleges
825 Fifteenth Street
Sacramento, CA 95814

Colorado: Supervisor, Agricultural Education
State Board for Community Colleges
 and Occupational Education
207 State Services Building
Denver, CO 80203

Connecticut: State Supervisor,
Agricultural Education
State Department of Education
P.O. Box 2219
Hartford, CT 06115

Delaware: State Supervisor,
Agricultural Education
University of Delaware
College of Agricultural Sciences
Agricultural Experiment Station
Newark, DE 19711

Florida: Administrator,
Agricultural Education
State Department of Education
Knott Building
Tallahassee, FL 32304

Georgia: State Supervisor,
Agricultural Education
State Department of Education
325 State Office Building
Atlanta, GA 30334

Hawaii: Program Specialist,
Agricultural Education Section
Office of Instructional Services
State Department of Education
Post Office Box 2360
Honolulu, HI 96804

Idaho: State Supervisor,
Agricultural Education
State Board for Vocational Education
518 Front Street
Boise, ID 83702

Illinois: Head Consultant,
Applied Biological and
 Agricultural Occupations
Board of Vocational Education
 and Rehabilitation
1035 Outer Park Drive
Springfield, IL 62706

Indiana: Chief Consultant,
Agribusiness Education
Division of Vocational Education
700 North High School Road
Indianapolis, IN 46224

Table 23-1 (Continued). Information on Careers and Jobs in the Environment and Related Subjects and Paraprofessional (Post-secondary) Training in the Environment and Allied Subjects

Iowa: Consultant,
State Department of Public Instruction
Area Schools and Career Education Branch
Des Moines, IA 50319

Kansas: State Supervisor,
Agricultural Education
Division of Vocational Education
Kansas State Department. of Education
Kansas State Education Building
120 East 10th Street
Topeka, KS 66612

Kentucky: Vice President,
University of Kentucky Community
  College System
Breckinridge Hall
Lexington, KY 40506

Director, Agricultural Education
Bureau of Vocational Education
Commonwealth of Kentucky
  Education Department
Frankfort, KY 40601

Louisiana: State Director,
Vocational Agriculture
State Department of Education
State Capitol Building
Baton Rouge, LA 70804

Maine: State Consultant,
Agricultural Education
State Department of Education
Augusta, ME 04330

Maryland: State Supervisor,
Agricultural Education
State Department of Education
600 Wyndhurst Avenue
Baltimore, MD 21210

Massachusetts: Supervisor,
Agricultural Education
Division of Occupational Education
The Commonwealth of Massachusetts
Department of Education
182 Tremont Street
Boston, MA 02111

Michigan: Supervisor,
Post-Secondary Unit
State Department of Education
Division of Voc Ed, Box 928
Lansing, MI 48904

Minnesota: State Supervisor,
Agricultural Education
State Department of Education
Capitol Square, 550 Cedar Street
St. Paul, MN 55101

Mississippi: State Supervisor,
Vocational Agriculture Depart.
State Department of Education
Vocational Education Division
Post Office Box 771
Jackson, MS 39205

Missouri: Director,
Agricultural Education
State Department of Education
Division of Public Schools
Post Office Box 480
Jefferson City, MO 65101

Montana: State Supervisor,
Agricultural Education
State Department of Public Inst.
State Capitol
Helena, MT 59601

Nebraska: State Director,
Agricultural Education
State Department of Education
223 South 10th Street
Lincoln, NE 68508

Nevada: State Supervisor,
Agricultural Education
State Department of Education
Heroes Memorial Building
Carson City, NV 89701

New Hampshire: Consultant,
Agricultural Education
State Department of Education
Stickney Avenue
Concord, NH 03301

Table 23-1 (Continued). Information on Careers and Jobs in the Environment and Related Subjects and Paraprofessional (Post-secondary) Training in the Environment and Allied Subjects

New Jersey: State Director,
Agricultural Education
State Department of Education
Vocational Division
225 West State Street
Trenton, NJ 08625

New Mexico: State Supervisor,
Agricultural Education
State Department of Vocational Education
200 DeVargas Street
Santa Fe, NM 87501

New York: Associate in Higher
  Occupational Education
University of the State of New York
The State Education Department
99 Washington Avenue
Albany, NY 12210

Vice Chancellor for Community Colleges
  and Provost for Vocational and
  Technical Education
State University of New York
Albany, NY 12201

North Carolina: Educational Consultant
Agricultural and Biological Education
Division of Occupational Education
Department of Community Colleges
State Board of Education
Raleigh, NC 27602

North Dakota: State Supervisor,
Agricultural Education
State Department of Education
State Office Building
900 East Boulevard
Bismarck, ND 58501

Ohio: State Supervisor,
Agricultural Education
Ag Administration Building
Ohio State University
2120 Fyffe Road
Columbus, OH 43210

Oklahoma: State Supervisor,
Oklahoma State Department of
  Vocational and Technical Education
1515 West Sixth Avenue
Stillwater, OK 74074

Oregon: Agribusiness Specialist
Oregon Board of Education
942 Lancaster Drive, N.E.
Salem, OR 97310

Pennsylvania: State Supervisor
Agricultural Education
Pennsylvania Department of Education
Post Office Box 911
Harrisburg, PA 17126

Puerto Rico: Director
Vocational Agriculture Program
Commonwealth of Puerto Rico
Area of Vocational and Technical Education
Post Office Box 759
Hato Rey, PR 00919

Rhode Island:
University of Rhode Island
Kingston, RI 02881

South Carolina: Coordinator,
Agricultural Technology Programs
State Committee for Technical Education
1429 Senate Street
Columbia, SC 29201

South Dakota: State Supervisor,
Agricultural Education
Agricultural Education Service
State Department of Public Inst.
Pierre, SD 57501

Tennessee: Supervisor,
Agricultural Occupations
State Department of Education
210 Cordell Hull Building
Nashville, TN 37219

Table 23-1 (Continued).   Information on Careers and Jobs in the Environment and Related Subjects and Paraprofessional (Post-secondary) Training in the Environment and Allied Subjects

Texas: Director,
Vocational Agricultural Education
Texas Education Agency
201 East Eleventh Street
Austin, TX 78701

Utah: State Specialist,
Agricultural Education
State Board for Vocational Education
1300 University Club Building
136 East South Temple
Salt Lake City, UT 84111

Vermont: State Consultant,
Agricultural Education
State Department of Education
State Office Building
Montpelier, VT 05602

Washington: Program Director,
Agricultural Education
Coordinating Council for
  Occupational Education
Post Office Box 248
Olympia, WA 98501

West Virginia: State Supervisor,
Vocational Agriculture
Department of Education
Building 6, Unit B, State Capitol
Charleston, WV 25305

Wisconsin: Supervisor,
Vocational Agriculture
State Board of Vocational,
  Technical and Adult Education
137 East Wilson Street
Madison, WI 53703

Wyoming: State Director,
Agricultural Education
State Department of Education
Capitol Building
Cheyenne, WY 82001

For further information on paraprofessional (post-secondary) training programs, write to:

Senior Educational Program Specialist, Agriculture,
Agribusiness and Renewable Resources
Division of Vocational and Technical Education
Office of Vocational and Adult Education
United States Department of Education
Washington, D.C. 20202

Please note that scientific truth has no geographic boundaries. What is true in New Orleans must also be true in Atlanta, New York City, Chicago, and Los Angeles.

As the nineteenth-century biologist, Thomas H. Huxley, wrote: the scientist must be, "as a little child—always asking, why? but continuously trying to answer his/her own question."

Further characteristics of a true **scientist** are:

- Scientists have had the best and most advanced formal education possible.

- Scientists try to learn from experienced scientists.

- Scientists fight against bias and preconceived ideas as these human failings tend to influence truth.

Table 23-2. These 130 Colleges and Universities Offer Special Courses and Degrees in the Environment and Related Subjects.

Alabama
Alabama A & M University
  Normal, AL 35762
Auburn University
  Auburn University, AL 36849
Tuskegee University
  Tuskegee Institute, AL 36088

Alaska
University of Alaska
  Fairbanks, AK 99701

Arizona
The University of Arizona
  Tucson, AZ 85721
Arizona State University
  Tempe, AZ 85281

Arkansas
Arkansas State University
  State University, AR 72467
Southern Arkansas University
  Magnolia, AR 71753
University of Arkansas
  Fayetteville, AR 72701
University of Arkansas
  Monticello, AR 71655
University of Arkansas
  Pine Bluff, AR 71601

California
California Polytechnic State University
  San Louis Obispo, CA 93407
California State Polytechnic University
  Pomona, CA 91768
California State University
  Chico, CA 95926
California State University
  Fresno, CA 93740
Humboldt State University
  Arcata, CA 95521
University of California
  Davis, CA 95616
University of California
  Riverside, CA 92521

Colorado
Colorado State University
  Fort Collins, CO 80523
Fort Lewis College
  Durango, CO 81301

Connecticut
The University of Connecticut
  Storrs, CT 06268

Delaware
Delaware State College
  Dover, DE 19901
University of Delaware
  Newark, DE 19711

Florida
Florida Southern College
  Lakeland, FL 33802
University of Florida
  Gainesville, FL 32611

Georgia
Abraham Baldwin Ag College
  Tifton, GA 31794
Berry College
  Mount Berry, GA 30149
Fort Valley State College
  Fort Valley, GA 31030
University of Georgia
  Athens, GA 30602

Hawaii
University of Hawaii
  Honolulu, HI 96822

Idaho
College of South Idaho
  Twin Falls, ID 83301
Ricks College
  Rexburg, ID 83440
University of Idaho
  Moscow, ID 83843

Illinois
Illinois State University
  Normal, IL 61761

Illinios (cont.)
Southern Illinois University
  Carbondale, IL 62901
University of Illinois
  Urbana, IL 61801
Western Illinois University
  Macomb, IL 61455

Indiana
Purdue University
  West Lafayette, IN 47907

Iowa
Iowa State University
  Ames, IA 50011

Kansas
Fort Hays State University
  Hays, KS 67601
McPherson College
  McPherson, KS 67460
Kansas State University
  Manhattan, KS 66506

Kentucky
Eastern Kentucky University
  Richmond, KY 40475
Morehead State University
  Morehead, KY 40351
Murray State University
  Murray, KY 42071
University of Kentucky
  Lexington, KY 40506
Western Kentucky University
  Bowling Green, KY 42101

Louisiana
Louisiana State University
  Baton Rouge, LA 70803
Louisiana Tech University
  Ruston, LA 71272
McNeese State University
  Lake Charles, LA 70609
Nicholls State University
  Thibodaux, LA 70310

Table 23-2 (Continued). These 130 Colleges and Universities Offer Special Courses and Degrees in the Environment and Related Subjects.

Louisiana (cont'd)
Northeast Louisiana University
  Monroe, LA 71209
Northwestern State University
  Natchitoches, LA 71497
Southeastern Louisiana
  University
  Hammond, LA 70402
Southern University
  Baton Rouge, LA 70813
University of Southwestern
  Louisiana
  Lafayette, LA 70504

Maine
University of Maine
  Orono, ME 04469

Maryland
University of Maryland
  College Park, MD 20742
University of Maryland
  Princess Anne, MD 21853

Massachusetts
University of Massachusetts
  Amherst, MA 01003

Michigan
Michigan State University
  East Lansing, MI 48824
Michigan Technological
  University
  Hancock, MI 49930
Northern Michigan University
  Marquette, MI 49855

Minnesota
University of Minnesota
  St. Paul, MN 55108
University of Minnesota
  Technical College
  Crookston, MN 56716
University of Minnesota
  Technical College
  Waseca, MN 56093

Mississippi
Alcorn State University
  Lorman, MS 39096
Mississippi State University
  Mississippi State, MS 39762

Missouri
Central Missouri State University
  Warrensburg, MO 64093
Lincoln University
  Jefferson City, MO 65101
Missouri Western State College
  St. Joseph, MO 64507
Northeast Missouri State University
  Kirksville, MO 63501
Northwest Missouri State University
  Maryville, MO 64468
Southwest Missouri State University
  Springfield, MO 65802
University of Missouri
  Columbia, MO 65211

Montana
Montana State University
  Bozeman, MT 59717
Northern Montana College
  Havre, MT 59501

Nebraska
University of Nebraska
  Lincoln, NE 68583

Nevada
University of Nevada
  Reno, NV 89557

New Hampshire
University of New Hampshire
  Durham, NH 03824

New Jersey
Rutgers University
  New Brunswick, NJ 08903

New Mexico
New Mexico State University
  Las Cruces, NM 88003

New York
Cornell University
  Ithaca, NY 14853

North Carolina
North Carolina A & T State University
  Greensboro, NC 27411
North Carolina State University
  Raleigh, NC 27650

North Dakota
North Dakota State University
  Fargo, ND 58105

Ohio
The Ohio State University
  Columbus, OH 43210
Wilmington College
  Wilmington, OH 45177

Oklahoma
Cameron University
  Lawton, OK 73505
Langston University
  Langston, OK 73050
Oklahoma Panhandle State University
  Goodwell, OK 73939
Oklahoma State University
  Stillwater, OK 74074

Oregon
Oregon State University
  Corvallis, OR 97331

Pennsylvania
Delaware Valley College of Science
  and Agriculture
  Doylestown, PA 18901
The Pennsylvania State University
  University Park, PA 16802
Temple University
  Ambler, PA 19002

Puerto Rico
University of Puerto Rico
  Mayaguez, PR 00708
University of Puerto Rico
  Rio Piedras, PR 00928

Table 23-2 (Continued). These 130 Colleges and Universities Offer Special Courses and Degrees in the Environment and Related Subjects.

Rhode Island
University of Rhode Island
Kingston, RI 02881

South Carolina
Clemson University
Clemson, SC 29631

South Dakota
South Dakota State University
Brookings, SD 57007

Tennessee
Austin Peay State University
Clarksville, TN 37040
Middle Tennessee State University
Murfreesboro, TN 37130
Tennessee Technological University
Cookeville, TN 38505
University of Tennessee
Knoxville, TN 37901
University of Tennessee
Martin, TN 38238

Texas
Abilene Christian University
Abilene, TX 79601
East Texas State University
Commerce, TX 75428
Prairie View A & M University
Prairie View, TX 77445

Texas (cont'd)
Sam Houston State University
Huntsville, TX 77341
Southwest Texas State University
San Marcos, TX 78666
Stephen F. Austin State University
Nacogdoches, TX 75962
Sul Ross State University
Alpine, TX 79832
Texas A & I University
Kingsville, TX 78363
Texas A & M University
College Station, TX 77843
Texas Tech University
Lubbock, TX 79409
West Texas State University
Canyon, TX 79015

Utah
Brigham Young University
Provo, UT 84602
Utah State University
Logan, UT 84322

Vermont
University of Vermont
Burlington, VT 05405

Virginia
Old Dominion University
Norfolk, VA 23508

Virginia (cont'd)
Virginia Polytechnic Institute
and State University
Blacksburg, VA 24061
Virginia State University
Petersburg, VA 23806

Washington
University of Washington
Seattle, WA 98195
Washington State University
Pullman, WA 99164

West Virginia
West Virginia University
Morgantown, WV 26506

Wisconsin
University of Wisconsin
Green Bay, WI 54302
University of Wisconsin
Madison, WI 53706
University of Wisconsin
Platteville, WI 53818
University of Wisconsin
River Falls, WI 54022
University of Wisconsin
Stevens Point, WI 54481

Wyoming
University of Wyoming
Laramie, WY 82071

■ Scientists always keep data from experiments to confirm or refute their own previous conclusions.

■ Scientists must always readily say, "I could be wrong but these data seem to indicate otherwise."

■ Scientists at any location must not be intimidated by a contrary opinion held by another scientist at a more prestigious location. The greatest truths often come from the "backwoods."

■ Scientists whose research is financed by commercial interests are usually forbidden to present their data to public scientific groups because public disclosure may release information to a competing company.

■ Scientists need recognition by their peers but never at the expense of "stealing data" or not giving proper recognition to coworkers. Priority of publication is dated from the time a scientific journal *receives* (not publishes) a research paper. Credit for the research goes to the person who sent the article to the scientific journal, not necessarily to the person who did the research. (Sometimes this seems unfair.)

Table 23-3. For Further Information on Environmental Occupations

Job Titles that Directly Relate to the Environment	Consult these Chapters in this Book for Further Information	Job Titles that Directly Relate to the Environment	Consult these Chapters in this Book for Further Information
Environmental Research Technician	All chapters	Soil Conservationist	1–3, 4, 5, 12, 18, 21
Environmental/Water/Soils/ Analyst/Scientist	All chapters	Biochemist	7, 8, 9, 11–14, 16
Environmental Control Technician	All chapters	Plant Botanist/Ecologist/ Taxonomist	1–3, 5, 6, 7, 18, 19, 21
Water Treatment Plant Engineer	8, 9, 13	Entomologist	7, 11, 19
Forest Engineer	3, 4, 6	Bacteriologist/Micro- biologist/Virologist	5, 7, 11, 21
Product Safety Engineer	11, 12, 14	Nematologist	4, 5, 7, 11
Quality Control Engineer	3, 9, 16	Parasitologist/ Helminthologist	5–7, 11, 12
Safety Manager	8, 9, 11, 14	Animal Ecologist	1–3, 6, 11, 18, 21
Quality Control Technician	3, 9, 16	Public Health Microbiologist	8, 9, 11, 12–14
Agricultural Engineer	All chapters	Park Naturalist	1–3, 6, 8, 16, 18, 21
Land Surveyor (Figure 23-11)	4, 5, 8, 18, 21	Biological Aide	1, 7, 11
Biomedical Engineer	9, 11, 12, 14, 16	Agricultural Economist	3, 7, 15, 22
Water Purification Chemist	4, 8, 9, 12	Veterinary Epidemiologist	7, 8, 12–14, 16
Geologist	4, 5, 8, 9	Veterinary Parasitologist	7, 8, 12-14, 16
Soils Engineer	4, 5, 21	Veterinary Laboratory Technician	7, 8, 12–14, 16
Meteorologist	8, 10, 16	Professor	All chapters
Soil Chemist	4, 5, 12, 13	Nutritionist	1, 7, 15, 22
Engine Emission/Pollution Control Technician (Figure 23-12)	12, 16, 20	Agricultural Extension Specialist	All chapters
Forest Ecologist	1–3, 6, 11, 16		
Horticulturist	7, 11, 12, 19		
Range Conservationist/ Manager/Specialist	1–3, 6, 11, 16, 18, 21		

■ Scientists are frequently told by their administrators: "Your advancement will depend on how many research papers you have published in first-class scientific journals and how many research grants you can bring to our university." This is a great temptation for scientists to practice fraud. (Source: "On Being a Scientist," 1989)

Summary

In truth, there are very few careers that could truly be said to be purely "environmental." At the same time, a great many (in fact most) careers are related to the environment in one way or another. In a very real sense, all of us are "environmentalists," regardless of our careers. Naturally, some careers are more closely oriented toward environmental science than others.

If you want to pursue a paraprofessional career that is heavily oriented toward environmental management or preservation, you can contact the persons in your state listed in Table 23-1. If you plan to attend a four-year college or university, those listed in Table 23-2 have programs that emphasize environmental specialties. Regardless of which route you take, you should consider your own characteristics and the nature of the potential careers. It is not necessary to make a firm career decision while you are in high school, but it is never too early to begin to make some tentative choices and to begin exploring your options.

DISCUSSION QUESTIONS

1. What are the differences between a job and a career?

2. How do paraprofessional and professional jobs compare? How do they differ from each other?

3. Who would you contact in your state and neighboring states for more information about paraprofessional training for environmental jobs and careers?

4. What colleges in your state and your neighboring states offer specialized education for environmental careers?

5. What are some of the important characteristics of scientists?

ADDITIONAL ACTIVITIES

1. Write to the person listed in your state and your neighboring states for information about paraprofessional training for environmental jobs and careers.

2. Contact the colleges in your state and your neighboring states and request information about their specialized education for environmental careers.

3. Visit a business in your community that specializes in some aspect of environmental management or environmental preservation. Find out what job titles are included in the business. Ask what sort of education would be appropriate for someone entering such a business.

4. Invite your local agricultural extension agent, soil conservation agent, biology teacher, or forester to speak to your class about career opportunities in the environment.

APPENDIX A

References and Bibliography

This appendix includes all references cited throughout the text and a bibliographic listing of works used in preparing this text. In addition, there are a great many books, magazines, and other publications that may be of interest to the student of environmental science. For that reason, many publications are listed here that are not cited in the text.

Altieri, M. A., & Schmidt, L. (1987, November-December). Mixing broccoli cultivars reduces cabbage aphid numbers. *California Agriculture*.

Association of American Plant Food Control Officials. (1990). *Official Publication No. 43*. West Lafayette, IN: Author.

Barnes, R. S. K., & Hughes, R. N. (1988). *An introduction to marine ecology*. Boston: Blackwell Scientific Publications.

Bluemle, J. P. (1975). *The prairie land and life*. Washington, D. C.: U. S. D. A. Forest Service and North Dakota Geological Survey.

Borgese, E. M., Ginsburg, N., & Morgan, J. R. (Eds.). (1989). *Ocean yearbook 8*. Chicago: The University of Chicago Press.

Borror, D. J., DeLong, D. M., & Triplehorn, C. A. (1981). *An introduction to the study of insects*. Philadelphia, PA: Saunders College Publishing.

Botkin, D. B. (1990). *Discordant harmonies: A new ecology for the twenty-first century*. New York: Oxford University Press.

Brown, L. R. (1974). *In the human interest: A strategy to stabilize world population*. New York: W. W. Norton & Company.

Brown, L. R., Brough, H., Durning, A., French, H., Jacobson, J., Lanssen, N., Lowe, M., Postel, S., Renner, M., Ryan, J., Starke, L., & Young, J. (1992). *State of the world 1992*. New York: W. W. Norton & Company.

Brown, L. R., Durning, A., Flavin, C., French, H., Jacobson, J., Lennsen, N., Lowe, M., Postel, S., Renner, M., Ryan, J., Starke, L., & Young, J. (1991). *State of the world 1991*. New York: W. W. Norton & Company.

Brown, L. R., Durning, A., Flavin, C., French, H., Jacobson, J., Lowe, M., Postel, S., Renner, M., Starke, L., & Young, J. (1990). *State of the world 1990*. New York: W. W. Norton & Company.

Burley, F. W., & Hazlewood, P. T. (1988). Tropical forests: A resource in jeopardy. In T. Kristensen & J. P. Paludan (Eds.). *The Earth's fragile systems: Perspectives on global change*. Boulder, CO: Westview Press.

Burton, I. (1988). Dimensions of a new vulnerability: The significance of large-scale urbanization in developing countries. In T. Kristensen & J. P. Paludan (Eds.). *The Earth's fragile systems: Perspectives on global change*. Boulder, CO: Westview Press.

Camp, W. G. & Daugherty, T. B. (1991). *Managing our natural resources*. Albany, NY: Delmar Publishers Inc.

Casagrande, R. A. (1990, January-February). You can learn to garden without pesticides. *In Touch*. Kingston, RI: University of Rhode Island.

Chandler, W. J., Labate, L., & Wille, C. (1988). *Audubon wildlife report 1988/1989*. New York: Academic Press, Inc.

Colinvaux, P. (1986). *Ecology*. New York: John Wiley & Sons.

Cooke, L. (1990). Ticked off about babesiosis. *Agricultural Research*, 38(3). Washington, D.C.: United States Department of Agriculture, Agricultural Research Service.

Council for Agricultural Science and Technology. (1987). *Science of Food and Agriculture*. 5(3). Ames: Iowa State University.

Cowardin, L. M., Carter, V., Golet, F. C., & LaRoe, E. T. (1979). *Classification of wetlands and deepwater habitats of the United States*. Washington, D. C.: United States Fish and Wildlife Service, Department of the Interior. United States Government Printing Office Stock Number GPO 024-010-00524-6.

DaDatta, S.K. (1981). *Principles and practices of rice production*. New York: John Wiley & Sons.

de Groot, P. (1990). Are we missing the grass for the trees? *New Scientist*, 125 (1698).

DeWitt, C. B., & Soloway, E. (Eds.) (1977). *Wetlands ecology, values, and impacts*. Madison, WI: Institute for Environmental Studies, University of Wisconsin-Madison.

Donahue, R. (1973). *Ecology and environment*. Kansas City, MO: Midwest Research Institute.

Dorland's illustrated medical dictionary, 26th ed. (1981). Philadelphia, PA: W. B. Saunders Co.

Duley, F.L., & Russell, J.C. (1942). Effect of stubble mulching on soil erosion and runoff. *Proceedings* (7-77-81) of the Soil Science Society of America. Madison, WI.

Edwards, C. A., & Lofty, J. R. (1972). *Biology of earthworms*. Ontario, CA: Bookworm Publishing Co.

Erlich, P. R., & Erlich, A. H. (1990). *The population explosion*. New York: Simon and Schuster.

Fein, G. G. (1984). Prenatal exposure to PCBs: Effects on birth size and gestational age. *Journal of Pediatrics*, pp. 105, 315-392.

Ferguson, F. A. (1968). A nonmyoptic approach to the problem of excess algal growths. *Environmental Science and Technology*, 2(3), pp. 188-193.

Firor, J. W. (1988). The heating up of the climate. In T. Kristensen & J. P. Paludan (Eds.). *The Earth's fragile systems: Perspectives on global change*. Boulder, CO: Westview Press.

Fitzharris, T. (1983). *The wild prairie: A natural history of the Western plains*. Toronto: Oxford University Press.

Follett, R. H., & Croissant, R. L. (1990). Use of manure in crop production. *Service in action*, No. 549. Boulder: Colorado State University.

Follett, R. H., & Self, J. R. (1989). *Domestic water quality criteria*, Cooperative Extension Service in Action, No. 513. Boulder: Colorado State University.

Grafton-Cardwell, E. E., & Hoy, M. A. (1986, September-October). Selection of the common green lacewing for resistance to carbaryl. *California Agriculture*.

Graham, F. (1990). Of broccoli and marshes. *Audubon*. 92(4).

Gribbin, J. (1989). The end of the ice ages? *New Scientist*, 122(1669).

Gribbin, J. (1990). Whatever happened to the mini-ice age? *New Scientist, 126*(1712).

Hays, S. M. (1990). Fungus found by accident protects lettuce. *Agricultural Research, 38*(10). Washington, D. C: United States Department of Agriculture.

Heimann, M. F., & Stevenson, W. R. (1981). *Walnut and butternut toxicity.* Madison: University of Wisconsin.

Henning, J.C., & Wheaton, H. (1987). White, ladino, and sweetclover. *Agricultural Guide* (No. G-4639). Columbus, MO: University of Missouri.

Herren, R. V., & Donahue, R. L. (1991). *The agriculture dictionary.* Albany, NY: Delmar Publishers, Inc.

Husband, T. P. (1987, September/October). Are ultrasonic pest repellers a wise investment? *In Touch.* Kingston: University of Rhode Island.

Jacobson, J. L. (1991). *Women's reproductive health: The silent emergency.* Worldwatch Paper 102. Washington, D. C.: Worldwatch Institute.

Juday, G. P., & Foster, N. R. (1990). A preliminary look at effects of the Exxon Valdez oil spill on Green Island Research Natural Area. *Agroborealis, 22*(1). Fairbanks: University of Alaska.

Juday, G. P., & Foster, N. R. (1991). A return to Green Island. *Agroborealis, 23*(1). Fairbanks: University of Alaska.

Kerr, R. A. (1989). Volcanos can muddle the greenhouse. *Science, 245*(4914).

Ketchum, B. H. (1983). Enclosed seas—Introduction. In B. H. Ketchum (Ed.). *Ecosystems of the world: Vol. 26. Estuaries and enclosed seas.* New York: Elsevier Scientific Publishing Company.

Ketchum, B. H. (1983). Estuarine characteristics. In B. H. Ketchum (Ed.). *Ecosystems of the world: Vol. 26. Estuaries and enclosed seas.* New York: Elsevier Scientific Publishing Company.

Klussman, W. *Improve your farm fish pond,* Bulletin B-213. College Station: Texas A&M University.

Kowal, N. E. (1985). *Health effects of land application of municipal sludge.* EPA/600/1-85/015. Triangle Park, NC: U.S. Environmental Protection Agency.

Kring, T. J., & Bush, L. (1991). Exotic aphid predator, European lady beetle, established in Arkansas. *Arkansas Farm Research. 40*(1).

Kristensen, T., & Paludan, J. P. (Eds.). (1988). *The Earth's fragile systems: Perspectives on global change.* Boulder, CO: Westview Press.

Lamb, H. H. (1972). *Climate: Present, past, and future* Vol 1. New York: Barnes & Noble Books.

Lamb, H. H. (1977). *Climate: Present, past, and future* Vol 2. New York: Barnes & Noble Books.

Lorius, C., Jougel, J., Aryan, D., Hansen, J., & LeTreut, H. (1990). The ice core record: Climate sensitivity and future warming. *Nature, 347*(6289).

Luoma, S. N. (1984). *Introduction to environmental issues.* New York: Macmillian Publishing Company.

Mackenthun, K. M. (1969). *The practice of water pollution biology.* Washington, D. C.: United States Department of the Interior, pp. 221-225.

Malthus, T. T. (1807). *An essay on the principle of population; Or, a view of its past and present effects on human happiness; With an inquiry into our prospects regarding the future removal or mitigation of the evils which it occasions.* London: Printed for J. Johnson by T. Bensley.

Matson, W. E. (1980). How we can double hydroelectric power. *The 1980 yearbook of agriculture.* Washington, D.C.: Author. pp. 367-377.

Matthews, S. W., & Sugar, J. A. (1991). Under the sun—Is our world warming? *National Geographic, 178*(4).

McNaughton, S. J. (1989). Diversity and stability. *Nature, 333*, pp. 204/205.

Merck and Co. (1983) *The Merck index: An encyclopedia of chemicals, drugs, and biochemicals,* 10th Ed. Rahway, NJ: Author.

Mertes, J. D. (1989). Trends in governmental control of erosion and sedimentation in urban developments. *Journal of Soil and Water Conservation, 44*(6).

Miller, P. (1991). A comeback for nuclear power? Our electric future. *National Geographic, 180*(2).

Miller, R.W., & Donahue, R.L. (1990). *Soils: An introduction to soils and plant growth.* Englewood Cliffs, NJ: Prentice Hall, Inc.

Milliman, J. D. (1989). Sea levels: Past, present, and future. *Oceanus, 32*(2).

Moore, J. W. (1986). *The changing environment.* New York: Springer-Verlag.

Morrison, J. (1987). Male insect's sex lure to trap females. *Agricultural Research, 35*(10).

Naderman, G.C., & Wagger, M.G. (1990). Tillage effects on infiltration and crop yields. *Agronomy Abstracts.* Madison, WI: American Society of Agronomy, Crop Science Society of America, and Soil Science Society of America.

National Research Council. (1986). *Soil conservation: Assessing the national resources inventory.* Vol. 1. Washington, D.C.: National Academy Press.

National Research Council Board on Agriculture. (1989). *Alternative agriculture.* Washington, D.C.: National Academy Press.

Nebel, B. J. (1990). *Environmental science: The world and how it works.* Englewood Cliffs, NJ: Prentice Hall, Inc.

Northeastern Regional Pesticide Coordinators. (undated). *Pesticide application training manual.* Ithaca, NY: Cornell University.

Norton, B. G. (Ed.). (1986). *The preservation of species.* Princeton, NJ: Princeton University Press.

Palmer, C. M. (1962). *Algae in water supplies.* Washington, D.C.: United States Department of Health, Education, and Welfare.

Pearse, F. (1989). Blowing hot and cold on the greenhouse. *New Scientist, 121* (1651).

Rathje, W. L., Wilson, D. C., Lambou, V. W., and Herndon, R. C. (1988). *Characterization of household hazardous waste from Marin County, California, and New Orleans, Louisiana,* EPA/600/S4-87/025. Triangle Park, NC: U.S. Environmental Protection Agency.

Ray, D. L. & Guzzo, L. (1990). *Trashing the planet: How science can help us deal with acid rain, depletion of the ozone, and nuclear waste (among other things).* Landham, MD: Regeny Gateway.

Rhoades, R. E., & Johnson, L. (1991). The world's food supply at risk. *National Geographic, 179*(4).

Ribaudo, M. O. (1989). *Water quality benefits from the conservation reserve program,* Agricultural Economic Report No. 606. Washington, D. C.: United States Department of Agriculture, Economic Research Service.

Schieder, S. (1990). Prudent planning for a warmer planet. *New Scientist. 128* (1743).

Seger, J. (1989). Diversity of little things. *Nature, 337,* pp. 305, 306.

Senft, D. (1990). Mighty mite takes on bindweed. *Agricultural Research, 38*(10). Washington, D. C.: United States Department of Agriculture.

Simon, J. L. (1981). *The ultimate resource.* Princeton, NJ: Princeton University Press.

Speidel, D., Ruedisili, L. C., & Agnew, A. F. (Eds). (1988). *Perspectives on water: Uses and abuses.* New York: Oxford University Press.

Staff. (1990, Fall). *Research perspectives.* Chapel Hill: University of North Carolina.

Steinhart, P. (1990). No net loss. *Audubon. 92*(4).

Stevens, G. C. (1989). The latitudinal gradient in geographical range: How so many species coexist in the tropics. *The American Naturalist. 133*(2).

Stork, N., & Gaston, K. (1990). Counting species one by one. *New Scientist, 127*(1729).

Tennessee Valley Authority. (1983). *The first fifty years: Changed land, changed lives.* Knoxville, TN: Author.

The Danube: Nature's resilience put to the test. (1991). *Calypso Log, 18*(6).

Thurman, H. V. (1985). *Introductory oceanography.* Columbus, OH: Charles E. Merrill Publishing Company.

Tierney, J. (1990, December 2). Betting the planet. *The New York Times Magazine.*

Tiner, R. W. (1988). *Field guide to nontidal wetland identification.* Annapolis: Maryland Department of Natural Resources, and Washington, D. C.: United States Fish and Wildlife Service, Department of the Interior.

Turner, B. (1990). Understanding streams. *Missouri Conservationist, 51*(12). Columbia: University of Missouri.

United States Department of Agriculture. (1980). *The 1980 yearbook of agriculture.* Washington, D.C.: Author.

United States Department of Agriculture. (1983). *Agricultural Research,* No. 30. Washington, D.C.: Author.

United States Department of Agriculture. (1984, June). *Agricultural Research,* No. 10. Washington, D.C.: Author.

United States Department of Agriculture. (1987). *The magnitude and costs of groundwater contamination from agricultural chemicals: A national perspective.* Staff Report AGES870318. Washington, D.C.: Economic Research Service.

United States Department of Agriculture. (1991). *Agriculture and the environment: The 1991 yearbook of agriculture.* Washington, D. C.: Author.

United States Department of Health and Human Services, Public Health Service. (1981). *Common poisonous and injurious plants,* Pub. No. (FDA) 81, 7006. Washington, D.C.: Author.

United States Environmental Protection Agency. (1992). *Another look: National survey of pesticides in drinking water wells, Phase II.* EPA/579/09/91/020. Washington, D. C.: Author.

United States Environmental Protection Agency. (1990). *Toxics in the community: National and local perspectives.* EPA/560/4-90/017. Washington, D. C.: Author.

United States Environmental Protection Agency. (1989). *Regulations and technology: Control of pathogens in municipal wastewater sludge.* EPA/625/10-89. Washington, D. C.: Author.

United States Environmental Protection Agency. (1988). *Toxics in the community: National and local perspectives.* Washington, D. C.: Author.

United States Environmental Protection Agency. (1985). *Health effects of land application of municipal sludge.* EPA/600/1-85/015. Washington, D. C.: Author.

United States Environmental Protection Agency. (1983). *Process design manual: Land application of municipal sludge.* EPA/625/1-83-016. Washington, D. C.: Author.

United States Department of the Interior. (1989). *National wetlands priority conservation plan.* Washington, D. C.: Author.

United States Department of the Interior. (1990). *Wetlands: Meeting the President's challenge..* Washington, D. C.: Author.

United States Soil Conservation Service. (1984). *1982 national resources inventory.* Washington, D.C.: United States Department of Agriculture.

United States Water Resources Council. (1978). *The Nation's Water Resources,* Vol., 1: Summary. Washington, D. C.: Author.

University of Arizona. (1991). *Arizona alumnus, 68*(2).

University of Illinois. (1971). *Agriculture's role in environmental quality,* Special Publication 21. Champaign: Author.

University of Rhode Island. (1991, May-June). *In Touch.* Kingston, RI: Author.

Vermeij, G. J. (1986). The biology of human-caused extinction. In B. G. Norton (Ed.). *The preservation of species.* Princeton, NJ: Princeton University Press.

Villee, C. A. (1985). *Fallout from the population explosion.* New York: Paragon House Publishers.

Wallace, D. (1982, July-August). Homemade pest control. *In Touch.* Kingston, RI: University of Rhode Island, Cooperative Extension Service.

Webb, F. J. (Ed.) (1986). *Proceedings of the thirteenth annual conference on wetlands restoration and creation.* Tampa, FL: Hillsboro Community College.

Webb, F. J. (Ed.) (1987). *Proceedings of the fourteenth annual conference on wetlands restoration and creation.* Tampa, FL: Hillsboro Community College.

Weil, R. R., Weismuller, R. A., & Turner, R. S. (1990). Nitrate contamination of ground water under irrigated coastal plain soils. *Journal of Environmental Quality, 19,* pp. 441-448.

Williams, W. M., Holden, P. W., Parsons, D. W., and Lorber, M. N. (1988). *Pesticides in ground water data base: 1988 interim report.* Washington, D.C.: United States Environmental Protection Agency.

Windholz, M. (Ed.) (1983). *The Merck index: An encyclopedia of chemicals, drugs, and biochemicals,* 10th Ed. Rahway, NJ: Merck and Co.

Winemiller, K. O. (1989). Musty connectedness decrease with species richness? *The American naturalist, 134*(6).

APPENDIX B

Environmental Conversion Factors

Conversion Factors for Weights and Measures*

To Convert From	To	Multiply By
Acres	Hectares	0.404 686
	Square Feet	43 560.000
	Square Kilometers	0.004
	Square Meters	4 046.856 422
	Square Miles	0.001 562
	Square Rods	160.000
	Square Yards	4 840.000
Acre-Feet	Acre Inches	12.000
	Cubic Feet	43 560.000
	Cubic Meters	1 233.482
	Cubic Yards	1 613.333
	Gallons (U.S.)	325 851.560
	Hectare-Centimeters	12.335
	Hectare-Meters	0.123 35
Acre-Feet/Day	Cubic Feet/Second	0.504 17
	Cubic Meters/Second	0.014 28
Acre/Inches	Acre/Feet	0.083 33
	Cubic Feet	3 630.000
	Cubic Meters	102.790 15
	Gallons (U.S.)	27 154.286
	Hectare-Centimeters	1.028
Board Feet	Cubic Centimeters	2 359.737 2
	Cubic Feet	0.083 333
	Cubic Inches	144.000
Bushels	Cubic Centimeters	35 239.070
	Cubic Feet	1.244 456
	Cubic Inches	2 150.420
	Cubic Meters	0.035 239
	Cubic Yards	0.046 091
	Gallons (U.S. dry)	8.000
	Gallons (U.S. liquid)	9.309 177
	Liters	35.239
	Pecks (U.S.)	4.000
	Pints (U.S. dry)	64.000

To Convert From	To	Multiply By
Bushels	Quarts (U.S. dry)	32.000
	Quarts (U.S. liquid)	37.236 71
	Gallons	7.481
Centimeters	Feet	0.032 808
	Inches	0.393 701
	Meters	0.010
	Millimeters	10.000
°Celsius	°Fahrenheit	1.8° + 32
Centimeters/Second	Feet/Minute	1.968 504
	Kilometers/Hour	0.036
	Meters/Minute	0.600
	Miles/Hour	0.022 369
Chains (Gunter's)	Feet	66.000
	Furlongs	0.100
	Links (Gunter's)	100.000
	Meters	20.116 8
	Miles	0.012 5
	Yards	22.000
Cubic Centimeters	Cubic Inches	0.061 024
	Cups	0.004 227
	Gallons (U.S. liquid)	0.000 264
	Liters	0.001
	Ounces	0.33 814
	Quarts (U.S. fluid)	0.001 057
Cubic Feet	Bushels (U.S.)	0.803 564
	Cords (of wood)	0.007 812
	Cubic Centimeters	28 316.847
	Cubic Inches	1 728.000
	Cubic Meters	0.028 317
	Cubic Yards	0.037 037
	Gallons (U.S. liquid)	7.480 520
	Liters	28.316 847
	Ounces (U.S. fluid)	957.506 49

To Convert From	To	Multiply By
Cubic Feet/Second	Acre Feet/Day	1.983 47
	Acre-Inches/Hour	0.991 73
	Cubic Centimeters/Second	28 316.847
	Cubic Meters/Second	0.028 317
	Hectare-Centimeters/Hour	1.019 4
	Liters/Minute	1699.011
	Liters/Second	28.316 85
	Million Gallons/Day	0.646 412
Cubic Inches	Board Feet	0.006 944
	Bushels (U.S.)	0.000 465
	Cubic Centimeters	16.387 064
	Cubic Feet	0.000 579
	Gallons (U.S. liquid)	0.004 329
	Liters	0.016 387
	Milliliters	16.387 064
	Ounces (U.S. fluid)	0.554 113
	Quarts (U.S. liquid)	0.017 316
Cubic Meters	Acre-Feet	0.000 811
	Cords (of wood)	0.384
	Cubic Feet	35.314 667
	Cubic Inches	61 023.740
	Cubic Yards	1.307 951
	Gallons (U.S. liquid)	264.172 05
	Liters	1 000.000
Cubic Yards	Cubic Centimeters	764 554.860
	Cubic Feet	27.000
	Cubic Inches	46 656.000
	Cubic Meters	0.764 555
	Gallons (U.S. liquid)	201.974 03
	Liters	764.554 86
Cups	Cubic Centimeters	236.588
	Liters	0.240
	Milliiters	240.000
	Ounces	8.000

To Convert From	To	Multiply By
°Fahrenheit	°Celius	0.556°F − 17.8
Feet	Centimeters	30.480 37
	Inches	12.000
	Meters	0.304 8
	Yards	0.333 333
Furlongs	Chain (Gunter's)	10.000
	Feet	660.000
	Meters	201.168
	Miles (statute)	0.125
	Rods	40.000
	Yards	220.000
Gallons (U.S. liquid)	Cubic Centimeters	3 785.411 8
	Cubic Feet	0.133 681
	Cubic Inches	231.000
	Cubic Meters	0.003 785
	Cubic Yards	0.004 951
	Gallons (British)	0.832 675
	Gallons (U.S. dry)	0.859 367
	Liters	3.785 412
	Ounces (U.S. fluid)	128.000
	Pints (U.S. liquid)	8.000
	Quarts (U.S. liquid)	4.000
Gallons (U.S. liquid)/Minute	Cubic Feet/Hour	8.020 8
	Cubic Meters/Hour	0.277
	Acre-Feet/Day	0.004 419
	Hectare-Centimeters/Hour	0.002 27
	Cubic Feet/Second	0.002 228
	Liters/Second	0.063 1
Grams	Kilograms	0.001
	Ounces (apothecary or troy)	0.032 151
	Ounces (avoirdupois)	0.035 274
	Pounds (apothecary or troy)	0.002 679
	Pounds (avoirdupois)	0.002 205

To Convert From	To	Multiply By
Grams/Cubic Centimeters	Pounds/Cubic Feet	62.426 961
	Pounds/Cubic Inches	0.036 127
	Pounds/Gallons (U.S. liquid)	8.345 404
Grams/Liters	Parts/Million	1 000.000
	Pounds/Cubic Feet	0.062 426
Hectare	Acres	2.471 054
	Square Feet	107 639.100
	Square Kilometers	0.010
	Square Meters	10 000.000
	Square Miles	0.003 861
Hectare/Centimeters	Acre-Feet	0.081 08
	Acre-Inches	0.972 76
Hectare/Meters	Acre-Feet	8.108
	Acre-Inches	97.276
Horsepower, mechanical	Foot-Pounds/Second	550.000
	Horsepower, boiler	0.076 018
	Horsepower, electrical	0.999 598
	Horsepower, metric	1.013 870
	Joules/Second	745.700
	Kilowatts	0.745 700
	Tons of Refrigeration	0.212 040
	Watts	745.700
Hundredweights (British long)	Kilograms	50.802 345
	Pounds (avoirdupois)	112.000
	Tons (long)	0.050
	Tons (metric)	0.50 817
	Tons (short)	0.056
Hundredweights (U.S. short)	Kilograms	45.359 237
	Pounds (avoirdupois)	100.000
	Tons (long)	0.044 643
	Tons (metric)	0.045 359
	Tons (short)	0.050
Inches	Centimeters	2.540
	Feet	0.083 333

To Convert From	To	Multiply By
Inches	Meters	0.025 4
	Yards	0.027 778
Kilograms	Ounces (apothecary or troy)	32.150 747
	Ounces (avoirdupois)	35.273 962
	Pounds (apothecary or troy)	2.679 229
	Pounds (avoirdupois)	2.204 623
	Quintals	0.010
Kilograms/Hectare	Pounds/Acre	0.892
Kilometers	Centimeters	100 000.000
	Feet	3 280.839 9
	Meters	1 000.000
	Miles (statute)	0.621 371
	Yards	1 093.613 3
Kilometers/Hour	Centimeters/Second	27.777 778
	Feet/Hour	3 280.839 9
	Meters/Second	0.277 778
	Miles (statute)/Hour	0.621 371
Liters	Bushels (U.S.)	0.028 378
	Cubic Centimeters	1000.000
	Cubic Feet	0.035 315
	Cubic Inches	31.023 744
	Cubic Meters	0.001 000
	Cubic Yards	0.001 308
	Gallons (U.S. liquid)	33.814 023
	Quarts (U.S. liquid)	1.056 688
Meters	Centimeters	100.000
	Feet	3.280 840
	Inches	39.370 079
	Kilometers	0.001 000
	Milliliters	1 000.000
	Rods	0.198 839
	Yards	1.093 613
Microns	Micrometers	1.000
	Centimeters	0.000 1

To Convert From	To	Multiply By
Microns	Milliliters	0.001
	Millimicrons	1 000.000
Miles (statute)	Chains (Gunter's)	80.000
	Feet	5 280.000
	Inches	63 360.000
	Kilometers	1.609 344
	Rods	320.000
	Yards	1 760.000
Milligrams/Liter	Grams/Liter	0.001
	Parts/Million	1.000
Milliliters	Centimeters	0.100
	Feet	0.003 281
	Inches	0.039 37
	Meters	0.001
Million Gallons/Day	Acre-Inches/Day	36.828
	Cubic Feet/Second	1.547
	Cubic Meters/Minute	2.629
Ounces (avoirdupois)	Grams	28.349 523
	Ounces (apothecary or troy)	0.911 458
	Pounds (apothecary or troy)	0.075 955
	Pounds (avoirdupois)	0.062 5
Ounces (U.S. fluid)	Cubic Centimeters	29.573 530
	Cubic Inches	1.804 688
	Cups	0.125
	Liters	0.029 574
	Quarts (U.S. liquid)	0.031 25
	Tablespoons	2.000
Parts/Million	Grams/Liter	0.001
	Milligrams/Liter	1.000
Pints (U.S. liquid)	Cubic Centimeters	473.176 47
	Cubic Feet	0.016 710
	Cubic Inches	28.875
	Cubic Yards	0.000 619

To Convert From	To	Multiply By
Pints (U.S. liquid)	Cups	2.000
	Gallons (U.S. liquid)	0.125
	Liters	0.473 177
	Ounces (U.S. fluid)	16.000
	Pints (U.S. dry)	0.859 367
Pounds (avoirdupois)	Grams	453.592 37
	Kilograms	0.453 592
	Ounces (apothecary or troy)	14.583 333
	Ounces (avoirdupois)	16.000
	Pounds (apothecary or troy)	1.215 277
Pounds/Acre	Kilograms/Hectare	1.121
	Megagrams/Hectare	0.001 12
	Metric Tons/Hectare	0.001 121
	Quintals/Hectare	0.011 21
Pounds/Cubic Feet	Grams/Cubic Centimeters	0.016 018
	Kilograms/Cubic Meter	16.018 463
Pounds/Cubic Inches	Grams/Cubic Centimeter	27.679 905
	Grams/Liter	27 679.905
	Kilograms/Cubic Meters	27 679.905
Pounds/Square Feet	Pascals	47.88
Pounds/Square Inch	Atmospheres	0.068 046
	Bars	0.068 948
	Grams/Square Centimeter	70.306 958
	Pascals	6 900.000
Quarts (U.S. liquid)	Cubic Centimeters	946.352 95
	Cubic Inches	57.750
	Cubic Feet	0.033 420
	Gallons (U.S. dry)	0.214 842
	Gallons (U.S. liquid)	0.250
	Liters	0.946 353
	Ounces (U.S. liquid)	32.000
	Pints (U.S. liquid)	2.000

To Convert From	To	Multiply By
Quintals	Kilograms	100.000
	Pounds (avoirdupois)	220.462 26
	Metric Tons	10.000
Quintals/Hectare	Kilograms/Hectare	100.000
	Metric Tons/Hectare	0.100
	Pounds/Acre	89.206 07
Rods	Feet	16.500
	Furlongs	0.025
	Inches	198.000
	Meters	5.029 2
	Miles (statute)	0.003 125
	Yards	5.500
Square Centimeters	Square Inches	0.155 000
	Square Meters	0.000 1
Square Chains (Gunter's)	Acres	0.100
	Square Meters	404.688
	Square Miles	0.000 156
	Square Rods	16.000
	Square Yards	484.000
Square Feet	Square Centimeters	929.030 4
	Square Inches	144.000
	Square Meters	0.092 903
	Square Yards	0.111 111
Square Inches	Square Centimeters	6.451 6
	Square Millimeters	645.16
Square Meters	Acres	247.105 38
	Hectares	100.000
	Square Miles (statute)	0.386 102
Square Miles (statute)	Acres	640.000
	Hectares	258.998 81
	Square Kilometers	2.589 988
	Square Meters	2 589 988.000
	Square Yards	3 097 587.500

To Convert From	To	Multiply By
Square Rods	Acres	0.006 25
	Hectares	0.002 529
	Square Centimeters	252 928.526 4
	Square Feet	272.250
	Square Inch	39 204.000
	Square Meters	25.292 853
	Square Yards	30.250
Square Yards	Square Centimeters	8 361.273 6
	Square Feet	9.000
	Square Inch	1 296.000
	Square Meters	0.836 127

*Adapted from: Herren, Ray V. and Donahue, Roy L. The Agriculture Dictionary. Delmar Publishers, Inc. 1991.

Glossary of Environmental Terms

Abiotic: The absence of life; nonliving. **Example**: the wind.

Abyssal Plain: The relatively flat surface of the ocean floor.

Acid Rain: The transfer of acidic or acidifying substances from the atmosphere to the Earth via precipitation.

Agronomy: The branch of agriculture dealing with field-crop production and soil management.

Albedo: The reflectivity of celestial bodies such as the Earth. **Example**: white snow has a higher albedo than black asphalt on a highway.

Algae: A group of one-celled green plants containing chlorophyll and having no true roots, stems, or leaves. They occur in wet places and sometimes appear as pond scum.

Algal Bloom: A rapid proliferation of algae caused by enrichment of waters.

Allergen: Any substance inducing an allergy such as ragweed pollen.

Amorality: Neither moral nor immoral; placing no meaning on the value judgments of right and wrong.

Anthracite Coal: The high-carbon, hardest kind of coal which gives great heat but little smoke and flame.

Anthropocentric: Literally "human-centered." A perspective that considers all aspects of life based on human perceptions and values.

Anthropogenic: Coming from the activities of humans.

Aphotic: Literally, without light. That part of ocean depth without light.

Arithmetic Increase: Simple linear increase. In a group of things that is arithmetically increasing, the number of new items added each time period would be the same. Example: 1, 2, 3, 4, 5, etc. is an example of 1 + n, where n is the number of time periods.

Art: The pursuit of beauty.

Atomic Energy: Energy released by nuclear reactions.

Avicide: A pesticide used to kill unwanted birds.

Axis of Spin: For the Earth, the angle between an imaginary line drawn between the poles and the plane of the ecliptic. At the present time, the axis of spin is approximately 23 1/2° from the vertical.

Backshore: That part of a beach that is above the level of the highest tides.

Bathymetry: The measurement of ocean depths and the charting of the ocean floor. It is similar to topography on the land.

Binomial: Having two parts. In this book, a scientific name consisting of genus and species is said to be binomial. Example: *Quercus alba*, white oak.

Biochemical Oxygen Demand: A measure of the amount of oxygen consumed in the biological decomposition of organic matter in water.

Biodiversity: Biological diversity; i.e., the number of species per unit area.

Biological Control: The control of pests by using plants and animals that are predators or that outcompete the pest.

Biomagnification: The concentration of a substance in the tissues of organisms as they go up a food chain or web.

Biome: Major land areas characterized by dominant life-forms, such as tundra and grassland.

Biosphere: The portion of the Earth where all life exists.

Biotic: Biological; pertaining to life.

Birth Rate: Number born per one thousand population, or percent born.

Bituminous Coal: Softer than anthracite coal; the most common kind of coal.

Boomsters (Cornucopians): People that believe that the more people there are, the more prosperity there will be.

Botany: A scientific study of plants.

Brackish: A mixture of fresh and salt water, containing between 1,000 and 10,000 milligrams per liter (parts per million) of salt.

Bristlecone Pine: Pinus aristata, a species of pine thought to represent the oldest living thing on Earth.

Calibration: The technique of setting a device to improve its accuracy, such as to deliver the desired amount of a fertilizer, pesticide, or any other substance.

Carbon Dioxide: (CO_2). A colorless, odorless, non-poisonous gas which results from fossil fuel use and all other organic decomposition.

Carbon Monoxide: (CO). A colorless, odorless, poisonous gas which results from fossil fuel use and certain types of organic decomposition.

Career: A life-time activity that matches one's ability and produces satisfaction (see *job*).

Carrying Capacity: The largest number of a species that can exist in a given ecosystem, without damaging the ecosystem.

Castor Bean: (*Ricinus communis*), family Euphorbiaceae. An annual herb often grown in the garden for beauty of foliage. However, foliage and seeds are poisonous to people.

Character Displacement: A process of evolution in which two closely related species gradually become more different.

Chemical Energy: Energy contained in the chemical bond between atoms. Example: the burning of wood releases heat energy.

Chemigation: The distribution of any chemical such as pesticides by mixing it with irrigation water.

Chlorofluorocarbons: A substance made by humans, used in refrigerators and as a propellant in spray cans. Some scientists believe they drift into the atmosphere where their chlorine breaks protective ozone down into oxygen gas.

Chlorosis: The yellowing of leaves caused by lack of nutrients such as iron, disease, or air pollution.

Clearcutting: The harvesting of all trees in an area.

Climax Vegetation: The ultimate association of plants in a specific habitat. The plant species that would eventually dominate a habitat, if it were to reach equilibrium.

Clods: Aggregates that form when soil is manipulated while it is too wet.

Coastline: An imaginary line at the highest point of land bordering a body of water, that is effectively reached by waves at high tide.

Coliform: A group of bacteria in the intestinal tract of humans and animals. The presence of coliform in the environment indicates the presence of possible disease organisms.

Common Properties: That which belongs to no one, but to everyone. Examples: air, oceans, parks.

Compartmentalization: In a diverse ecosystem, smaller groups of species (subsets) associated with each other. The process by which the subsets form is *compartmentalization.*

Competition: A situation when two or more organisms are struggling for the same resource or niche.

Competitive Exclusion: When two or more species compete for a necessary resource, one will always win and the other will lose, eventually. The losing species must adapt, move, or die. This principle is called competitive exclusion.

Composting: A technique of piling organic materials in a favorable environment for their decomposition. The finished product is then used to make plants grow better.

Computer Simulation: A technique of putting a series of assumptions in a computer to obtain an educated guess as to the results of some activity.

Conservation: The wise use of resources.

Consumers: Organisms such as people and animals that eat what green plants produce by photosynthesis.

Contact Dermatitis: (dermal toxicity). The ability of a toxic substance such as poison ivy to produce a rash on the skin of a sensitive person.

Continental Rise: A smooth-surface accumulation of sediment formed at the base of the ocean floor toward land at a gradient between 1:100 and 1:700.

Continental Shelf: The gently seaward-sloping surface from the shoreline to an ocean depth of about five hundred feet.

Continental Slope: The relatively steeply sloping surface that extends from the outer edge of the continental shelf to the continental rise.

Copper Sulfate: $CuSO_4$.

Coriolis Effect: The tendency for any moving object in the northern hemisphere to be deflected to the *right* and to the *left* in the southern hemisphere.

Corn: Zea mays, family Gramineae (grass family). One of the most important of the world's cereals.

Cotton: Gossypium hirsutum—upland cotton, and American Egyptian cotton (*G. barbadense*) both of family mabreaceae. The fiber is used principally for clothing.

Crop Science: Research, teaching, and extension about mostly annual plants grown for food and fiber.

Crownvetch: (*Coronilla varia,* family Leguminosae). A perennial herb planted extensively to stabilize disturbed soil.

Cultivar: Any *culti*vated *vari*ety of a plant.

Cultural Pest Control: A technique of management to control pests. Example: Using disease-resistant varieties when available and rotating crops.

Death Rate: The number of deaths, usually expressed as deaths per one thousand population, or expressed as a percentage of population.

Decomposer: Organisms in nature that help organic materials to degrade. They include bacteria, fungi, actinomyces, and insects.

Decreaser: Plants in a range or pasture that usually become fewer when overgrazing occurs.

Deforestation: The conversion of a forest to barren land because of excess cutting, insects, diseases, or excess pollution.

Desiccation: Drying of an object.

Disease Resistance: The ability of an organism to avoid a given disease or category of diseases. Example: some tomato cultivars have been bred to resist certain diseases.

Divine Order: An outlook on life that believes that "what will be – will be" because the gods have so ordered.

Doomsters (Malthusians): People who believe that zero population growth is the only way to "save the earth" – if it is not already too late.

Drain Field: A series of buried horizontal drain tile or perforated pipe surrounded by gravel that is used to slowly inject septage into soil from a septic tank.

Draining: The process of reducing the amount of water in an area such as a wetland.

Dumbcane: (*Dieffenbachia seguine,* family Araceae). A foliage plant grown in greenhouses. Eating a part of a leaf causes loss of speech for a few hours.

Ecliptic Plane: An imaginary line traced by a planet as it revolves around the sun.

Ecological Erosion Control: The control of soil erosion by planting adapted vegetation.

Ecology: The study of interrelationships among organisms and their environment.

Economic Threshold: As used in this book, the point at which insect control measures should be started. The point at which the cost of applying some management technique becomes less than the damage caused by not applying it.

Ecosystem: The entire system of living organisms and nonliving environmental and geographic factors in a given area.

Effluent: The flowing out, such as a discharge of a pollutant.

El Niño: A warm-water ocean current which periodically flows southward along the coast of Ecuador and modifies world climate. Literally refers to the Christ-Child—"The Boy"—and traditionally capitalized.

Electrical Energy: A property of particles of matter as electrons (negative charges) and protons (positive charges) to create a field to do work. The energy can be generated by duplicate friction, induction, or chemical change.

Emergency Episode: Climatic intensifications such as heat or cold waves or storms. Extreme air pollution build-ups resulting from lack of air movement are one example.

Emergent Hydrophyte: A water-loving plant that has a part of its structure above the surface of the water. Example: cattail. May or may not grow in standing water.

Emissions: The discharge of a pollutant.

Employability Qualities: Characteristics of persons that affect (hopefully favorably) the employer and fellow employees.

Endophyte: A plant that grows within another plant. Example: an alga within a fungus. Example: species of lichens.

Energy: The capacity to do work.

Energy Flow (also energy transfer): The transfer of energy from the sun to green plants by means of photosynthesis, then to other plants and animals in the ecosystem.

Environment: The combination of all factors external to and affecting an organism or a population.

Environmentalism: Social or political activities undertaken for the purpose of affecting the environment, normally in a way that the individual perceives as positive.

Environmental Diseases: Diseases caused by environmental conditions such as temperature extremes, drought, or acidification.

Environmental Science: All those branches of science which deal with the study of any aspects of the environment.

Equator: An imaginary line around the earth halfway between the true poles and designated as zero latitude.

Equilibrium: A temporary or permanent harmony (stability) in any system. If equilibrium were reached in an ecosystem, there would be no more changes until some outside force caused a change.

Essential Elements: For seed-bearing plants there are seventeen elements necessary for growth and reproduction — carbon, hydrogen, oxygen, nitrogen, phosphorus, potassium, sulfur, calcium, iron, manganese, magnesium, boron, copper, zinc, molybdenum, chlorine and lead.

Estuarine System: Deepwater and tidal wetlands beneath and associated with estuaries.

Estuary: "A semi-enclosed, tidal, coastal body of water open to the sea in which fresh and salt water mix" (Source: Chesapeake Bay Foundation).

Ethanol: (C_2H_5OH) (Ethyl alcohol; grain alcohol). The alcohol product of fermentation. When blended with gasoline, it is known as **gasohol** and is used as a motor fuel.

Eutrophication: A process of waters becoming richer in nutrients. The typical result of this process is a heavy growth of aquatic plants and is often accompanied by a decreased level of dissolved oxygen.

Evaporation: The change from liquid water to water vapor.

Evapotranspiration: The process of evaporation of water plus the water loss from plants through the openings (stomata) in the leaves.

Exhaustible: As used in this book, exhaustible refers to the property of a natural resource as being capable of being "used up."

Exploitation: Use of a resource.

Exponential Growth: As used in this book, the tendency of a population to increase at an increasingly rapid rate.

Exponential Increase: An increase in numbers that occurs in an exponential way rather than an additive way. Example: 2, 4, 8, 16, 32, 64 is an example of 2^n, where n is the number of time periods.

Fertigation: The mixing of fertilizers in irrigation water and the spreading of fertilizer while irrigating. A specialized type of chemigation.

Fertilizer: Any chemical that contains one or more of the seventeen essential plant nutrients in available form.

Fescue Foot: A fungus disease transmitted to cattle on fescuegrass.

Filling: The use of soil and debris to raise the level of depressions.

Flux: A coming in (flow) or going out (ebb) of the tide; any continental movement; any change. As used in this book, particularly Chapter 8, the movement of water into or out of a specific reservoir.

Food Chain: A sequence of organisms, each of which uses the next lower member of the sequence as food (see *Food Web*).

Food Web: The dependence of organisms upon others in a series for food. A food web may be made up of many interconnected food chains (see *Food Chain*).

Forb: A broad-leaved, herbaceous flowering, seed-bearing plant.

Foreshore: That portion of the shore between normal low tide and normal high tide.

Forest Type: A classification of forest stands based on dominant tree species present. Example: oak-hickory forest type.

Forestry: The science, art, and business practices of studying and managing lands designated as forests.

Foxglove: (Genus *Digitalis*, family Scrophulariaceae). Animals may be poisoned by grazing the biennial herb.

Fuelwood: Woody plants used for heating and cooking.

Fungicides: Pesticides designed to kill specific fungi that cause disease.

Gasohol: A motor fuel made from ten percent ethanol (ethyl alcohol, C_2H_5OH), and ninety percent unleaded gasoline.

Gause Curves: Population graphs that resulted from an experiment in which two species, one a predator and the other its prey, were introduced into a closed ecosystem with only the two species present. The end result was that the predator consumed all of the prey, then it died from lack of food (see *Lotka-Volterra Curves*).

General Circulation Model (GCM): A very large computer simulation of the world's atmosphere. GCMs are used to predict the movements of weather formations, like hurricane Andrew. They are also used to predict the results of atmospheric changes.

Genetic Control: The development of plants resistant to specific diseases and some insect pests by gene transfer and other manipulations.

Genetics: The science of plant, animal, and human inheritance. Applied genetics is called breeding.

Geologic Materials: Materials, such as loose sands or clays, that always touch soils at the soil's lower extremities and extend through the earth's crust.

Geology: Science of the earth, including water, air, minerals, rocks, and earth movements.

Geothermal Energy: The use of heat generated within the Earth. Example: use of heat from hot springs for heating a home.

Global Warming: A theory, sometimes related to the "greenhouse effect," which proposes that the Earth's mean temperature is rising rapidly because of the generation of emissions resulting from human activity.

Grass Tetany: A disease of cattle and sheep resulting from a deficiency of magnesium. Also known as "grass staggers."

Grassland: Land in uncultivated grass. It may also include legumes and a few trees and shrubs.

Green (Pond) Scum: A local term referring to a small body of eutrophic (enriched) water in which many species of algae are growing.

Green Revolution: A term coined by the international news media about two years after the release of the rice cultivar 1R-8 by the International Rice Research Institute in the Philippines. The so-called green revolution resulted in very rapid increases in the yields of food production, mainly rice, because of new varieties.

Greenhouse Effect: A measurable warming of the Earth's atmosphere by the absorption of solar energy.

Greenhouse Gasses: In general, all atmospheric gasses that tend to capture energy from sunlight and release it as heat. Carbon dioxide, nitrous oxide, methane, ozone, water vapor, and chlorofluorocarbons 11 and 12 are generally considered the important greenhouse gasses.

Groundwater: Water present below the water table stored in rock cracks and between mineral and rock particles.

Growth Rate: An increase in numbers, height, girth, or weight of plants, animals, or people.

Guild: (1) The members of any association with a designated purpose; (2) a group of closely associated plants or animals that depend on each other.

Hardwood: (1) Any timber of broad-leaved tree species (angiosperms) whose wood has vessels; (2) a tree whose wood is hard.

Hazardous Waste: Any throw-away substance that poses a threat to human health and/or the environment.

Herbicides: A chemical pesticide designed to kill weeds (unwanted plants).

Hermaphrodite: An animal having *both* male and female sex organs. Example: earthworms.

High-Level Nuclear Waste: Radioactive waste that emits more than one hundred nanocuries per gram. (see *Low-Level Nuclear Waste*).

Homeothermic: A characteristic of people and animals whose bodies must remain within a constant temperature range, regardless of the ambient temperature of their environment (warm blooded).

Hybrid: A cell or organism formed from a cross of two closely related species.

Hybridization: The process of producing a hybrid plant or animal.

Hydric Soil: A poorly-drained soil formed in a wet environment such as in a wetland.

Hydrologic Cycle: The movement of water from air (atmosphere) to land to water bodies back to the atmosphere.

Hydrophyte: A plant that either tolerates wetness or must have wetness to survive.

Ice Core Records: An analysis of the composition of ice and air entrapped in glaciers to help to determine environmental conditions at the time the glacial ice was formed.

Igneous Rock: Rock formed by cooling and solidification of molten silicate minerals within deep layers of the Earth, i.e., granite.

Incinerating: Burning to reduce the volume of wastes.

Increaser: Plants that reproduce rapidly. Particularly in an area that is being grazed, increasers tend to reproduce rapidly when moderate grazing occurs.

Inexhaustible: Resources that have no practical limit over an indefinite period of time.

Infiltration: Movement of a liquid *into* a substance. Example: rain moves downward into soil.

Insect Resistance: A characteristic of plants that are seldom or never damaged by insects or by specific insects.

Insecticide: A pesticide that kills insects.

Integrated Pest Management: The use of *all* means possible to control pests.

Intermittent Subsystem: A wetland that is covered by flowing water only part of the year. Example: an intermittent stream will be dry part of the year.

Interplanting: The planting of crops between other crops for the purpose of "confusing" insects which find plants by odor.

Intertidal Subsystem: The part of the coastline that lies between the average low tide and the average high tide. Intertidal plants and animals are submerged part of the time and exposed to the air part of the time.

Invader: Plants that reproduce very rapidly. Particularly in an area that is being grazed, invaders tend to become dominant when overgrazing occurs.

IPM: (see *Integrated Pest Management*).

Jimson Weed: (*Datura stramonium*, family Solanaceae). A weedy annual herb that is poisonous to people and animals.

Job: (see *Career*). A means of earning a living.

Juglone: The substance in all parts of black walnut and butternut plants that is toxic to most garden plants.

Kinetic Energy: The ability to do work produced by a mass in motion. Example: flowing water.

La niña: Literally "the girl." A *cold* ocean current in the Pacific Ocean (see *El Niño*). Because la nina does not refer to a specific person, it is not traditionally capitalized.

Lacustrine System: An area dominated by lakes.

Lagoon: (1) A body of shallow water established for anaerobic decomposition of human or animal effluent; (2) any body of shallow water.

Land Use Planning: An official technique of guiding orderly development to assure a more healthful and more artful environment.

Landfill: A trench dug to bury solid wastes.

Latent Heat of Evaporation: The energy required by liquid water to cause it to evaporate.

Latent Heat of Melting: The energy required by ice to change its state from solid to liquid (melt).

Latitude Gradient: A change in distance from the equator. As used in this book, the difference between the northernmost and southernmost ranges of a species.

LD_{50}: A pesticide dose lethal to fifty percent of the target species.

Leaching: The movement of substances dissolved or suspended in water as the water moves in the soil.

Lead: (Pb). A heavy metal hazardous to health if breathed or swallowed. Its use has been sharply restricted. It is essential in small amounts for plants, animals, and humans.

Legume: A family of plants (Leguminosae) including clovers and alfalfa that have the capability of aiding symbiotic bacteria to manufacture protein nitrogen from the air. By that process, legumes are said to "fix" nitrogen (take it from the atmosphere, form nitrogen compounds that are usable to plants, then release those compounds into the soil).

Lignite Coal: The lowest grade of coal between peat and bituminous coal.

Lily of the Valley: (*Convallaria*). A perennial, white-flowered, low-growing plant that is widely grown but is toxic to humans.

Linear Function: A relationship between two variables. The addition of a unit of one variable produces the same change in the value of the other variable, every time. Example: for the linear function Y = 3X, for every additional X, Y becomes 3 larger.

Linnaeus: Carl von Linne (1707–1778). A Swedish botanist, the originator of the binomial system of latinizing plant and animal scientific names. He Latinized his name to Carolus Linnaeus.

LISA: Low-Input Sustainable Agriculture. An acronym for programs designed to encourage less use of chemicals that may pollute the environment.

Lithosphere: (Greek *lithos*, [stone]). The more or less solid, rocky crust of the Earth's surface below the soil.

Littoral: Pertaining to the seashore.

Livable: An environment suitable for people to live.

Logistic Curve: A graph that shows what would likely happen to the numbers of a single species if it were introduced into a sealed environment. At first the species would reproduce rapidly, then it would overcrowd its environment. Finally, the species would become extinct in the closed system.

Longleaf Pine: (*Pinus palustris*). One of the principal southern pines valued for its long, straight bole and its heavy, hard wood.

Lotka-Volterra Curves: A theory that proposed what might happen if two species, one predator and one prey, were enclosed in a sealed environment with no other species present. According to this theory, the predator and prey would take turns increasing and decreasing in numbers around some "equilibrium" level (see *Gause Curves*).

Low-level Nuclear Waste: Radioactive waste that emits *less than* 100 nanocuries per gram (see *High-Level Nuclear Waste*).

Lower Perennial Subsystem: A wetland that is nearly level, but which is covered by slowly flowing water year round.

Maize: (see *Corn*).

Marine Science: The branches of science dealing with the biology, hydrology, bathymetry, and other aspects of the seas and estuaries.

Marine System: Wetlands underlying shallow ocean water, including intertidal regions.

Maximum Sustainable Growth Rate: The maximum rate of population increase that occurs at the point where the rate of increase changes from increasing to decreasing.

Maximum Sustainable Harvest: The highest rate at which a species can be taken for use by humans without decreasing the total population of the species in a given ecosystem. In theory, the maximum sustainable harvest would occur at half the carrying capacity for a species.

Mechanical Erosion Control: The use of dams, terraces, and diversion ditches to reduce soil erosion and sedimentation.

Metamorphic Rock: An igneous rock such as granite changed to agneiss or a sedimentary rock such as limestone changed to marble by the Earth's heat and pressure.

Meteoric Precipitation: Water, snow, or ice falling from the sky.

Meteorology: The science of the atmosphere, including climate and weather.

Methane: (CH_4). A gas created by anaerobic decomposition. It is classed among the greenhouse gases.

Methanol: (CH_3OH) (methyl alcohol). A colorless, volatile, flammable gas that is poisonous. It is made from the distillation of wood.

Methemoglobinemia: A serious disease of young children caused by the drinking of water high in nitrates. Also known as "blue-baby disease."

Milankovich Model: A theory that describes large shifts in the tilt of the Earth's axis of spin, smaller "wobbles" in the tilt of the Earth's axis of spin, and "bulges" in the plane of the Ecliptic.

Mineral Soil Sediments: Inorganic soil particles moved by water or wind erosion from one place to another. Mineral sediments make up the largest source of pollution (by weight) in the environment.

Mixed-Grass Prairie: An ecosystem with a combination of tall grasses such as big bluestem grass and short grasses such as buffalo grass.

Monoculture: The growing of large areas of a single species year after year. Example: corn on a large field in any one year.

Mores: Habits, customs, traditions, instinctive behaviors of humans.

Multiple Use: The simultaneous or alternative use of land for many purposes. Example: a tract of forest used for wood products, recreational trails, and hunting.

Mycorrhizae: A group of fungi whose hyphae (roots) function as root hairs which assist growth of many vascular plants.

Natural Gas: A mixture of gaseous hydrocarbons, chiefly methane (CH_4).

Natural Resource: Any naturally occurring substance or property that humans use. Examples: coal, water power, arable soil.

Natural Succession: The crowding out of one species by another because the aggressor is better adapted.

Naturalist: A person who studies nature as a science, especially by direct observation and enjoyment.

Nearshore: The region of shore between the line where breakers form and normal lowtide shoreline.

Nematicide: A pesticide to kill nematodes.

Neritic: The zone of ocean waters located outside the foreshore but still above the continental shelf.

Niche: A specific set of environmental conditions (habitat) favorable for one specific species of plant or animal.

Niche Overlap: A situation in which two or more plant or animal species that require similar resources can exist without one species being crowded out.

Nightshade: Any species of a large plant genus (*Solanum* sp.) with narcotic qualities. Some species are poisonous and some are not.

Nitrate: (NO_3).

Nitrate Nitrogen: (NO_3-N). A nitrogenous compound that is capable of moving into groundwater and polluting it. (see *Methemoglobinimia*).

Nitrite: (NO_2) or a salt or ester of nitrous acid.

Nitrogen: (N). A colorless, tasteless, odorless gas comprising nearly eighty percent of the ambient atmosphere and a component of all living things.

Nitrogen Fixation: (see *Legume*).

No-till: A soil conservation technique of planting a current crop in last year's crop residue without soil preparation.

Noise: Unwanted sound.

Nonpoint Source: Pollution that originates from an area such as a farm yard in contrast to point sources such as from a sewer pipe.

Nonrenewable Resource: All natural resources such as coal, oil, and other minerals are not capable of being renewed in a human lifetime.

Nuclear Waste: (see *Low-Level Nuclear Waste*).

Obligate Hydrophyte: A water-loving plant that *must* grow in water, in contrast to a plant that grows on land as well as in water.

Ocean Dumping Act: A federal law that authorizes the U.S. Environmental Protection Agency and the Corps of Engineers to regulate ocean dumping by means of a permit system. A 1988 amendment prohibits dumping of sewage sludge, industrial waste, and infectious medical wastes into the ocean. In 1992 it phased out dumping of sludge.

Offshore: The region beyond the line of breakers.

Oncogenicity: The property of a substance that causes tumors, including cancers.

OPEC: Organization of Petroleum Exporting Countries.

Opportunity Cost: Whenever a choice must be made between two or more possible courses of action, that which you must give up to select the other course of action is said to be "opportunity cost."

Organic Gardening: A system of gardening that proposes to use only organic fertilizers, no synthetic chemical fertilizers, and no synthetic chemical pesticides. Insects are removed from vegetables by hand. Careful management is necessary to hold down plant diseases. Animal manure or plant residue is used to provide the necessary plant nutrients.

Organic Sediments: Pollutants in water consisting of plant, animal, and human effluents.

Overdraft: Using more than that which is replaced. For example, using more irrigation water from groundwater than that which is replaced by rainfall infiltration.

Overgrazing: Allowing grazing animals to eat more forage than is produced by regrowth.

Overpopulation: A condition in which there are more individuals of a species than the ecosystem can support.

Palustrine System: Wetlands that are dominated by trees, shrubs, and emergent hydrophytes and which are not associated with standing water over 6.6 feet deep. We might describe such an area as a bog or swamp.

Paraprofessional: (Greek *para* [alongside of]). A person whose career requires somewhat less specialized education than a professional.

Pathogen: A disease-producing organism.

Peds: Soil aggregates that form from natural sources.

Percolation: A liquid moving through something. *Example:* rainwater moving through the soil and rock to groundwater.

Permafrost: Permanently frozen soil.

Permeability: The characteristic of a substance to permit percolation.

Pesticide: Any substance, usually a chemical, designed to kill any pest whether a weed, a pathogen, rodent, a bird, or other unwanted species.

Philodendron: Any plant of genus *Philodendron* whose leaves are toxic to humans.

Phosphorus Pentoxide: P_2O_5. This is the chemical equivalent of all phosphorus-bearing fertilizers. The conversions are: $P \times 2.291 = P_2O_5$; $P_2O_5 \times 0.436 = P$.

Photic: (1) The effect of light upon organisms or the production of light by organisms. *Example:* lightning bug (firefly). (2) That zone in the ocean in which sunlight penetrates in adequate amounts to support photosynthesis.

Photoperiodicity: The response of plants to the length of the lighted period (day). Some plants produce seed when the light is longer each day, others when it is shorter each day, while other plants, "don't care" either way.

Photosynthesis: The manufacture of food by chlorophyll in the green plant, light, and carbon dioxide. Other nutrients are also necessary (see *Essential Elements*).

Phytoplankton: Plankton that photosynthesizes glucose in the presence of light, such as algae.

Poikilothermic: Animals whose body temperatures vary with the outside environment (cold-blooded animals such as snakes).

Point Source: Pollution source such as from a sewer pipe (see *Nonpoint Source*).

Poison Ivy/Poison Oak: (Family Anacardiaceae). Poison ivy is *Toxicodendron radicans* and Pacific poison oak is *Toxicodendron diversiloba*. Both species cause a rash on the skin of sensitive humans.

Pokeweed: (Pokeberry). Genus *Phytolacca* sp., family Phytolaccaceae. Leaves are often eaten by humans but seeds and roots are poisonous.

Population Density: The number of people per unit area, usually per square mile (United States) or square kilometer (elsewere).

Pores(Soil): The spaces (voids) in a soil *not* occupied by mineral or organic soil particles. A good garden soil has fifty percent voids and fifty percent solids. Voids are filled with water or air.

Porosity (soil): The state of being porous—the characteristic of a soil by which free water can move through the soil.

Potable: Drinkable; such as potable water. The EPA has set quality standards for potable water.

Potassium Oxide: (K_2O). In a fertilizer, the water-soluble potassium reported as percent potassium converted to its chemical equivalent of K_2O. Conversions: $K \times 1.204 = K_2O$; $K_2O \times 0.83 = K$.

Potential Energy: Stored energy. *Example:* water behind a dam.

Prescription Burn (prescribed burn): The intentional setting on fire of the vegetation on a land area such as a forest, to kill unwanted vegetation.

Preservation: Retaining an item or an area in its present state. It differs from conservation which is *wise use*; preservation means *no use* (lock it up).

Primary Succession: The natural colonization of plants on bare soil such as on mine spoils.

Producers: Green plants that manufacture food through photosynthesis.

Professional: Decision-makers in any subject; people who engage in a sport for pay; people who receive money for writing; people who do something with great skill.

Radioactive Waste: (see *Nuclear Waste*).

Rappoport's Rule: The closer one gets to the equator, the more biodiversity will be found. More scientifically, there is a direct relationship between nearness to the equator and biodiversity.

Reclamation: The recovery of waste into a useful resource; irrigating drylands; draining wetlands; making productive farmland from wasteland.

Recycling: The conversion of waste into useful products. *Example:* recycled paper.

Regolith: Mantle rock; surface rock; the R horizon in soil science.

Renewable Resource: A thing useful to humans that can be replaced. Renewable resources are living things that can reproduce themselves. Example: trees that produce wood.

Reservoir: (1) A place where anything can be stored for future use; (2) a pond or lake where water is stored for use by a community of people; (3) a category of similar locations which hold water. Example: an ocean, the tissues in a human body, or a soft drink can.

Residence Time: The length of time required for the movement of a full volume into or out of a reservoir. If one gallon of water per week evaporates from a five-gallon aquarium, then the residence time of water in the tank would be five weeks.

Respiration: An exchange of oxygen by all forms of life, both plants and animals. Respiration takes in O_2 and releases food energy, CO_2, and H_2O. The net result of respiration is the opposite of photosynthesis.

Responsible Environmentalism: Practices that are based on scientifically accurate information, and that will lead to a livable, sustainable ecosystem for humans.

Rhododendron: A genus of plants with beautiful flowers but whose leaves are toxic to humans.

Riverine System: Streams and rivers that move through fairly well-defined channels which periodically or continuously contain moving water, or which form connecting links between two bodies of standing water.

Rodenticide: Any substance used to kill rodents such as rats and mice.

Saline: Salty, mainly waters high in NaCl (sodium chloride).

Salinity: (see *Saline*)

Salinization: The build-up of salt in soil, often as a result of irrigation with water containing dissolved salts.

Salt Tolerance: The ability of plants to survive in saline waters or on salty soil on ocean beaches.

Sanitary Landfill: A land-burial site for solid wastes intended to be environmentally sound.

Science: The pursuit of truth.

Scientist: A person engaged in the pursuit of truth.

Secondary Succession: The natural migration of plants into an area where other plants were formerly the dominant vegetation. Example: grasses and forbs moving into a forested area after a serious fire.

Section 404, Clean Water Act of 1972: Federal legislation that charged the Army Corps of Engineers with responsibility for issuing permits for "filling" of wetlands on both public and private lands in the United States.

Sediment Core Record: A boring into the sediment of the ocean floor or the floor of some other current or previous body of standing water to determine former vegetation and animal life and as a means of determining former climates.

Sedimentary Rock: Rock made of solidified sediments such as calcium carbonate that hardened into limestone.

Selection Cutting: Harvesting only mature trees in a forest.

Septage: Human effluent such as that pumped from a septic tank or that which flows from the septic tank into a drain field.

Septic Tank: A container into which human effluent flows and in which anaerobic fermentation takes place before the septage flows into a drain field.

Sewage Sludge: The solids resulting from a municipal treatment plant.

Sex Attractants: A chemical substance either secreted by an insect or artificially developed to attract mostly female, but occasionally male, insects.

Shelterwood Harvesting : A system of forest harvest in which only part of the merchantable trees are removed. The system is adapted to a stand of heavy-seeded species such as oaks.

Shifting Cultivation: A system of agriculture in which a patch of trees are clearcut and burned and crops are planted among the stumps until crop yields decline, then a new patch is clearcut, burned, and planted.

Shoreline: The edge of a given body of water. In a large body of water like an ocean, the shoreline changes continuously with the tides.

Shortgrass Prairie: Areas of low-growing grasses such as buffalograss. Precipitation is too small for midgrasses and tall grasses and forests. Any less rainfall and desert conditions would prevail.

Slash: Tree limbs and defective trees left in a forest after a tree harvest.

Slash and Burn: A primitive farming practice in which a forested area is cleared and the materials burned to allow for planting crops.

Sodbuster Legislation: Laws restricting the use of sloping lands for cultivated crops for the purpose of reducing erosion and sedimentation.

Softwood: Trees such as pines and spruces in contrast to hardwoods such as oaks, hickories, and maples.

Soil (Agriculture): The surface of the Earth to a depth of plant roots.

Soil (Engineer): All soil and nonbedrock geological material. Soil must be removed in constructing a foundation.

Soil (Geologist): All earthy material, including soil, overlying bedrock.

Soil Conservation District: A legal, nongovernmental organization that establishes guidelines for work by the U.S.D.A. Soil Conservation Service.

Soil Science: The organized study of the mineral and organic matter on the surface of the Earth capable of supporting plants.

Soil Survey: In the United States, the National Cooperative Soil Survey. It consists of making maps of soil series on an aerial map base. Interpretations of the use of each mapping unit for production agriculture and for environmental uses.

Solar Constant: The amount of energy reaching the upper atmosphere from sunlight. The solar constant (not actually constant) is usually about 2 gram calories per square centimeter of area.

Solid Wastes: All "throw-away" paper products, plastics, metals, and other non-liquid things not wanted.

Sorghum: A cereal grain used in the United States primarily for livestock feed. (*Sorghum* sp., family Gramineal - grass family).

Soybeans: (*Glycine max*, family Leguminaceae; some recent publications have listed the family as Fabaceae). A principal legume grown in humid U.S. primarily for oil and high-protein feed.

Speciation: The process by which new species evolve.

Specific Gravity: The ratio of the mass (weight) of a substance in comparison with the mass (weight) of an equal volume of water.

Stability: In an ecosystem, a (theoretical) condition in which the populations of all the organisms would be at equilibrium. Numbers of all the species would be relatively constant. There would be no changes in the environment. Such a condition has never existed on Earth.

Sub-bituminous Coal: Coal between lignite and bituminous.

Subtidal Subsystem: Marine wetlands that are permanently submerged.

Suspended Particulates: Particles such as clay and humus floating or moving about in a body of water or in the air.

Sustainable: The ability to maintain yields of plants and animals over a period of time.

Sustainable Harvest: The harvest of a crop with almost the same yield year after year.

Swamp Acts: Laws passed between 1849 and 1860 granting tracts of wetlands to the states and requiring the states to reclaim the areas for productive and heathful use.

Swamp-Buster Act: Laws forbidding wetlands to be drained for agricultural use.

Swampbuster Legislation: (see *Swamp Buster Act*).

Sweetclover: Any herb of genus *Melilotus*, family Leguminosae, used as a bee pasture, cattle pasture, and soil-building crop.

Tallgrass Prairie: Regions of land dominated by tall grasses such as big bluestem growing naturally in an area where rainfall is not enough to support trees.

Thomas Malthus: (1766–1834). An English clergyman who proposed that if human population is not controlled, poverty and famine are inevitable.

Tidal Subsystem: The zone of water near the mouths of rivers flowing into the ocean. In the tidal subsystem, the water level rises and falls with the tides.

TINSTAAFL: There *Is No Such Thing As a Free Lunch.* The TINSTAAFL principle means that we have to give up something if we want something else. Example: In dealing with the environment, we cannot have cheap lumber and preserve all of our forests at the same time.

Toxin: A poison.

Transpiration: The movement of liquid water from the soil, through a plant, and out of the stomata as a gas into the atmosphere.

Ultrasonic Pest Repellers: A mechanical device that makes sound waves above 20,000 vibrations per second (above the sound people can hear). The purpose is to scare away birds or fleas, or whatever. The authors have found no scientific evidence that any of these devices work.

Undergrazing: Animal grazing less than forage growth.

Upper Perennial Subsystem: A wetland that is sloping somewhat, and which is covered by flowing water year-round.

Utility: That which satisfies a human need or desire.

Volatile: A substance which, upon exposure, changes from a liquid to a gas at ambient temperature, i.e., gasoline.

Water Budget: The total amount of water that moves through a given system (like North America) in a given time period (usually a year).

Water Hemlock: (*Cicuta* sp., family Umbelliferae). Grows in moist places and is very poisonous to humans and animals when eaten.

Water Table: The surface of the zone of saturation. Example: the level of the water in a well.

Wheat: (*Triticum* sp., family Gramineae). A cereal that is the principal food and feed grain of the world.

White Clover: (*Trifolium repens,* family Leguminosae). A perennial herb which spreads rapidly by stolons (underground runners). An excellent pasture and lawn legume.

Wisteria: (*Wisteria* sp., family Leguminosae). A twining woody plant toxic to humans.

Zero Population Growth: A proposition that deaths should equal births for humans.

Zoology: The branch of biology that studies the science of animals.

Index

417